NEUTRON OPTICS

OXFORD SERIES ON NEUTRON SCATTERING IN CONDENSED MATTER

SERIES EDITORS

S. W. LOVESEY
E. W. J. MITCHELL

1. G. Williams: *Polarized Neutrons*
2. E. Balcar and S. W. Lovesey: *Theory of Magnetic Neutron and Photon Scattering*
3. V. F. Sears: *Neutron Optics*

NEUTRON OPTICS

An Introduction to the Theory of Neutron Optical Phenomena and Their Applications

VARLEY F. SEARS

Atomic Energy of Canada Limited

New York Oxford
OXFORD UNIVERSITY PRESS
1989

Oxford University Press

Oxford New York Toronto
Delhi Bombay Calcutta Madras Karachi
Petaling Jaya Singapore Hong Kong Tokyo
Nairobi Dar es Salaam Cape Town
Melbourne Auckland

and associated companies in
Berlin Ibadan

Published by Oxford University Press, Inc.,
200 Madison Avenue, New York, New York 10016

Library of Congress Cataloging-in-Publication Data
Sears, Varley F.
Neutron optics: by Varley F. Sears.
p. cm.
Bibliography: p. Includes index.
ISBN 0-19-504601-3
1. Neutron optics. 2. Neutrons—Diffraction. I. Title.
QC793.5.N4628S43 1989
539.7'213—dc 19 88-10009 CIP

1 3 5 7 9 8 6 4 2

Printed in the United States of America
on acid-free paper

PREFACE

The past 15 years have witnessed a rapid and continuing growth in the use of optical techniques in neutron beam research. Optical devices such as multilayer monochromators and polarizers are now commonplace components in neutron spectrometers. In addition, neutron optics has developed into an important field of research in its own right, centered on instruments such as the perfect-crystal neutron interferometer, and with applications ranging from the accurate measurement of neutron scattering lengths to sensitive tests of the fundamental principles of quantum mechanics.

This book is an attempt to integrate all these developments into a coherent story. The main emphasis is on a systematic development of the fundamental theory of neutron-optical phenomena and their applications, but experimental results are displayed and discussed wherever possible.

On the other hand, instrumentation and experimental technique have not been given the attention they would deserve in a truly balanced account of neutron optics. In addition, it is clear to me, as it will be to others, that some subjects, most notably ultracold neutrons and dynamical diffraction in crystals, have not been treated in sufficient detail to really do them justice. These omissions and weaknesses are due partly to space limitations and partly to the priorities dictated by my own interests and knowledge. I emphasize these deficiencies not by way of apology but simply to indicate that this book is not intended to be a definitive treatise on all aspects of neutron optics. However, within the restricted context of the fundamental theory of the subject, this book is reasonably balanced and complete, and the treatment is as thorough as the subject demands.

In writing this book I have tried to keep the interests of many groups of potential readers in mind: theorists and experimenters, students and professionals, those who are interested in the subject as a whole and those who are mainly interested in one particular area or application. Thus, for example, I have tried to make each chapter as self-contained as possible. No prior knowledge of optics is assumed beyond what would normally be acquired and retained from undergraduate physics courses. I do, however, assume that the reader has a good working knowledge of quantum mechanics.

I am most grateful to Oxford University Press and to Stephen Lovesey and Sam Werner for encouraging me to write this book. I am equally grateful to Atomic Energy of Canada Limited, and especially Gerald

Dolling, for granting me the time and resources necessary to complete it. Special thanks are due to Margaret Carey who typed most of the manuscript.

Deep River, Ontario V. F. S.
May, 1988

CONTENTS

NEUTRON OPTICS

1

INTRODUCTION

Neutron optics

The optical phenomena of reflection, refraction, diffraction, and interference are characteristic of waves in general and are not in any way unique properties of light. Under suitable experimental conditions such phenomena can be produced with any kind of radiation including, for example, neutrons.

The subject of neutron optics began on July 14, 1944, when the mirror reflection of thermal neutrons was first observed by Enrico Fermi and Walter Zinn using a collimated beam from the thermal column of the CP-3 reactor at what is now the Argonne National Laboratory. According to their colleague, Herbert Anderson (Segrè, 1962):

> There was a certain fascination associated with experiments which depended so clearly on the wave rather than the corpuscular properties of these particles. Their wavelengths and scattering cross sections were comparable with those of X-rays, and it seemed possible that various materials would have an index of refraction for neutrons . . . This was the beginning of a series of experiments on neutron optics . . . .

The purpose of these experiments was not simply to demonstrate the mirror reflection of neutrons but to determine from the measured index of refraction the scattering length that characterizes the neutron–nucleus interaction. The accurate measurement of the neutron scattering lengths of the elements and their isotopes is still one of the main practical applications of neutron optics.

In subsequent years virtually all of the classical optical phenomena that can be produced with light or X-rays have also been demonstrated with neutrons. This has led to the construction of many neutron-optical instruments that represent valuable tools for studies in physics, chemistry, metallurgy, and biology.

Figure 1.1 illustrates a typical neutron optics experiment, namely, the diffraction of slow neutrons by apertures (Klein and Zeilinger, 1984). Results are shown for (a) a straight edge, (b) a single slit, and (c) a double slit. The last can also be regarded as a two-beam interferometer that operates on the division-of-wavefront principle. It will be noted that absorbing slits are also used to collimate the beams, and that a prism monochromator is used to produce incident neutrons of wavelength $\lambda = 20$ Å.

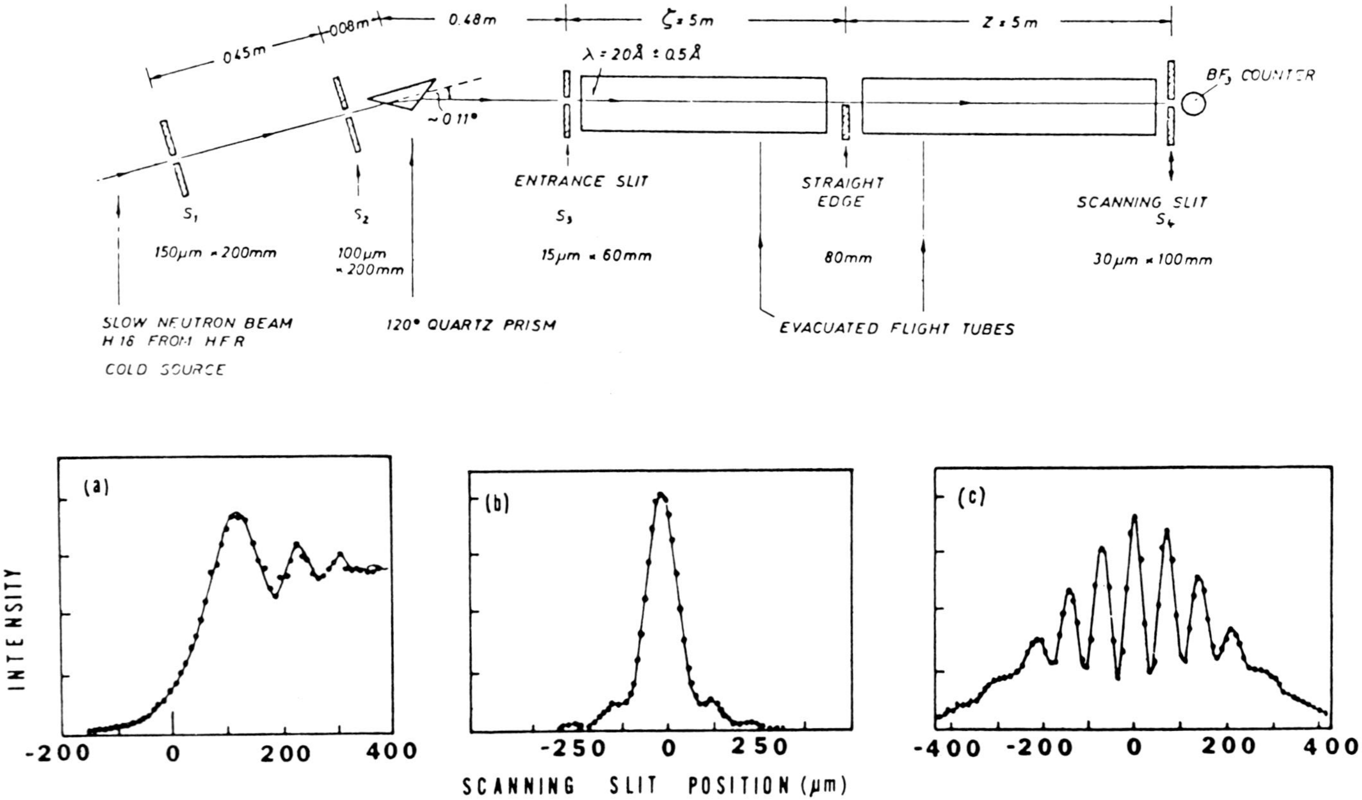

Fig. 1.1 Neutron diffraction by (a) a straight edge, (b) a single slit, and (c) a double slit (from Klein and Zeilinger, 1984).

Casual readers might easily mistake Fig. 1.1 for a textbook demonstration of the diffraction and interference of light. On closer examination, however, they would notice a few peculiarities. First, the prism is inverted with respect to its usual orientation for light because, for neutrons, the index of refraction has a value $n < 1$ while for light $n > 1$. Second, the overall size of the instrument ($\simeq 10$ m) is an order of magnitude larger than one would expect in ordinary optics with visible light. This is due to the fact that the neutron wavelength in these experiments is two orders of magnitude smaller than that of visible light. Finally, the resolution and statistical precision of the measurements are not as good as one would expect in similar experiments with light. This is because even the most intense thermal neutron source is still many orders of magnitude weaker than conventional light sources.

Neutron optics experiments such as those in Fig. 1.1 demonstrate quantum-mechanical phenomena on a truly macroscopic scale and are clearly of great pedagogical value. Nevertheless, a critical reader might well ask what such experiments tell us that we do not already know. One answer is that they provide very sensitive tests of the validity of quantum mechanics. For example, a fundamental assumption of quantum mechanics is that the Schrödinger equation is linear. The results shown in Fig. 1.1a put an upper limit of 3.3×10^{-15} eV on the magnitude of the coupling constant in a possible nonlinear term in the Schrödinger equation, as we shall see later in Chap. 7. This value is five orders of magnitude smaller than an earlier limit that was obtained from the measured Lamb shift in hydrogen.

Applications of neutron optics

The principal neutron optical phenomena and their applications are listed in Table 1.1. These applications fall into two general classes, which we have labeled *applied* and *fundamental.* In the applied class, existing knowledge is used to produce some optical instrument or device. In the fundamental class, an existing optical instrument is used to gain new knowledge.

For example, the Bragg scattering of neutrons from a crystal of known structure can be used to produce a collimated monoenergetic beam of neutrons with a specified wavelength. Conversely, the Bragg scattering of a collimated monoenergetic beam of neutrons of known wavelength from a crystal can be used to determine its structure. In the former (applied) case the crystal is a *device* and in the latter (fundamental) case it is a *sample.*

The purpose of most of the neutron-optical devices in Table 1.1 is to transport, collimate, focus, monochromate, filter, polarize, or otherwise manipulate neutron beams for applications in basic research. These applications include the study of the microscopic structure and dynamics

Table 1.1 Neutron-optical phenomena and their applications

Phenomena	Applications	
	Applied	Fundamental
Transmission	Resonance filters	Total cross sections Neutron radiography
Reflection	Polarizing mirrors Curved mirrors for focusing Guide tubes Bottles for ultracold neutrons	Scattering lengths
Refraction	Prism monochromators Lenses for focusing	Scattering lengths
Diffraction		
by crystals (Bragg scattering)	Monochromators Polarizers Filters Perfect-crystal interferometer	Scattering lengths Crystal structures Magnetic structures Diffraction topography Residual stress measurements Crystal texture
by inhomogeneities (small-angle scattering)		Scattering lengths Materials science
by apertures	Zone plates for focusing	Tests of quantum mechanics
Interference		
two-beam	Interferometers	Scattering lengths Tests of quantum mechanics Materials science
multiple-beam	Multilayer monochromaters Polarizing supermirrors	

of materials, the accurate measurement of neutron scattering lengths, and fundamental tests of quantum-mechanical principles.

An exception is the perfect-crystal neutron interferometer, which is an instrument that enables one to control and measure the relative phase between two plane-wave states. It has found a wide range of applications because the large separation of the beams provides easy access for material phase-shifting devices and magnetic fields. It was the development of this remarkable instrument, more than anything else, that has led to the great resurgence of activity in neutron optics in the past 10 years or so.

Theory of neutron optics

Neutron-optical phenomena arise from *coherent elastic scattering* in condensed matter. In fact, these two terms are synonymous. The characteristic property of coherent elastic scattering is that it is not accompanied by any change in the microscopic state of the system and, hence, can be described by a one-body Schrödinger equation,

$$\left[-\frac{\hbar^2}{2m}\Delta + v(\mathbf{r})\right]\psi(\mathbf{r}) = E\psi(\mathbf{r}) \tag{1.1}$$

in which E is the incident-neutron energy, m the neutron mass, and $v(\mathbf{r})$ the optical potential that represents the effective interaction of the neutron with the system. The neutron wave function $\psi(\mathbf{r})$, which then provides a formally exact description of the coherent elastic scattering, is called the *coherent wave*.

Coherent elastic scattering is, of course, not the only collision process that can occur. In optics, all other scattering processes are referred to collectively as *diffuse scattering*. The incident neutron might also be absorbed by a nucleus by means of, for example, radiative capture. The presence of diffuse scattering and absorption causes an attenuation of the coherent wave $\psi(\mathbf{r})$ in the medium, so that in a self-consistent theory the optical potential $v(\mathbf{r})$ is complex.

It is important to recognize that $v(\mathbf{r})$ and $\psi(\mathbf{r})$ are macroscopic thermodynamic quantities. That is, they depend only on the thermodynamic variables (temperature, pressure, etc.) that define the macroscopic state of the system and are independent of the instantaneous positions of the atoms of which the system is composed. In other words, the system behaves like a continuous macroscopic medium as far as the coherent elastic scattering is concerned. Thus, in a homogeneous medium, such as a liquid or amorphous solid, $v(\mathbf{r})$ is constant, independent of $\mathbf{r}$; in a crystal, $v(\mathbf{r})$ is a periodic function of $\mathbf{r}$ with the same periodicity as the lattice. Outside the medium, on the other hand, $v(\mathbf{r}) = 0$. Thus, the Schrödinger equation (1.1) is a macroscopic equation that describes the coherent elastic scattering, and hence all neutron optical phenomena, in terms of the collision of the neutron with a potential barrier.

From a theoretical point of view, neutron optics divides naturally into two quite distinct parts:

1. the theory of dispersion;
2. the theory of reflection, refraction, and diffraction.

Part (1) is concerned with the derivation of the macroscopic one-body Schrödinger equation (1.1) from the underlying microscopic many-body Schrödinger equation, and with the calculation of the complex optical potential $v(\mathbf{r})$ for the particular material of interest. In other words, the theory of dispersion is concerned with the proof that as far as neutron optics is concerned a system of atoms behaves like a continuous macroscopic medium. This is analogous to the problem of deriving the macroscopic Maxwell equations for ordinary optics from the microscopic electrodynamic equations of Lorentz. Part (2) is concerned with the solution of the Schrödinger equation (1.1) subject to boundary conditions appropriate to the experimental arrangement of interest.

Scope of this work

It would be impossible to discuss in detail all aspects of neutron-optical phenomena and their applications in a single volume. In the present book we emphasize the basic theoretical principles of neutron optics rather than the practical techniques required to produce and study neutron-optical phenomena in the laboratory. In particular, Chapters 2 and 4 are concerned with the theory of dispersion and the remaining chapters with the theory of reflection, refraction, and diffraction.

In addition, we emphasize the optical principles that underlie the applications of neutron optics rather than the end-products of these applications. Thus, for example, in the case of neutron diffraction by crystals we discuss in considerable detail the intrinsic structure of the Bragg peaks in various kinds of crystalline materials. On the other hand, we make no attempt to survey the vast literature on the application of neutron diffraction to crystallography and materials science.

We do, however, include a fairly complete review of the current applications of neutron-optical techniques to the accurate measurement of coherent and incoherent scattering lengths and to fundamental studies of quantum mechanics, since these applications are central to the subject of neutron optics as it is understood by those working in the field.

2

GENERAL THEORY OF THERMAL NEUTRON SCATTERING

Many readers will be familiar with the elementary theory of neutron scattering (Squires, 1978; Lovesey, 1984) that is based on the Born approximation and uses the Fermi pseudopotential to represent the effective interaction between the neutron and the nuclei. This approach is valid for neutron scattering in condensed matter if the sample is sufficiently small that multiple scattering is negligible.

Neutron optics requires a different and more rigorous approach for two reasons. First, the multiple-scattering effects that are neglected in the Born approximation must be taken properly into account in order to understand the performance of neutron-optical devices, because optical phenomena are inherently multiple-scattering effects of one kind or another. Second, one of the main practical applications of neutron optics is in the accurate measurement of the "scattering length" that occurs as a phenomenological constant in the Fermi pseudopotential. The theoretical framework used to interpret measurements of such high accuracy (typically 0.05 percent) demands a more rigorous justification than is provided by the heuristic arguments that motivate the elementary theory. There are also situations, such as when the imaginary part of the scattering length is important, in which corrections to the Fermi pseudopotential itself must be taken into account.

The purpose of the present chapter is to develop a general theoretical foundation for the discussion of neutron-optical phenomena that is presented in the succeeding chapters. We must abandon the naive conception of the Fermi pseudopotential as an effective "potential" and recognize that it is, in fact, the impulse approximation for the T-matrix. Since this approach involves concepts and techniques that may not be well known to some readers, we begin by deriving a number of basic results in scattering theory that will be needed later. We then consider the general theory of neutron scattering by a macroscopic body and show, without approximation, that the optical component of the scattering is governed by a one-body Schrödinger equation containing a complex optical potential that describes the effective interaction of the neutron with the medium. The treatment given here is a more detailed version of one presented (Sears, 1978) in a review article some years ago.

2.1. Formal theory of scattering

We begin by discussing the formal theory of the scattering of a neutron by an arbitrary macroscopic system. The total Hamiltonian can be expressed as

$$H = H_0 + V \tag{2.1.1}$$

where H_0 denotes the Hamiltonian in the absence of any interaction between the neutron and the system, and V is the interaction energy.

Lippmann–Schwinger equation

Let Φ denote a solution of the equation

$$H_0\Phi = E\Phi \tag{2.1.2}$$

corresponding to a possible initial state for the collision when the total energy is E. Given Φ, we can construct a second state

$$\Psi = (1 - GV)^{-1}\Phi \tag{2.1.3}$$

in which G is the retarded Green function,

$$G = (E + i\varepsilon - H_0)^{-1} \tag{2.1.4}$$

The purpose of the quantity ε will become apparent in Sec. 2.2. In the meantime we simply assert that ε approaches zero through positive values ($\varepsilon \to 0^+$). According to Eq. 2.1.3, Ψ satisfies the equation (Lippmann and Schwinger, 1950)

$$\Psi = \Phi + GV\Psi \tag{2.1.5}$$

Operating on both sides of this equation with G^{-1}, we then get

$$(H - E)\Psi = i\varepsilon(\Psi - \Phi) \to 0 \tag{2.1.6}$$

so that

$$H\Psi = E\Psi \tag{2.1.7}$$

Thus, if Φ is an eigenstate of H_0 with energy E, then Ψ is (formally, at least) an eigenstate of H with the same energy.

The usefulness of the Lippmann–Schwinger equation (2.1.5) lies in the fact that its solution satisfies not only the Schrödinger equation (2.1.7) but also the correct boundary conditions for the scattering problem. We shall see in Sec. 2.2 that, in the coordinate representation, Φ corresponds to a plane wave and G to an outgoing spherical wave. Thus, the first term in Eq. 2.1.5 represents the incident wave and the second term the scattered wave. More generally, it can be shown that Ψ is the *scattering state* that evolves out of the *channel state* Φ when the interaction energy V is switched on adiabatically with a time constant $\hbar/\varepsilon$.

Transition operator

It follows from Eq. 2.1.3 that

$$V\Psi = T\Phi \tag{2.1.8}$$

where T is called the *transition operator*, or *T-matrix*, and is given by

$$T = V(1 - GV)^{-1} \tag{2.1.9}$$

The Lippmann–Schwinger equation (2.1.5) then becomes

$$\Psi = (1 + GT)\Phi \tag{2.1.10}$$

This equation provides a formal solution of the scattering problem in terms of T. In other words, the problem of solving the Schrödinger equation (2.1.7), or the equivalent Lippmann–Schwinger equation (2.1.5), is reduced to finding the T-matrix.

According to Eq. 2.1.9, T satisfies the equation

$$T = V + TGV \tag{2.1.11}$$

the iterated solution of which gives the *Born series* (Born, 1926),

$$T = V + VGV + VGVGV + \cdots \tag{2.1.12}$$

This can be interpreted as a multiple scattering expansion in which V represents the contribution from single scattering, VGV the contribution from double scattering, and so on. The *Born approximation*

$$T = V \tag{2.1.13}$$

is then equivalent to neglect of multiple scattering.

Ewald equations

The macroscopic system is composed of particles (nuclei and electrons), so that we can write

$$V = \sum_i V_i \tag{2.1.14}$$

where V_i denotes the energy of interaction of the neutron with the ith particle. The Born series (2.1.12) then becomes

$$T = \sum_i V_i + \sum_{ij} V_i G V_j + \sum_{ijk} V_i G V_j G V_k + \cdots \tag{2.1.15}$$

and includes contributions from both intraparticle multiple scattering (e.g., V_1GV_1) and interparticle multiple scattering (e.g., V_1GV_2).

The first important point that must be recognized is that, at least as far as the nuclei are concerned, intraparticle multiple scattering cannot be neglected. In other words, V_i cannot be treated in the Born approximation. This approximation would be valid for thermal neutrons only if the scattering cross section of a nucleus were much smaller than its geometrical cross section (Messiah, 1964). However, it is known from experiment that this is not the case (Mughabghab et al., 1981; Mughabghab, 1984). The importance of intranuclear multiple scattering becomes especially clear when one recognizes that it is responsible for the formation of the compound nucleus in an actual physical collision.

To discuss the intraparticle multiple scattering it is convenient to introduce a quantity T_i, which is defined such that

$$T_i = V_i + T_i G V_i \tag{2.1.16}$$

Thus, T_i is the transition operator for the hypothetical situation in which the neutron interacts only with the ith particle. The iterated solution of Eq. 2.1.16,

$$T_i = V_i + V_i G V_i + V_i G V_i G V_i + \cdots \tag{2.1.17}$$

then generates all the intraparticle multiple-scattering processes for the ith particle. The inverse of Eq. 2.1.16,

$$V_i = T_i - T_i G V_i \tag{2.1.18}$$

has the interated solution

$$V_i = T_i - T_i G T_i + T_i G T_i G T_i - \cdots \tag{2.1.19}$$

On substituting Eq. 2.1.19 into Eq. 2.1.15, we get

$$T = \sum_i T_i + \sum_{i \neq j} T_i G T_j + \sum_{i \neq j \neq k} T_i G T_j G T_k + \cdots \tag{2.1.20}$$

By expressing T in terms of T_i, rather than V_i, all intraparticle multiple-scattering processes are implicitly summed to infinite order, and we are left with an explicit interparticle multiple-scattering expansion. This latter expansion can now be summed to give

$$T = \sum_i T_i(1 + GZ_i) \tag{2.1.21}$$

in which Z_i is the solution of the following set of inhomogeneous linear equations:

$$Z_i = \sum_{j(\neq i)} T_j(1 + GZ_j) \tag{2.1.22}$$

The physical significance of these results can be seen by constructing the states

$$\chi_i = (1 + GZ_i)\Phi \tag{2.1.23}$$

It then follows from Eq. 2.1.10 that

$$\Psi = \Phi + \sum_i G T_i \chi_i \tag{2.1.24}$$

and

$$\chi_i = \Phi + \sum_{j(\neq i)} G T_j \chi_j \tag{2.1.25}$$

These are the fundamental equations of the theory of multiple scattering (Goldberger and Watson, 1964), and are the quantum-mechanical analog of equations first introduced by Ewald (1916a, b, 1917a, b) to describe the multiple scattering of electromagnetic radiation. In the coordinate representation, Eq. 2.1.24 states that the total wave function Ψ is the sum of the incident wave Φ plus a scattered wave $GT_i\chi_i$ from each particle in the system. Here χ_i is the *local field* at the ith particle, which, according to Eq. 2.1.25, is the sum of the incident wave Φ plus a scattered wave $GT_j\chi_j$ from every other particle in the system. The term GZ_i in Eq. 2.1.21 therefore represents the local field correction for the transition operator.

Kinematical theory

The kinematical theory of neutron scattering is based on the approximation

$$T = \sum_i T_i \tag{2.1.26}$$

which can be regarded as coming either from Eq. 2.1.20 by neglecting interparticle multiple scattering, or from Eq. 2.1.21 by neglecting the local field correction. In the kinematical theory the scattering cross section has a simple interpretation in terms of the atomic structure and dynamics of the system. Consequently, in applications of thermal neutron scattering to the study of the microscopic properties of condensed matter, the linear dimensions of the sample are chosen sufficiently small that multiple scattering is negligible and, hence, that the kinematical theory is valid.

The preceding kinematical expression for T is formally similar to the Born approximation,

$$T - \sum_i V_i \tag{2.1.27}$$

so that, in elementary treatments of thermal neutron scattering, T_i is introduced heuristically as an effective potential that is chosen to give the "right answer" in the "Born approximation." Instead of introducing a model for the potential V_i, and then solving Eq. 2.1.16 for the transition operator T_i, one simply introduces a model for T_i itself. This model is the *Fermi pseudopotential* and provides a simple phenomenological treatment of thermal neutron scattering that is adequate for most applications in condensed-matter physics.

In neutron optics, on the other hand, one is not using neutrons to study the properties of condensed matter. Rather, one is using the known properties of condensed matter to study the quantum-mechanical properties of neutrons. In other words, the scattering system is not a "sample," it is a "device" that is designed such that the quantum-mechanical interference of the scattered waves with each other and with the incident wave will produce the desired optical effect. Depending on its size, shape, and composition, the device will then act as a mirror, lens, monochromator, interferometer, etc. To understand the behavior of such devices it is necessary to go beyond the kinematical theory and take multiple scattering properly into account. The only optical effect that is present in the kinematical theory is Fraunhofer diffraction and even this is not described correctly, since extinction effects are absent.

Before we can discuss the theory and application of neutron-optical devices we must first demonstrate how the formal identities that we have derived in the present section can be used as the basis of a rigorous theory of thermal neutron scattering in general and of neutron optics in particular. We begin in the next section with a discussion of the scattering of a neutron by a single atom.

2.2. Scattering amplitude

Complex optical potential

In general, the interaction of a neutron with an atom includes contributions from the strong and weak interactions of the neutron with the nucleus, and from various electromagnetic interactions of the neutron with the nucleus and the electrons. These neutron–atom interactions have been discussed in detail in a recent review article (Sears, 1986a). In the present chapter we shall, for simplicity, confine our attention to the strong interaction between the neutron and the nucleus, since this is by far the most important interaction in nonmagnetic materials. The theory will later be generalized to include other interactions when this becomes necessary.

When a thermal neutron collides with a nucleus it may be either scattered or absorbed. By *absorption* we mean reactions such as (n, γ), (n, p), or (n, α) in which there is no neutron in the final state. We shall show in Sec. 2.4 that the effect of absorption can be included without going to a multichannel formalism by using a complex optical potential to represent the effective interaction of the neutron with the nucleus. Thus, the scattering and absorption of a thermal neutron by a single fixed nucleus can be described by a Hamiltonian of the form

$$H = \frac{1}{2m}\mathbf{p}^2 + V(\mathbf{r}) \tag{2.2.1}$$

in which m is the neutron mass, $\mathbf{p}$ is the neutron momentum, and $V(\mathbf{r})$ is the complex optical potential, $\mathbf{r}$ being the position of the neutron relative to the nucleus.

In Sec. 2.3 we shall show that only s-wave scattering is appreciable at thermal-neutron energies. Thus, there is no loss of generality in taking $V(\mathbf{r})$ to be a central potential. Any noncentral component in the actual potential will affect only the p-wave and higher-order terms in the partial wave expansion that are negligible for thermal neutrons. For simplicity, we shall also assume that $V(\mathbf{r})$ is spin independent. The effect of the neutron and nuclear spins on the scattering and absorption of thermal neutrons will be discussed later in Sec. 2.7.

It will not be necessary to introduce a specific model for $V(\mathbf{r})$, since the thermal neutron scattering and absorption cross sections do not depend explicitly on the detailed behavior of $V(\mathbf{r})$ but only on a single complex parameter, the s-wave scattering length, which can (and in practice must) be treated as a phenomenological constant.

Momentum representation

Let $|\mathbf{k}\rangle$ denote the state in which the neutron has momentum $\hbar\mathbf{k}$,

$$\mathbf{p}\,|\mathbf{k}\rangle = \hbar\mathbf{k}\,|\mathbf{k}\rangle \tag{2.2.2}$$

These momentum eigenstates form a complete orthonormal set, so that

$$\langle \mathbf{k} \mid \mathbf{k}' \rangle = \delta(\mathbf{k}-\mathbf{k}'), \qquad \int d\mathbf{k}\,|\mathbf{k}\rangle\,\langle\mathbf{k}| = 1 \tag{2.2.3}$$

The free-particle Hamiltonian

$$H_0 = \frac{1}{2m}\,\mathbf{p}^2 \tag{2.2.4}$$

is also diagonal in the momentum representation,

$$H_0\,|\mathbf{k}\rangle = E_{\mathbf{k}}\,|\mathbf{k}\rangle \tag{2.2.5}$$

and the energy is given by

$$E_{\mathbf{k}} = \frac{(\hbar k)^2}{2m} \tag{2.2.6}$$

Finally, the momentum representation provides a useful spectral decomposition of the Green function (2.1.4),

$$\begin{aligned} G &= \frac{1}{E_{\mathbf{k}} + i\varepsilon - H_0}\int d\mathbf{k}'\,|\mathbf{k}'\rangle\,\langle\mathbf{k}'| \\ &= \int d\mathbf{k}'\,\frac{|\mathbf{k}'\rangle\,\langle\mathbf{k}'|}{E_{\mathbf{k}} + i\varepsilon - E_{\mathbf{k}'}} \end{aligned} \tag{2.2.7}$$

Coordinate representation

Let $|\mathbf{r}\rangle$ denote the state in which the neutron has position $\mathbf{r}$. These states also form a complete orthonormal set so that

$$\langle \mathbf{r} \mid \mathbf{r}' \rangle = \delta(\mathbf{r}-\mathbf{r}'), \qquad \int d\mathbf{r}\,|\mathbf{r}\rangle\,\langle\mathbf{r}| = 1 \tag{2.2.8}$$

The wave function of the state $|\mathbf{k}\rangle$ in the coordinate representation is, by definition, the probability amplitude

$$\phi_{\mathbf{k}}(\mathbf{r}) = \langle \mathbf{r} \mid \mathbf{k} \rangle = \langle \mathbf{k} \mid \mathbf{r} \rangle^* \tag{2.2.9}$$

and this quantity is a plane wave with wave vector **k**,

$$\phi_{\mathbf{k}}(\mathbf{r}) = (2\pi)^{-3/2} \exp(i\mathbf{k} \cdot \mathbf{r}) \tag{2.2.10}$$

The normalization factor in the wave function is determined by Eqs. 2.2.3 and 2.2.8. Thus, since

$$\int d\mathbf{r}\, \langle \mathbf{k} \mid \mathbf{r} \rangle \langle \mathbf{r} \mid \mathbf{k}' \rangle = \langle \mathbf{k} \mid \mathbf{k}' \rangle \tag{2.2.11}$$

$\phi_{\mathbf{k}}(\mathbf{r})$ must be normalized such that

$$\int d\mathbf{r}\, \phi_{\mathbf{k}}(\mathbf{r})^* \phi_{\mathbf{k}'}(\mathbf{r}) = \delta(\mathbf{k} - \mathbf{k}') \tag{2.2.12}$$

The *concrete* momentum operator is given in the coordinate representation by

$$\mathbf{p} = \frac{\hbar}{i} \nabla \tag{2.2.13}$$

so that

$$\mathbf{p}\phi_{\mathbf{k}}(\mathbf{r}) = \hbar\mathbf{k}\phi_{\mathbf{k}}(\mathbf{r}) \tag{2.2.14}$$

Similarly, the concrete free-particle Hamiltonian is

$$H_0 = -\frac{\hbar^2}{2m} \Delta, \tag{2.2.15}$$

so that

$$H_0\phi_{\mathbf{k}}(\mathbf{r}) = E_{\mathbf{k}}\phi_{\mathbf{k}}(\mathbf{r}) \tag{2.2.16}$$

Finally, the Green function can be calculated in the coordinate representation with the help of the spectral decomposition (2.2.7),

$$\begin{aligned} \langle \mathbf{r}|\, G\, |\mathbf{r}'\rangle &= \int d\mathbf{k}' \frac{\langle \mathbf{r} \mid \mathbf{k}' \rangle \langle \mathbf{k}' \mid \mathbf{r}' \rangle}{E_{\mathbf{k}} + i\varepsilon - E_{\mathbf{k}'}} \\ &= \int d\mathbf{k}' \frac{\phi_{\mathbf{k}'}(\mathbf{r})\phi_{\mathbf{k}'}(\mathbf{r}')^*}{E_{\mathbf{k}} + i\varepsilon - E_{\mathbf{k}'}} \\ &= \frac{m}{4\pi^3\hbar^2} \int d\mathbf{k}' \frac{\exp[i\mathbf{k}' \cdot (\mathbf{r} - \mathbf{r}')]}{k^2 + i\eta - k'^2} \end{aligned} \tag{2.2.17}$$

where $\eta = 2m\varepsilon/\hbar^2$. This integral can be evaluated using Cauchy's theorem, and one finds that in the limit $\eta \to 0^+$,

$$\langle \mathbf{r}|\, G\, |\mathbf{r}'\rangle = -\frac{m}{2\pi\hbar^2} \frac{\exp[ik\, |\mathbf{r} - \mathbf{r}'|]}{|\mathbf{r} - \mathbf{r}'|} \tag{2.2.18}$$

If we regard $\langle \mathbf{r}|G\,|\mathbf{r}'\rangle$ as a function of the complex energy $E_\mathbf{k}+i\varepsilon$, then this function has a branch cut along the positive real axis. Immediately above the cut ($\varepsilon\to 0^+$), the function is given by Eq. 2.2.18 and represents an outgoing spherical wave, the so-called *retarded* wave. Immediately below the cut ($\varepsilon\to 0^-$), the function is given by an expression similar to Eq. 2.2.18 except that the exponent has the opposite sign. This represents an incoming or *advanced* wave. The boundary condition for the scattering problem requires that we use the retarded Green function (2.2.18).

Scattering state

The scattering state $\psi_\mathbf{k}$ that evolves out of the channel state $|\mathbf{k}\rangle$ when the interaction energy V is switched on adiabatically is given according to Eq. 2.1.10 by

$$\psi_\mathbf{k}=(1+GT)\,|\mathbf{k}\rangle \tag{2.2.19}$$

and satisfies the Schrödinger equation

$$H\psi_\mathbf{k}=E_\mathbf{k}\psi_\mathbf{k} \tag{2.2.20}$$

The wave function of the state $\psi_\mathbf{k}$ in the coordinate representation is

$$\begin{aligned}\psi_\mathbf{k}(\mathbf{r})&=\langle \mathbf{r}\mid\psi_\mathbf{k}\rangle\\&=\langle \mathbf{r}|\left\{1+G\int d\mathbf{r}'\,|\mathbf{r}'\rangle\langle\mathbf{r}'|\,T\right\}|\mathbf{k}\rangle\\&=\langle\mathbf{r}\mid\mathbf{k}\rangle+\int d\mathbf{r}'\,\langle\mathbf{r}|\,G\,|\mathbf{r}'\rangle\langle\mathbf{r}'|\,T\,|\mathbf{k}\rangle\\&=\phi_\mathbf{k}(\mathbf{r})+\int d\mathbf{r}'\,\langle\mathbf{r}|\,G\,|\mathbf{r}'\rangle\langle\mathbf{r}'|\,T\,|\mathbf{k}\rangle\end{aligned} \tag{2.2.21}$$

Thus, the wave function of the scattering state is a coherent superposition of the incident plane wave $\phi_\mathbf{k}(\mathbf{r})$ plus a spherical scattered wave $\langle\mathbf{r}|\,G\,|\mathbf{r}'\rangle$ from every point $\mathbf{r}'$ where $\langle\mathbf{r}'|\,T\,|\mathbf{k}\rangle\neq 0$.

All we really need to know is the asymptotic behavior of $\psi_\mathbf{k}(\mathbf{r})$ as $r\to\infty$, because that is where the neutron detector is located. In the present context "$r\to\infty$" means $r\gg r'$ wherever $V(\mathbf{r}')$ and, hence, $\langle\mathbf{r}'|\,T\,|\mathbf{k}\rangle$ are different from zero. In other words $r\gg 10^{-12}$ cm, say.

We note first that

$$|\mathbf{r}-\mathbf{r}'|=r-\hat{\mathbf{r}}\cdot\mathbf{r}'+O(r^{-1}) \tag{2.2.22}$$

where $\hat{\mathbf{r}}$ denotes a unit vector in the direction of $\mathbf{r}$. Hence,

$$k\,|\mathbf{r}-\mathbf{r}'|=kr-\mathbf{k}'\cdot\mathbf{r}'+O(r^{-1}) \tag{2.2.23}$$

where $\mathbf{k}'=k\hat{\mathbf{r}}$ and is called the *scattered neutron wave vector* (see Fig.

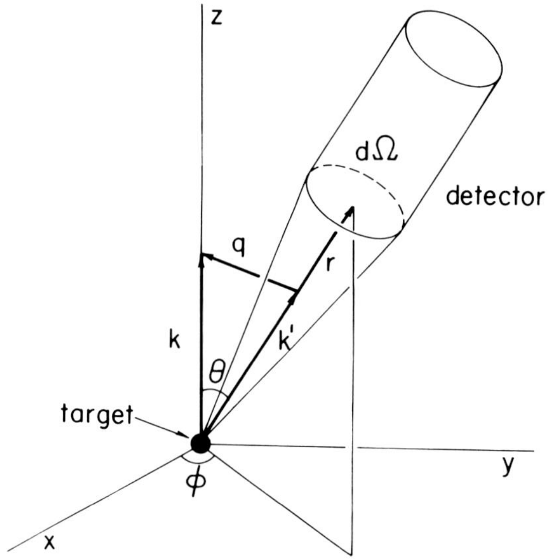

Fig. 2.1 Scattering geometry.

2.1). It then follows from Eq. 2.2.18 that as $r \to \infty$,

$$\langle \mathbf{r} | G | \mathbf{r}' \rangle \to -\frac{m}{2\pi\hbar^2} \frac{\exp[i(kr - \mathbf{k}' \cdot \mathbf{r}')]}{r}$$

$$= -\frac{\sqrt{2\pi}\, m}{\hbar^2} \frac{\exp(ikr)}{r} \phi_{\mathbf{k}'}(\mathbf{r}')^* \tag{2.2.24}$$

$$= -\frac{\sqrt{2\pi}\, m}{\hbar^2} \frac{\exp(ikr)}{r} \langle \mathbf{k}' | \mathbf{r}' \rangle$$

and hence, from Eq. 2.2.21, that

$$\psi_{\mathbf{k}}(\mathbf{r}) \to \phi_{\mathbf{k}}(\mathbf{r}) - \frac{\sqrt{2\pi}\, m}{\hbar^2} \frac{\exp(ikr)}{r} \int d\mathbf{r}' \langle \mathbf{k}' | \mathbf{r}' \rangle \langle \mathbf{r}' | T | \mathbf{k} \rangle$$
$$= \phi_{\mathbf{k}}(\mathbf{r}) - \frac{\sqrt{2\pi}\, m}{\hbar^2} \frac{\exp(ikr)}{r} \langle \mathbf{k}' | T | \mathbf{k} \rangle \tag{2.2.25}$$

Thus, in the limit $r \to \infty$ the scattered waves $\langle \mathbf{r} | G | \mathbf{r}' \rangle$ in Eq. 2.2.21 reduce to a single spherical scattered wave. The result, Eq. 2.2.25, can be expressed as

$$\psi_{\mathbf{k}}(\mathbf{r}) \to (2\pi)^{-3/2} \left\{ \exp(i\mathbf{k} \cdot \mathbf{r}) + f(\theta) \frac{\exp(ikr)}{r} \right\} \tag{2.2.26}$$

where

$$f(\theta) = -\frac{4\pi^2 m}{\hbar^2} \langle \mathbf{k}' | T | \mathbf{k} \rangle \tag{2.2.27}$$

The quantity $f(\theta)$ gives the amplitude of the scattered wave relative to the incident wave and is accordingly called the *scattering amplitude.* In general, $f(\theta)$ depends on $\mathbf{k}$ and $\mathbf{k}'$ separately and is not simply a function of the angle of scattering θ. However, it is conventional to denote the scattering smplitude as $f(\theta)$. Apart from a constant factor, the scattering amplitude is simply the T-matrix element between the initial and final states. Alternatively,

$$f(\theta) = \langle \mathbf{k}' | \, F \, | \mathbf{k} \rangle \tag{2.2.28}$$

where F is the *scattering amplitude operator,*

$$F = -\frac{4\pi^2 m}{\hbar^2} T \tag{2.2.29}$$

2.3. Partial waves

In the coordinate representation, the Schrödinger equation (2.2.20) becomes

$$\left\{-\frac{\hbar^2}{2m}\Delta + V(r)\right\}\psi_{\mathbf{k}}(\mathbf{r}) = E_{\mathbf{k}}\psi_{\mathbf{k}}(\mathbf{r}) \tag{2.3.1}$$

We must find the solution of this equation that has the asymptotic behavior of Eq. 2.2.26. Thus, omitting the arbitrary normalization factor, we require that as $r \to \infty$,

$$\psi_{\mathbf{k}}(\mathbf{r}) \to \exp(ikr\cos\theta) + f(\theta)\frac{\exp(ikr)}{r} \tag{2.3.2}$$

Since $V(r)$ is assumed to be a central potential and the incident wave is independent of the azimuthal angle ϕ in Fig. 2.1, it is evident that $\psi_{\mathbf{k}}(\mathbf{r})$ and, hence, $f(\theta)$ must also be independent of ϕ. Thus, $\psi_{\mathbf{k}}(\mathbf{r})$ can be expanded in terms of Legendre polynomials,

$$\psi_{\mathbf{k}}(\mathbf{r}) = \sum_{l=0}^{\infty} R_l(r) P_l(\cos\theta) \tag{2.3.3}$$

and the radial coefficients $R_l(r)$ satisfy the equation

$$\left[\frac{1}{r^2}\frac{d}{dr}\left(r^2\frac{d}{dr}\right) + k^2 - \frac{l(l+1)}{r^2} - \frac{2m}{\hbar^2}V(r)\right]R_l(r) = 0 \tag{2.3.4}$$

The expression (2.3.3) is called a partial wave expansion, and the spectroscopic nomenclature s, p, d, f, . . . is used to denote the $l =$ 0, 1, 2, 3, . . . partial waves, respectively. The method of partial waves was originally developed by Rayleigh (1878) to describe the scattering of sound waves by a sphere. It was later applied to the quantum-mechanical theory of scattering by Faxén and Holtsmark (1927).

Exterior wave function

Since the optical potential $V(r)$ represents the short-range strong interaction between the neutron and the nucleus, we can assume that

$$V(r) = 0, \qquad r > r_0 \tag{2.3.5}$$

where r_0 is a measure of the nuclear radius. Hence,

$$\left[\frac{1}{r^2}\frac{d}{dr}\left(r^2\frac{d}{dr}\right) + k^2 - \frac{l(l+1)}{r^2}\right]R_l(r) = 0, \qquad r > r_0 \tag{2.3.6}$$

Since this is a second-order differential equation, it has two linearly independent solutions. The first is the spherical Bessel function $j_l(kr)$, which is regular at $r = 0$, and the second is the spherical Neumann function $n_l(kr)$, which is singular at $r = 0$. For example,

$$j_0(x) = \frac{\sin x}{x}, \qquad n_0(x) = -\frac{\cos x}{x} \tag{2.3.7}$$

The general exterior solution can therefore be expressed in the form

$$R_l(r) = A_l[\cos(\delta_l)j_l(kr) - \sin(\delta_l)n_l(kr)], \qquad r > r_0 \tag{2.3.8}$$

The coefficients A_l and δ_l are determined by the asymptotic condition (2.3.2) and by the boundary condition at $r = r_0$.

Free-particle solution

If $V(r) = 0$ for all r then there is no scattered wave and

$$\psi_{\mathbf{k}}(\mathbf{r}) = \exp(ikr\cos\theta) \tag{2.3.9}$$

Since $\psi_{\mathbf{k}}(0) = 1$ while $n_l(kr)$ diverges as $r \to 0$, it follows that $\delta_l = 0$ for a free particle and

$$R_l(r) = A_l j_l(kr) \tag{2.3.10}$$

Thus, the wave function (2.3.9) can be expressed equivalently as

$$\psi_{\mathbf{k}}(\mathbf{r}) = \sum_{l=0}^{\infty} A_l j_l(kr) P_l(\cos\theta) \tag{2.3.11}$$

It then follows, with the help of the orthonormality relation for the Legendre polynomials,

$$\int_{-1}^{1} P_l(x)P_{l'}(x)\,\mathrm{d}x = \frac{2}{2l+1}\delta_{ll'} \tag{2.3.12}$$

and Gegenbauer's theorem (Abramowitz and Stegun, 1965),

$$\int_{-1}^{1} P_l(x)\exp(ixy)\,dx = 2i^l j_l(y) \tag{2.3.13}$$

that

$$A_l = (2l+1)i^l \tag{2.3.14}$$

Thus, for a free particle,

$$\psi_{\mathbf{k}}(\mathbf{r}) = \exp(ikr\cos\theta) = \sum_{l=0}^{\infty} (2l+1)i^l j_l(kr)P_l(\cos\theta) \tag{2.3.15}$$

Scattering amplitude

In general, as $r \to \infty$,

$$j_l(kr) \to \frac{1}{kr}\cos(kr+\beta_l), \qquad n_l(kr) \to \frac{1}{kr}\sin(kr+\beta_l) \tag{2.3.16}$$

where

$$\beta_l = -(l+1)\frac{\pi}{2} \tag{2.3.17}$$

Hence, the radial wave function (2.3.8) has the asymptotic behavior

$$R_l(r) \to \frac{A_l}{kr}\cos(kr+\alpha_l), \qquad \text{as } r \to \infty \tag{2.3.18}$$

in which

$$\alpha_l = \beta_l + \delta_l \tag{2.3.19}$$

Here α_l is the phase of the lth partial wave, β_l is the corresponding phase if $V(r) = 0$, and δ_l is the *phase shift* caused by the potential $V(r)$.

It now follows from Eq. 2.3.3 that as $r \to \infty$,

$$\psi_{\mathbf{k}}(\mathbf{r}) \to \frac{\exp(ikr)}{2kr}\left[\sum_{l=0}^{\infty} A_l \exp(i\alpha_l)P_l(\cos\theta)\right] + \frac{\exp(-ikr)}{2kr}\left[\sum_{l=0}^{\infty} A_l \exp(-i\alpha_l)P_l(\cos\theta)\right] \tag{2.3.20}$$

On the other hand, Eqs. 2.3.2 and 2.3.15 imply that as $r \to \infty$,

$$\psi_{\mathbf{k}}(\mathbf{r}) \to \frac{\exp(ikr)}{2kr}\left[\sum_{l=0}^{\infty} (2l+1)i^l \exp(i\beta_l)P_l(\cos\theta) + 2kf(\theta)\right] + \frac{\exp(-ikr)}{2kr}\left[\sum_{l=0}^{\infty} (2l+1)i^l \exp(-i\beta_l)P_l(\cos\theta)\right] \tag{2.3.21}$$

Comparing the preceding two relations, we then get

$$\sum_{l=0}^{\infty} A_l \exp(i\alpha_l)P_l(\cos\theta) = \sum_{l=0}^{\infty} (2l+1)i^l \exp(i\beta_l)P_l(\cos\theta) + 2kf(\theta) \tag{2.3.22}$$

and

$$\sum_{l=0}^{\infty} A_l \exp(-i\alpha_l)P_l(\cos\theta) = \sum_{l=0}^{\infty} (2l+1)i^l \exp(-i\beta_l)P_l(\cos\theta) \tag{2.3.23}$$

It then follows from Eq. 2.3.23 that

$$A_l = (2l+1)i^l \exp(i\delta_l) \tag{2.3.24}$$

and, hence, from Eq. 2.3.22 that

$$f(\theta) = \frac{1}{2ik}\sum_{l=0}^{\infty}(2l+1)[\exp(2i\delta_l)-1]P_l(\cos\theta) \tag{2.3.25}$$

Thus, the scattering amplitude is uniquely determined by the phase shift δ_l.

Phase shift

To determine the phase shift δ_l we must match the exterior solution (2.3.8) to the corresponding interior solution at $r = r_0$, and the interior solution depends on $V(r)$. Thus, for example, in the case of a hard-sphere potential,

$$V(r) = \begin{cases} \infty, & r < r_0 \\ 0, & r > r_0 \end{cases} \tag{2.3.26}$$

the interior solution of Eq. 2.3.4 is

$$R_l(r) = 0, \qquad r < r_0 \tag{2.3.27}$$

Hence,

$$R_l(r_0) = A_l[\cos(\delta_l)j_l(kr_0) - \sin(\delta_l)n_l(kr_0)] = 0 \tag{2.3.28}$$

so that

$$\tan\delta_l = \frac{j_l(kr_0)}{n_l(kr_0)} \tag{2.3.29}$$

In particular, it follows from Eq. 2.3.7 that for s waves,

$$\tan\delta_0 = -\tan kr_0 \tag{2.3.30}$$

and, hence,

$$\delta_0 = -kr_0 \tag{2.3.31}$$

In general, as $x \to 0$, $j_l(x) \sim x^l$ and $n_l(x) \sim x^{-(l+1)}$. Hence,

$$\delta_l \sim (kr_0)^{2l+1} \qquad \text{as } k \to 0 \tag{2.3.32}$$

Since $r_0 \simeq 5$ fm and $k \simeq 2$ Å^{-1} for thermal neutrons, it is evident that $kr_0 \simeq 10^{-4}$ and, hence, that

$$\frac{\delta_l}{\delta_0} \sim (kr_0)^{2l} \simeq 10^{-8l} \tag{2.3.33}$$

Thus, we conclude that for thermal neutrons the terms in Eq. 2.3.25 with $l \geq 1$ are completely negligible for all practical purposes and we need only retain the $l = 0$ (s-wave) term. Although this conclusion is based on the hard-sphere potential, it would also follow from a more realistic model for the optical potential $V(r)$, since it can be shown (Landau and Lifshitz, 1958) that for any short-range potential $\delta_l \sim k^{2l+1}$ as $k \to 0$.

Since $P_0(\cos\theta) = 1$, the s-wave scattering amplitude is therefore given by

$$f(\theta) = \frac{1}{2ik}[\exp(2i\delta_0) - 1] \tag{2.3.34}$$

or, equivalently,

$$f(\theta) = [k \cot \delta_0 - ik]^{-1} \tag{2.3.35}$$

Scattering length

Since δ_0 is an odd function of k, it follows that $k \cot \delta_0$ is an even function of k and, hence, has an expansion of the form (Blatt and Jackson, 1949; Bethe, 1949),

$$k \cot \delta_0 = -\frac{1}{a} + \frac{1}{2} r_e k^2 + O(k^4) \tag{2.3.36}$$

in which a is called the *scattering length* and r_e is the *effective range*. The parameters a and r_e are characteristic properties of the optical potential $V(r)$. Thus, for example, in the case of a hard-sphere potential it follows from Eq. 2.3.31 that

$$a = r_0, \qquad r_e = \tfrac{2}{3} r_0 \tag{2.3.37}$$

The scattering amplitude (2.3.35) has the corresponding expansion

$$f(\theta) = -a + ika^2 + O(k^2) \tag{2.3.38}$$

in which the k^2 term contains the effective range correction. When higher-order partial waves are included, the k^2 term also contains an anisotropic p-wave contribution.

In practice, the scattering length a is treated as a phenomenological constant to be determined by experiment. In fact, one of the main applications of neutron optics is in the accurate measurement of a. It is found that $|a|$ is typically of the order of 5 fm and, since $k \simeq 2$ Å^{-1} for thermal neutrons, we see that $|ka| \simeq 10^{-4}$.

Fermi pseudopotential

For most practical purposes we need only retain the first term in Eq. 2.3.38, in which case

$$f(\theta) = -a \tag{2.3.39}$$

On the other hand, Eq. 2.2.28 gives

$$\begin{aligned} f(\theta) &= \langle \mathbf{k}' | F | \mathbf{k} \rangle \\ &= \int d\mathbf{r}\, \phi_{\mathbf{k}'}(\mathbf{r})^* F(\mathbf{r}) \phi_{\mathbf{k}}(\mathbf{r}) \\ &= \frac{1}{(2\pi)^3} \int d\mathbf{r} \exp[i(\mathbf{k} - \mathbf{k}') \cdot \mathbf{r}] F(\mathbf{r}) \end{aligned} \tag{2.3.40}$$

where $F(\mathbf{r})$ is the scattering amplitude operator in the coordinate representation. Comparing Eqs. 2.3.39 and 2.3.40, we see that

$$F(\mathbf{r}) = -(2\pi)^3 a\,\delta(\mathbf{r}) \tag{2.3.41}$$

and hence, from Eq. 2.2.29, that the corresponding transition operator is given by

$$T(\mathbf{r}) = \frac{2\pi\hbar^2}{m} a\,\delta(\mathbf{r}) \tag{2.3.42}$$

This is the *Fermi pseudopotential* that was referred to at the end of Sec. 2.1. In his original expression Fermi (1936)[1] included a minus sign that is absent in Eq. 2.3.42, and that arose from the fact that, at that time, he defined the scattering length a simply as the low-energy limit of the scattering amplitude. Following the development of nuclear reactors, when it became possible to measure a directly by neutron-optical methods, it was found (Fermi and Zinn, 1946; Fermi and Marshall, 1946) that the scattering length of most nuclei was negative. Fermi then changed the sign convention to that of Eq. 2.3.39 in order that the scattering length would be a predominantly positive quantity (Fermi and Marshall, 1947). This sign convention is now universally employed in neutron scattering work.

The scattering length a is, in general, complex and we shall see in the next section that its imaginary part is usually of the same order of magnitude as the second term in Eq. 2.3.38. Thus, in situations where the imaginary part of $f(\theta)$ is important, the second term in Eq. 2.3.38 cannot be neglected and the Fermi pseudopotential, which is based on Eq. 2.3.39, is no longer valid. It is therefore evident that, while the thermal neutron scattering amplitude is for all practical purposes uniquely determined by the scattering length, it is not always adequately described by the Fermi pseudopotential.

Some people use the terms *scattering amplitude* and *scattering length* as if they were synonymous. This is bad practice since, quite apart from the conventional minus sign in Eq. 2.3.39, there are conceptual differences between these two quantities that become especially important when the second term in Eq. 2.3.38 is not negligible. For example, we shall show in the next section that the imaginary part of the scattering length determines the absorption cross section, while the imaginary part of the scattering amplitude determines the total collision cross section, which includes contributions from both scattering and absorption.

2.4. Scattering and absorption cross sections

Equation of continuity

We shall first demonstrate that the complex optical potential,

$$V(\mathbf{r}) = V'(\mathbf{r}) - iV''(\mathbf{r}) \tag{2.4.1}$$

[1] An English translation of this historic paper can be found in Segrè (1962).

does, as asserted in Sec. 2.2, describe both the scattering and absorption of neutrons. We consider the time-dependent Schrödinger equation,

$$i\hbar\dot{\Psi}(\mathbf{r}, t) = \left\{ -\frac{\hbar^2}{2m}\Delta + V(\mathbf{r}) \right\} \Psi(\mathbf{r}, t) \tag{2.4.2}$$

where the dot denotes differentiation with respect to t, and introduce the quantities

$$\begin{aligned} \rho &= |\Psi|^2 \\ \mathbf{J} &= \mathrm{Re}\{\Psi^* \mathbf{v} \Psi\} \\ s &= \frac{2}{\hbar} \rho V'' \end{aligned} \tag{2.4.3}$$

in which $\mathbf{v}$ is the group velocity of the neutron,

$$\mathbf{v} = \frac{1}{m}\mathbf{p} = \frac{\hbar}{im}\nabla \tag{2.4.4}$$

It then follows from (2.4.2) that

$$\dot{\rho} + \operatorname{div} \mathbf{J} = -s \tag{2.4.5}$$

We can interpret this as an equation of continuity in which ρ is[2] the neutron *number density*, $\mathbf{J}$ is the neutron *current density*, and s is the neutron *sink density*.

With this interpretation, the total number of neutrons in any volume V (not to be confused with the potential $V(\mathbf{r})$) is given by

$$N = \int_V \rho \, d\mathbf{r} \tag{2.4.6}$$

so that

$$\dot{N} = -\int_V (\operatorname{div} \mathbf{J} + s) \, d\mathbf{r} \tag{2.4.7}$$

According to Gauss' theorem,

$$\int_V \operatorname{div} \mathbf{J} \, d\mathbf{r} = \int_S \mathbf{J} \cdot \hat{\mathbf{n}} \, dS \tag{2.4.8}$$

where $\hat{\mathbf{n}}$ is a unit vector in the direction of the outward normal to the surface S that bounds V, and dS is an element of surface area. Hence,

$$\dot{N} = \int_S \mathbf{J} \cdot (-\hat{\mathbf{n}}) \, dS - \int_V s \, d\mathbf{r} \tag{2.4.9}$$

This equation states that the rate of increase in the number of neutrons in V is equal to the rate at which neutrons are flowing into V through S minus the rate at which neutrons are being absorbed in V.

[2] More precisely, ρ is the *average* neutron number density, etc.

When V represents all space, the first term in Eq. 2.4.9 vanishes and

$$\dot{N} = -\int s\, d\mathbf{r} \tag{2.4.10}$$

If the optical potential is real then $s = 0$ and $\dot{N} = 0$, so that N is constant in time and the total number of neutrons is conserved. In this case we can, if we wish, take $N = 1$, whereupon $\int \rho\, d\mathbf{r} = 1$ and ρ has its usual interpretation as a probability density. However, when the optical potential is complex, neutrons are not conserved and it is necessary to regard ρ as the neutron number density.

Differential scattering cross section

Since the scattering state $\psi_{\mathbf{k}}$ is a stationary state with energy $E_{\mathbf{k}}$, we get

$$\Psi_{\mathbf{k}}(\mathbf{r}, t) = \exp(-iE_{\mathbf{k}}t/\hbar)\psi_{\mathbf{k}}(\mathbf{r}) \tag{2.4.11}$$

in which $\psi_{\mathbf{k}}(\mathbf{r})$ satisfies the time-independent Schrödinger equation (2.3.1). In this case ρ and $\mathbf{J}$ are independent of t:

$$\rho = |\psi_{\mathbf{k}}(\mathbf{r})|^2, \qquad \mathbf{J} = \mathrm{Re}\{\psi_{\mathbf{k}}(\mathbf{r})^* \mathbf{v} \psi_{\mathbf{k}}(\mathbf{r})\} \tag{2.4.12}$$

As $r \to \infty$,

$$\psi_{\mathbf{k}}(\mathbf{r}) \to \sqrt{\rho_0}\left\{\exp(i\mathbf{k}\cdot\mathbf{r}) + f(\theta)\frac{\exp(ikr)}{r}\right\} \tag{2.4.13}$$

where the wave function is now normalized to represent an incident neutron beam with number density ρ_0.

In practice, the incident neutron beam is collimated such that the detector in Fig. 2.1 sees only the scattered wave. Consequently, the incident and scattered neutron current densities can be measured separately and there is no quantum-mechanical interference between them.

The incident neutron current density is

$$\mathbf{J} = \rho_0 \mathbf{v} \tag{2.4.14}$$

where $\mathbf{v} = \hbar\mathbf{k}/m$ is the incident neutron velocity. Let $\mathbf{v} = v\hat{\mathbf{k}}$ where $v = \hbar k/m$ and $\hat{\mathbf{k}}$ is a unit vector in the direction of $\mathbf{k}$. Then

$$\mathbf{J} = J\hat{\mathbf{k}} \tag{2.4.15}$$

where J is called the *incident flux* and is given by

$$J = \rho_0 v \tag{2.4.16}$$

The *scattered flux* J' is defined as the radial component of the neutron current density in the scattered wave. Thus,

$$J' = \mathbf{J}' \cdot \hat{\mathbf{r}} \tag{2.4.17}$$

where $\hat{\mathbf{r}}$ denotes a unit vector in the direction of $\mathbf{r}$. As $r \to \infty$,

$$J' = J\frac{|f(\theta)|^2}{r^2} \tag{2.4.18}$$

In the limit $r \to \infty$ the detector in Fig. 2.1 makes an infinitesimal solid angle with the nucleus that is given by

$$d\Omega = \frac{S}{r^2} \tag{2.4.19}$$

where S is the area of the face of the detector. The *differential scattering cross section* is denoted by $d\sigma$, and is defined as the average number of incident neutrons that are scattered into $d\Omega$ per unit time per unit incident flux. Thus,

$$d\sigma = \frac{J'S}{J} = |f(\theta)|^2\, d\Omega \tag{2.4.20}$$

and

$$\frac{d\sigma}{d\Omega} = |f(\theta)|^2 \tag{2.4.21}$$

Total scattering cross section

The *total scattering cross section* σ_s is defined as the average number of incident neutrons that are scattered per unit time per unit incident flux. Thus,

$$\sigma_s = \int d\sigma = \int_{4\pi} |f(\theta)|^2\, d\Omega \tag{2.4.22}$$

Since $d\Omega = \sin\theta\, d\theta\, d\phi$, and $f(\theta)$ is independent of ϕ for a central potential, we get

$$\sigma_s = 2\pi \int_0^\pi |f(\theta)|^2 \sin\theta\, d\theta \tag{2.4.23}$$

It then follows from the partial wave expansion, Eq. 2.3.25, and the orthonormality relation, Eq. 2.3.12, for the Legendre polynomials, that

$$\sigma_s = \frac{\pi}{k^2}\sum_{l=0}^{\infty}(2l+1)\,|\exp(2i\delta_l) - 1|^2 \tag{2.4.24}$$

Absorption cross section

The *absorption cross section* σ_a is defined as the average number of incident neutrons that are absorbed per unit time per unit incident flux. Thus,

$$\sigma_a = \frac{1}{J}\int_V s\, d\mathbf{r} \tag{2.4.25}$$

where V now denotes any volume that contains the nucleus. Since ρ is independent of t it follows that $\dot{N}=0$ and, hence, from Eq. 2.4.9 that

$$\sigma_a = \frac{1}{J}\int_S \mathbf{J}' \cdot (-\hat{\mathbf{n}})\, dS \tag{2.4.26}$$

where $\mathbf{J}'$ is the total neutron current density including both the incident and scattered waves.

To calculate σ_a it is convenient to take V to be a sphere of radius r, where $r \to \infty$. Then,

$$\sigma_a = -\frac{r^2}{J}\int_{4\pi} \mathbf{J}' \cdot \hat{\mathbf{r}}\, d\Omega \tag{2.4.27}$$

It then follows with the help of the partial wave expansion (2.3.25) that

$$\sigma_a = \frac{\pi}{k^2}\sum_{l=0}^{\infty} (2l+1)[1 - |\exp(2i\delta_l)|^2] \tag{2.4.28}$$

It will be noted that for a real potential δ_l is real and $\sigma_a = 0$.

Total collision cross section

The *total collision cross section* σ_t is defined as the average number of incident neutrons that are scattered or absorbed per unit time per unit incident flux. Thus,

$$\sigma_t = \sigma_s + \sigma_a \tag{2.4.29}$$

The phase shift is in general complex,

$$\delta_l = \delta_l' + i\delta_l'' \tag{2.4.30}$$

so that

$$\sigma_s = \frac{\pi}{k^2}\sum_{l=0}^{\infty} (2l+1)[1 + \exp(-4\delta_l'') - 2\exp(-2\delta_l'')\cos(2\delta_l')] \tag{2.4.31}$$

and

$$\sigma_a = \frac{\pi}{k^2}\sum_{l=0}^{\infty} (2l+1)[1 - \exp(-4\delta_l'')] \tag{2.4.32}$$

Hence,

$$\sigma_t = \frac{2\pi}{k^2}\sum_{l=0}^{\infty} (2l+1)[1 - \exp(-2\delta_l'')\cos(2\delta_l')] \tag{2.4.33}$$

An alternative expression for σ_t can be obtained from the partial wave expansion (2.3.25) by using the fact that $P_l(1) = 1$ for all l. One then finds that Eq. 2.4.33 becomes simply

$$\sigma_t = \frac{4\pi}{k}\,\mathrm{Im}[f(0)] \tag{2.4.34}$$

This is the well-known *optical theorem* (Feenberg, 1932; Newton, 1976) that relates the total collision cross section to the imaginary part of the forward scattering amplitude.

Thermal neutron cross sections

For thermal neutrons one can use the expansion (2.3.38) for the scattering amplitude. Since the scattering length is in general complex, i.e.,

$$a = a' - ia'' \tag{2.4.35}$$

it follows from Eq. 2.4.21 that

$$\begin{aligned}\frac{d\sigma}{d\Omega} &= (a'^2 + a''^2)[1 - 2ka'' + O(k^2)] \\ &= |a|^2[1 - 2ka'' + O(k^2)]\end{aligned} \tag{2.4.36}$$

and, hence, that

$$\sigma_s = 4\pi\, |a|^2\, [1 - 2ka'' + O(k^2)] \tag{2.4.37}$$

The optical theorem (2.4.34) gives

$$\sigma_t = \frac{4\pi}{k} a'' + 4\pi(a'^2 - a''^2) + O(k) \tag{2.4.38}$$

so that

$$\sigma_a = \frac{4\pi}{k} a''[1 - 2ka'' + O(k^2)] \tag{2.4.39}$$

In general, $|a| \simeq 5$ fm and $k \simeq 2$ Å^{-1} for thermal neutrons, so that $|ka| \simeq 10^{-4}$. Also, for most nuclides used in neutron scattering research, $\sigma_a \lesssim \sigma_s$ so that

$$\frac{a''}{a'} \lesssim ka' \simeq 10^{-4} \tag{2.4.40}$$

and, hence,

$$ka'' \lesssim (ka')^2 \simeq 10^{-8} \tag{2.4.41}$$

For a strongly absorbing nuclide, where $\sigma_a \simeq 10^4 \sigma_s$ say, we get $a'' \simeq a'$ so that $ka'' \simeq 10^{-4}$, which is still much less than unity. Thus, for all practical purposes,

$$\sigma_s = 4\pi\, |a|^2 \tag{2.4.42}$$

and

$$\sigma_a = \frac{4\pi}{k} a'' \tag{2.4.43}$$

For thermal neutrons σ_s is therefore independent of k, while σ_a is inversely proportional to k and, hence, to the incident neutron velocity v.

This "$1/v$ law" was first obtained by means of a simple heuristic argument (Amaldi et al., 1935). At that time it was not realized that a is complex and that the constant of proportionality in the $1/v$ law is determined by the imaginary part of a.

The expression (2.3.39) for the scattering amplitude gives σ_s correctly but not σ_t. The contribution of the scattering to σ_t comes from the second term in Eq. 2.3.38, which is neglected in Eq. 2.3.39. Since the Fermi pseudopotential is based on Eq. 2.3.39, it therefore violates the optical theorem (Lippmann and Schwinger, 1950; Sears, 1982a). This defect in the Fermi pseudopotential is not important in most applications, but it has serious consequences in the theory of the neutron refractive index, as we shall see later in Chap. 4.

Since $\sigma_a \geq 0$ by definition, it follows from Eq. 2.4.43 that $a'' \geq 0$. This is the reason for the conventional minus sign in Eq. 2.4.35. On the other hand, a' may be either positive or negative. If the optical potential $V(r)$ is repulsive then a' is positive. If $V(r)$ is attractive then a' is, again, usually positive; it is negative only if $V(r)$ is a weak potential in the sense that it has no bound states, or in the neighborhood of a size resonance. Since $V(r)$ is, in fact, a strong attractive potential we therefore expect that a' will be positive for most nuclides. This is what is found by experiment; there are only a few nuclides, such as ^{1}H, ^{7}Li, ^{48}Ti, ^{55}Mn, and a few others, for which a' is negative. Thus, the Fermi pseudopotential Eq. 2.3.42 is usually repulsive even though the underlying optical potential is attractive. This point has been discussed in detail by Peshkin and Ringo (1971).

These results are modified for a few heavy nuclides, such as ^{113}Cd, which have an (n, γ) resonance at thermal neutron energies. In this case the discussion of the scattering and absorption in terms of a complex optical potential, while still possible in principle, is very artificial because $V(r)$ is then strongly dependent on the incident neutron energy. In the neighborhood of an (n, γ) resonance, $f(\theta)$ is given by the Breit–Wigner formula (Breit and Wigner, 1936; Landau and Lifshitz, 1958; Shore, 1967; Lynn, 1968), and σ_s and σ_a are then strongly varying functions of k. For the time being we shall, for simplicity, ignore this situation. When necessary, the theory given below can easily be generalized to include resonance scattering and absorption.

2.5. Neutron scattering by a free atom

Two-body problem

Consider the collision of a neutron with a single atom in otherwise free space. Let the neutron have mass m and position $\mathbf{r}$ relative to axes fixed in space (the laboratory system), while the atom has mass M and position

R. Then the Hamiltonian is

$$H = \frac{1}{2M}\mathbf{P}^2 + \frac{1}{2m}\mathbf{p}^2 + V(\mathbf{r} - \mathbf{R}) \tag{2.5.1}$$

where $\mathbf{p} = m\dot{\mathbf{r}}$ is the momentum of the neutron, $\mathbf{P} = M\dot{\mathbf{R}}$ is the momentum of the atom, and $V(\mathbf{r} - \mathbf{R})$ is the complex optical potential that describes the interaction of the neutron with the nucleus of the atom.

With the definitions,

$$\begin{aligned} M_x &= M + m, & m_0 &= \frac{Mm}{M+m} \\ \mathbf{R}_x &= \frac{M\mathbf{R} + m\mathbf{r}}{M+m}, & \mathbf{r}_0 &= \mathbf{r} - \mathbf{R} \\ \mathbf{P}_x &= M_x\dot{\mathbf{R}}_x, & \mathbf{p}_0 &= m_0\dot{\mathbf{r}}_0 \end{aligned} \tag{2.5.2}$$

the preceding Hamiltonian can be expressed equivalently as

$$H = \frac{1}{2M_x}\mathbf{P}_x^2 + \frac{1}{2m_0}\mathbf{p}_0^2 + V(\mathbf{r}_0) \tag{2.5.3}$$

Thus, the two-body problem is dynamically equivalent to two one-body problems. The first particle (called the *extracule*) is located at the center of mass $\mathbf{R}_x$ and moves like a free particle of mass M_x. The second particle (called the *intracule*) has reduced mass m_0 and position $\mathbf{r}_0$ relative to the nucleus, and moves as if the latter were fixed in space. The Hamiltonian of the intracule alone is the same as Eq. 2.2.1, except that m is replaced by m_0. Thus, all the results obtained in the preceding three sections apply to the intracule when m is replaced by m_0.

Kinematics

Suppose that in the laboratory system the neutron has momentum $\mathbf{p} = \hbar\mathbf{k}$ and the atom has momentum $\mathbf{P} = \hbar\mathbf{K}$. If unprimed quantities refer to the initial state and primed quantities to the final state (see Fig. 2.2a) then momentum conservation requires that

$$\mathbf{k} + \mathbf{K} = \mathbf{k}' + \mathbf{K}' \tag{2.5.4}$$

The momentum transferred from the neutron to the atom in the collision is therefore given (in units of $\hbar$) by

$$\mathbf{q} = \mathbf{k} - \mathbf{k}' = \mathbf{K}' - \mathbf{K} \tag{2.5.5}$$

Similarly, energy conservation requires that

$$\varepsilon + \frac{\hbar K^2}{2M} = \varepsilon' + \frac{\hbar K'^2}{2M} \tag{2.5.6}$$

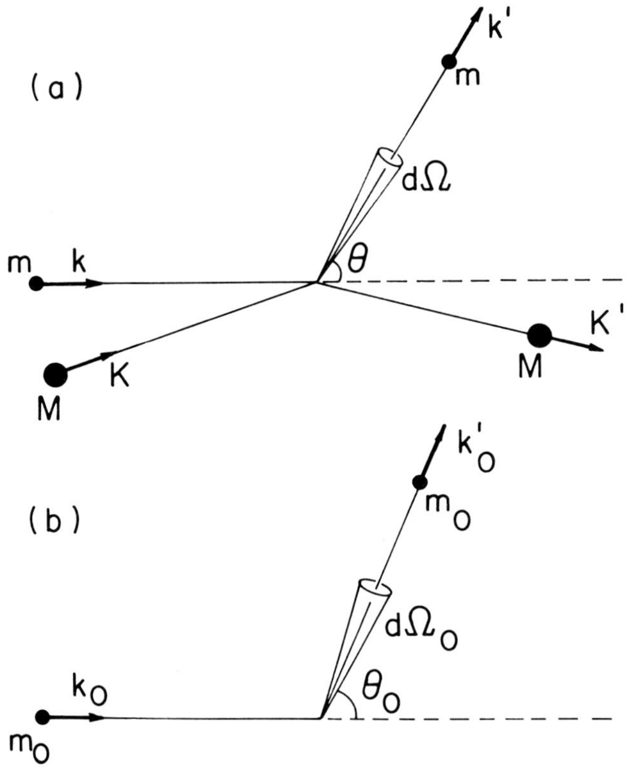

Fig. 2.2 Kinematics for the scattering of a neutron by a free atom as viewed in (a) the laboratory system and (b) the centre-of-mass system.

where $\hbar\varepsilon$ is the incident neutron energy,

$$\hbar\varepsilon = E_{\mathbf{k}} = \frac{(\hbar k)^2}{2m} \tag{2.5.7}$$

The energy transferred from the neutron to the atom in the collision is therefore given (in units of $\hbar$) by

$$\omega = \varepsilon - \varepsilon' = \frac{\hbar}{2M}(K'^2 - K^2) \tag{2.5.8}$$

Combining Eqs. 2.5.5 and 2.5.8, we get

$$\omega = \omega_r + \mathbf{q} \cdot \mathbf{V} \tag{2.5.9}$$

where $\mathbf{V} = \hbar\mathbf{K}/M$ is the initial velocity of the atom and $\hbar\omega_r = (\hbar q)^2/2M$ is the energy transfer when the atom is initially at rest. The quantity $\hbar\omega_r$ is usually referred to simply as the recoil energy. Equation (2.5.9) applies to any scattering process. In the case of X-ray scattering by electrons, for example, ω is the total frequency shift, ω_r is the Compton shift, and $\mathbf{q} \cdot \mathbf{V}$ is the Doppler shift.

Quantities in the center-of-mass system are indicated by the subscript 0 (see Fig. 2.2b), so that the momentum and energy transfers are given (in units of $\hbar$) by

$$\mathbf{q}_0 = \mathbf{k}_0 - \mathbf{k}_0', \qquad \omega_0 = \varepsilon_0 - \varepsilon_0' \tag{2.5.10}$$

Momentum is not conserved in the center-of-mass system (i.e., $\mathbf{q}_0 \neq 0$), but energy is conserved (i.e., $k_0' = k_0$ and $\omega_0 = 0$).

The transformation between the laboratory and the center-of-mass systems is provided by the relative velocity, $\dot{\mathbf{r}}_0 = \dot{\mathbf{r}} - \dot{\mathbf{R}}$, so that

$$\frac{\hbar \mathbf{k}_0}{m_0} = \frac{\hbar \mathbf{k}}{m} - \frac{\hbar \mathbf{K}}{M} \tag{2.5.11}$$

Hence,

$$\mathbf{k}_0 = \frac{A\mathbf{k} - \mathbf{K}}{A+1} \tag{2.5.12}$$

where $A = M/m$ is the atom/neutron mass ratio that is approximately equal to the mass number of the nucleus. Similarly,

$$\mathbf{k}_0' = \frac{A\mathbf{k}' - \mathbf{K}'}{A+1} \tag{2.5.13}$$

It then follows that

$$\mathbf{q}_0 = \mathbf{q} \tag{2.5.14}$$

so that the momentum transfer is the same in both systems. Also,

$$\omega_0 = \omega - \omega_r - \mathbf{q} \cdot \mathbf{V} = 0 \tag{2.5.15}$$

The relations (2.5.4) and (2.5.13) imply that

$$\mathbf{k}' = \mathbf{k}_0' + \frac{\mathbf{k} + \mathbf{K}}{A+1} \tag{2.5.16}$$

so that for given values of $\mathbf{k}$ and $\mathbf{K}$,

$$d\mathbf{k}' = d\mathbf{k}_0' \tag{2.5.17}$$

If $\mathbf{k}'$ has polar coordinates (k', θ, ϕ) as in Fig. 2.1, then

$$d\mathbf{k}' = k'^2\, dk' \sin\theta\, d\theta\, d\phi = \frac{mk'}{\hbar}\, d\Omega\, d\varepsilon' \tag{2.5.18}$$

where $d\Omega = \sin\theta\, d\theta\, d\phi$. Similarly,

$$d\mathbf{k}_0' = \frac{m_0 k_0'}{\hbar}\, d\Omega_0\, d\varepsilon_0' \tag{2.5.19}$$

so that from Eq. 2.5.17,

$$k'\, d\Omega\, d\varepsilon' = \left(\frac{A}{A+1}\right) k_0'\, d\Omega_0\, d\varepsilon_0' \tag{2.5.20}$$

Double differential scattering cross section

The *double differential scattering cross section* $d^2\sigma$ is defined as the average number of incident neutrons that are scattered into $d\mathbf{k}'$ per unit time per incident flux. It is therefore evident from Eq. 2.5.17 that

$$J\, d^2\sigma = J_0\, d^2\sigma_0 \tag{2.5.21}$$

where J is the incident flux in the laboratory system,

$$J = \rho \frac{\hbar k}{m} \tag{2.5.22}$$

in which ρ is the incident neutron number density (cf. Eq. 2.4.16), and J_0 is the incident flux in the center-of-mass system,

$$J_0 = \rho \frac{\hbar k_0}{m_0} \tag{2.5.23}$$

Let us define

$$\eta = \frac{J_0}{J} = \frac{m k_0}{m_0 k} = \left(\frac{A+1}{A}\right)\frac{k_0}{k} \tag{2.5.24}$$

Then

$$d^2\sigma = \eta\, d^2\sigma_0 = \left(\frac{A+1}{A}\right)\frac{k_0}{k}\, d^2\sigma_0 \tag{2.5.25}$$

and since $k_0' = k_0$ it then follows, with the help of Eq. 2.5.20, that

$$\frac{d^2\sigma}{d\Omega\, d\varepsilon'} = \left(\frac{A+1}{A}\right)^2 \frac{k'}{k} \frac{d^2\sigma_0}{d\Omega_0\, d\varepsilon_0'} \tag{2.5.26}$$

In general,

$$\frac{d\sigma_0}{d\Omega_0} = \int_0^\infty \frac{d^2\sigma_0}{d\Omega_0\, d\varepsilon_0'}\, d\varepsilon_0' \tag{2.5.27}$$

and for thermal neutrons (see Eq. 2.4.36),

$$\frac{d\sigma_0}{d\Omega_0} = |a|^2 \tag{2.5.28}$$

Since the scattering is elastic in the center-of-mass system, it follows that $d^2\sigma_0 = 0$ unless $\omega_0 = 0$. Hence,

$$\frac{d^2\sigma_0}{d\Omega_0\, d\varepsilon_0'} = |a|^2\, \delta(\omega_0) \tag{2.5.29}$$

so that from Eqs. 2.5.15 and 2.5.26,

$$\frac{d^2\sigma}{d\Omega\, d\varepsilon'} = \frac{k'}{k} |b|^2\, \delta(\omega - \omega_r - \mathbf{q}\cdot\mathbf{V}) \tag{2.5.30}$$

in which

$$b = b' - ib'' = \left(\frac{A+1}{A}\right)a \tag{2.5.31}$$

The quantity b is called the *bound scattering length,* for reasons that will become clear later. In case it seems paradoxical that the double differential scattering cross section for a free atom (Eq. 2.5.30) should depend on the "bound" scattering length, we should perhaps emphasize that Eq. 2.5.31 is simply a transformation from the center-of-mass to the laboratory system. The quantity b is as much a property of a free atom as a is and, in spite of its name, has nothing to do with chemical binding per se.

For an ideal gas of N atoms,

$$\frac{d^2\sigma}{d\Omega\, d\varepsilon'} = N\frac{k'}{k}|b|^2 S(\mathbf{q}, \omega) \tag{2.5.32}$$

where $S(\mathbf{q}, \omega)$ is called the *scattering function* and is given by

$$S(\mathbf{q}, \omega) = \langle \delta(\omega - \omega_r - \mathbf{q}\cdot\mathbf{V})\rangle \tag{2.5.33}$$

where the brackets denote an average over the initial velocity $\mathbf{V}$ of the scattering atom. Thus,

$$\langle\cdots\rangle = \int \cdots f(\mathbf{V})\, d\mathbf{V} \tag{2.5.34}$$

in which $f(\mathbf{V})$ is the normalized velocity distribution function so that

$$\int f(\mathbf{V})\, d\mathbf{V} = 1 \tag{2.5.35}$$

If the gas is in thermodynamic equilibrium at temperature T, then $f(\mathbf{V})$ is given by the Maxwell–Boltzmann distribution,

$$f(\mathbf{V}) = \left(\frac{M}{2\pi k_B T}\right)^{3/2} \exp\left(-\frac{MV^2}{2k_B T}\right) \tag{2.5.36}$$

where k_B is Boltzmann's constant, and

$$S(\mathbf{q}, \omega) = \left(\frac{M}{2\pi k_B T}\right)^{1/2} \frac{1}{q}\exp\left[-\frac{M}{2k_B T}\left(\frac{\omega - \omega_r}{q}\right)^2\right] \tag{2.5.37}$$

Thus, the scattering function of an ideal gas is a Gaussian centered at the recoil energy ω_r, and with a width proportional to q.

Total scattering cross section

The total scattering cross section is given by

$$\sigma_s = \iint d^2\sigma = \iint \eta\, d^2\sigma_0 \tag{2.5.38}$$

where η is given by Eq. 2.5.24, and can also be expressed as

$$\eta = \frac{v_0}{v} = \left(1 + \frac{V^2 - 2\mathbf{v} \cdot \mathbf{V}}{v^2}\right)^{1/2} \tag{2.5.39}$$

Here $\mathbf{v}$ is the incident neutron velocity in the laboratory system and $\mathbf{v}_0 = \mathbf{v} - \mathbf{V}$ is the incident neutron velocity in the center-of-mass system. since η is independent of $\mathbf{k}_0'$, it follows that

$$\sigma_s = \eta \iint d^2\sigma_0 = \eta\sigma_{s0} \tag{2.5.40}$$

where σ_{s0} is the total scattering cross section in the center-of-mass system and is given, according to Eq. 2.4.42, by

$$\sigma_{s0} = 4\pi\ |a|^2 \tag{2.5.41}$$

Thus,

$$\sigma_s = 4\pi\eta\ |a|^2 \tag{2.5.42}$$

If the atom is initially at rest, so that $\mathbf{V} = 0$, then $\eta = 1$ and $\sigma_s = \sigma_{s0}$. However, in general $\mathbf{V} \neq 0$ and $\sigma_s \neq \sigma_{s0}$.

For an ideal gas of N atoms,

$$\sigma_s = 4\pi N \langle \eta \rangle\ |a|^2 \tag{2.5.43}$$

If the gas is in thermodynamic equilibrium at temperature T, so that $f(\mathbf{V})$ is given by Eq. 2.5.36, then

$$\langle \eta \rangle = 1 + \frac{k_B T}{2AE_\mathbf{k}} \tag{2.5.44}$$

where $E_\mathbf{k}$ is, as before, the incident neutron energy in the laboratory system.

Absorption cross section

The average number of incident neutrons that are absorbed per unit time is given by

$$J\sigma_a = J_0\sigma_{a0} \tag{2.5.45}$$

Hence,

$$\sigma_a = \eta\sigma_{a0} \tag{2.5.46}$$

which is formally the same as Eq. 2.5.40. The absorption cross section in the center-of-mass system is given, according to Eq. 2.4.43, by

$$\sigma_{a0} = \frac{4\pi}{k_0} a'' \tag{2.5.47}$$

Hence, it follows from Eq. 2.5.24 that the absorption cross section is given in the laboratory system by

$$\sigma_a = \frac{4\pi}{k} b'' \tag{2.5.48}$$

where b'' is the imaginary part of the "bound" scattering length (Eq. 2.5.31), and k is the incident neutron wave vector in the laboratory system.

Finally, for an ideal gas of N atoms,

$$\sigma_a = \frac{4\pi N}{k} b'' \tag{2.5.49}$$

which, unlike σ_s, is independent of temperature.

2.6. Neutron scattering by a bound system of atoms

Internal versus external forces

We now consider the scattering of a neutron by a bound system of atoms. The first point that must be recognized is that there are two distinct kinds of binding forces. First, there are the internal interatomic forces that, depending on the nature of the atoms, may bind the atoms together to form molecules or crystals. Second, there are the external forces that constrain the atoms to remain within a fixed volume of space so that, for example, the system as a whole does not recoil to infinity following the collision, as the neutron does. In practice, the external constraints are imposed by rigidly mounting the sample on the specimen table of the neutron spectrometer. If the system is a gas or liquid, the external constraints are imposed by the walls of the container.

The effect of the internal interatomic forces will be considered in the next section. In the meantime we shall focus our attention on the behavior that is imposed on the scattering cross section by the presence of the external constraints. Although these constraints serve a number of purposes in practice, their role in the present context is largely a matter of convenience. We invoke the external constraints simply to ensure that the elastic scattering, which contains all the neutron-optical effects, will occur in the laboratory system rather than in the center-of-mass system.

Channel state

The channel Hamiltonian H_0, which was introduced in Sec. 2.1, is of the form

$$H_0 = h + H_s \tag{2.6.1}$$

where h is the Hamiltonian of the neutron,

$$h = \frac{1}{2m} \mathbf{p}^2 \tag{2.6.2}$$

and H_s is the Hamiltonian of the atomic system including the external constraints. As before (cf. Eq. 2.2.5),

$$h\,|\mathbf{k}\rangle = E_{\mathbf{k}}\,|\mathbf{k}\rangle \tag{2.6.3}$$

The effect of the external constraints is that the eigenstates of H_s are labeled by a discrete index α. More precisely, α represents a set of discrete quantum numbers, one for each degree of freedom of the system. To the extent that the atoms remain permanently in their ground electronic states, these degrees of freedom are then the positions of the atoms and their nuclear spins, if any. Thus,

$$H_s\,|\alpha\rangle = E_\alpha\,|\alpha\rangle \tag{2.6.4}$$

and the eigenstates $|\alpha\rangle$ form a complete orthonormal set so that

$$\langle \alpha \mid \alpha' \rangle = \delta_{\alpha\alpha'}, \qquad \sum_\alpha |\alpha\rangle\langle\alpha| = 1 \tag{2.6.5}$$

The essential point to note there is that, as a result of the assumed presence of the external constraints, α is a discrete rather than a continuous variable, and $\delta_{\alpha\alpha'}$ is the Kronecker delta rather than the Dirac delta function. This is important, because we shall later make use of the relation

$$\delta^2_{\alpha\alpha'} = \delta_{\alpha\alpha'} \tag{2.6.6}$$

which is a property of the Kronecker delta but not of the Dirac delta function.

The channel state Φ, which is defined by Eq. 2.1.2, is therefore the product state

$$\Phi = |\mathbf{k}\rangle\,|\alpha\rangle \equiv |\mathbf{k}\alpha\rangle \tag{2.6.7}$$

and its energy is

$$E = E_{\mathbf{k}} + E_\alpha \tag{2.6.8}$$

At this point we can, if we wish, include the neutron spin quantum number in the "lumped" index α.

Scattering state

The scattering state Ψ that evolves out of the channel state Φ when the interaction energy V is switched on adiabatically can be expressed as a superposition of product states,

$$\Psi = \sum_{\alpha'} \psi_{\alpha'\alpha}\,|\alpha'\rangle \tag{2.6.9}$$

in which, according to Eq. 2.6.5,

$$\psi_{\alpha'\alpha} = \langle \alpha' \mid \Psi \rangle \tag{2.6.10}$$

The quantity $\psi_{\alpha'\alpha}$ is a vector in the Hilbert space of the neutron and can be interpreted as the one-body scattering state corresponding to the transition $\alpha \to \alpha'$.

It follows from Eq. 2.1.10 that

$$\begin{aligned}\psi_{\alpha'\alpha} &= \langle \alpha' | (1 + GT) |\mathbf{k}\alpha\rangle \\ &= \left\{ \langle \alpha' \mid \alpha \rangle + \sum_{\alpha''} \langle \alpha' | G |\alpha''\rangle \langle \alpha'' | T |\alpha\rangle \right\} |\mathbf{k}\rangle \end{aligned} \quad (2.6.11)$$

According to Eq. 2.1.4,

$$G = (E_{\mathbf{k}} + E_\alpha + i\varepsilon - h - H_s)^{-1} \quad (2.6.12)$$

Hence,

$$\langle \alpha' | G |\alpha''\rangle = g'\, \delta_{\alpha'\alpha''} \quad (2.6.13)$$

where g' is the one-body Green function,

$$g' = (E_{\mathbf{k}'} + i\varepsilon - h)^{-1} \quad (2.6.14)$$

in which, by definition,

$$E_{\mathbf{k}'} = E_{\mathbf{k}} + E_\alpha - E_{\alpha'} \quad (2.6.15)$$

The quantity $\mathbf{k}'$ will be called the *scattered neutron wave vector,* and Eq. 2.6.15 expresses conservation of energy in the collision in which the neutron is scattered from the state $|\mathbf{k}\rangle$ to the state $|\mathbf{k}'\rangle$ while the system makes the transition $\alpha \to \alpha'$. It will be noted that Eq. 2.6.15 defines only the magnitude of $\mathbf{k}'$. Its direction will be defined later. Hence,

$$\psi_{\alpha'\alpha} = \{\delta_{\alpha'\alpha} + g'\langle \alpha' | T |\alpha\rangle\} |\mathbf{k}\rangle \quad (2.6.16)$$

This equation is analogous to Eq. 2.2.19 and the remaining development proceeds as before. Thus, the wave function of the state $\psi_{\alpha'\alpha}$ in the coordinate representation is

$$\begin{aligned}\psi_{\alpha'\alpha}(\mathbf{r}) &= \langle \mathbf{r} \mid \psi_{\alpha'\alpha}\rangle \\ &= \delta_{\alpha'\alpha}\langle \mathbf{r} \mid \mathbf{k}\rangle + \int d\mathbf{r}'\, \langle \mathbf{r} | g' |\mathbf{r}'\rangle \langle \mathbf{r}' | \, \langle \alpha' | \mathrm{T} |\alpha\rangle \, |\mathbf{k}\rangle \end{aligned} \quad (2.6.17)$$

As $r \to \infty$ the Green function is given, according to Eq. 2.2.24, by

$$\langle \mathbf{r} | g' |\mathbf{r}'\rangle \to -\frac{\sqrt{2\pi}\, m}{\hbar^2} \frac{\exp(ik'r)}{r} \langle \mathbf{k}' \mid \mathbf{r}'\rangle \quad (2.6.18)$$

where $\mathbf{k}' = k'\hat{\mathbf{r}}$. Hence,

$$\psi_{\alpha'\alpha}(\mathbf{r}) \to \delta_{\alpha'\alpha}\phi_{\mathbf{k}}(\mathbf{r}) - \frac{\sqrt{2\pi}\, m}{\hbar^2} \frac{\exp(ik'r)}{r} \langle \mathbf{k}'\alpha' | T |\mathbf{k}\alpha\rangle \quad (2.6.19)$$

where $\phi_{\mathbf{k}}(\mathbf{r})$ is given by Eq. 2.2.10 as before. Omitting the normalization factor $(2\pi)^{-3/2}$ we get finally that, as $r \to \infty$,

$$\psi_{\alpha'\alpha}(\mathbf{r}) \to \delta_{\alpha'\alpha} \exp(i\mathbf{k} \cdot \mathbf{r}) + \langle \mathbf{k}'\alpha' \mid F |\mathbf{k}\alpha\rangle \frac{\exp(ik'r)}{r} \quad (2.6.20)$$

in which F is again the scattering amplitude operator (Eq. 2.2.29),

$$F = -\frac{4\pi^2 m}{\hbar^2} T \tag{2.6.21}$$

Microscopic scattering cross section

The *microscopic differential scattering cross section* is denoted by $d\sigma_{\alpha'\alpha}$ and is defined as the average number of incident neutrons that are scattered into $d\Omega$ (see Fig. 2.1) as a result of the transition $\alpha \to \alpha'$ per unit time per unit incident flux. Thus, as in Eq. 2.4.20,

$$d\sigma_{\alpha'\alpha} = \frac{J'}{J} r^2 \, d\Omega \tag{2.6.22}$$

The incident flux is given by

$$J = \frac{\hbar k}{m} \tag{2.6.23}$$

and the scattered flux is given in the limit $r \to \infty$ by

$$J' = \frac{\hbar k'}{mr^2} |\langle \mathbf{k}'\alpha' | F | \mathbf{k}\alpha \rangle|^2 \tag{2.6.24}$$

Hence,

$$\frac{d\sigma_{\alpha'\alpha}}{d\Omega} = \frac{k'}{k} |\langle \mathbf{k}'\alpha' | F | \mathbf{k}\alpha \rangle|^2 \tag{2.6.25}$$

The *microscopic double differential scattering cross section* is denoted by $d^2\sigma_{\alpha'\alpha}$ and is defined as the average number of incident neutrons that are scattered into $d\mathbf{k}'$ (see Eq. 2.5.18) as a result of the transition $\alpha \to \alpha'$ per unit time per unit incident flux. In general,

$$\frac{d\sigma_{\alpha'\alpha}}{d\Omega} = \int_0^\infty \frac{d^2\sigma_{\alpha'\alpha}}{d\Omega \, d\varepsilon'} \, d\varepsilon' \tag{2.6.26}$$

where $\hbar\varepsilon' = E_{\mathbf{k}'}$ is the scattered neutron energy as before (see Sec. 2.5). Also, $d^2\sigma_{\alpha'\alpha} = 0$ unless the energy conservation condition (Eq. 2.6.15) is satisfied. With

$$\hbar\omega = E_{\mathbf{k}} - E_{\mathbf{k}'} \tag{2.6.27}$$

and

$$\hbar\omega_{\alpha'\alpha} = E_{\alpha'} - E_{\alpha} \tag{2.6.28}$$

this condition is $\omega = \omega_{\alpha'\alpha}$. Hence, we get

$$\begin{aligned} \frac{d^2\sigma_{\alpha'\alpha}}{d\Omega \, d\varepsilon'} &= \frac{d\sigma_{\alpha'\alpha}}{d\Omega} \delta(\omega - \omega_{\alpha'\alpha}) \\ &= \frac{k'}{k} |\langle \mathbf{k}'\alpha' | F | \mathbf{k}\alpha \rangle|^2 \, \delta(\omega - \omega_{\alpha'\alpha}) \end{aligned} \tag{2.6.29}$$

Macroscopic scattering cross section

In practice, the system is initially in a mixed (thermodynamic) state rather than in a pure state $|\alpha\rangle$, and the final state $|\alpha'\rangle$ is not measured. In this case, the relevant quantity for describing neutron scattering experiments is the *macroscopic double differential scattering cross section* $d^2\sigma$. This quantity is defined as the average number of incident neutrons that are scattered into $d\mathbf{k}'$ per unit time per unit incident flux when the system is in thermodynamic equilibrium at temperature T. Thus,

$$d^2\sigma = \sum_{\alpha'\alpha} P_\alpha \, d^2\sigma_{\alpha'\alpha} \tag{2.6.30}$$

where P_α is the normalized probability that the system is initially in state $|\alpha\rangle$,

$$\sum_\alpha P_\alpha = 1 \tag{2.6.31}$$

This quantity is the normalized Boltzmann factor,

$$P_\alpha = Z^{-1} \exp\left(\frac{-E_\alpha}{k_B T}\right) \tag{2.6.32}$$

in which the partition function is given by

$$Z = \sum_\alpha \exp\left(\frac{-E_\alpha}{k_B T}\right) \tag{2.6.33}$$

Hence, it follows from Eq. 2.6.29 that

$$\frac{d^2\sigma}{d\Omega\, d\varepsilon'} = \frac{k'}{k} \sum_{\alpha'\alpha} P_\alpha \, |\langle \mathbf{k}'\alpha'|\, F \,|\mathbf{k}\alpha\rangle|^2 \, \delta(\omega - \omega_{\alpha'\alpha}) \tag{2.6.34}$$

The delta function in this expression automatically ensures that the energy conservation condition (Eq. 2.6.15) is satisfied. Consequently, $\mathbf{k}$, α, $\mathbf{k}'$, and α' are all effectively independent variables, so that the factor k'/k in Eq. 2.6.34 can be taken outside the summation sign.

The total scattering cross section is given by

$$\sigma_s = \iint d^2\sigma = \sum_{\alpha'\alpha} P_\alpha \frac{k'}{k} \int_{4\pi} |\langle \mathbf{k}'\alpha'|\, F \,|\mathbf{k}\alpha\rangle|^2 \, d\Omega \tag{2.6.35}$$

Here, the factor k'/k must be put back inside the summation sign, since k and k' are once more restricted to values such that Eq. 2.6.15 is satisfied.

In Eq. 2.6.30 it is tacitly assumed that the energy that is transferred from the incident neutron beam to the system by inelastic scattering and absorption is conducted out of the system into a heat bath sufficiently quickly that the system remains in thermodynamic equilibrium. This assumption is certainly valid for the neutron fluxes that are available in practice. The only exception we are aware of is in neutron diffraction in solid ^{3}He at temperatures $T \simeq 1$ mK, in which the heating of the sample due to (n, p) reactions is a nonnegligible effect.

Optical theorem

The *macroscopic absorption cross section* σ_a is defined as the average number of incident neutrons that are absorbed per unit time per unit incident flux when the system is in thermodynamic equilibrium at temperature T. It is evident from Eq. 2.4.27 that

$$\sigma_a = \sum_{\alpha'\alpha} P_\alpha\left(-\frac{r^2}{J}\right)\int_{4\pi} \mathbf{J}'_{\alpha'\alpha} \cdot \hat{\mathbf{r}}\, d\Omega \tag{2.6.36}$$

in which $\mathbf{J}'_{\alpha'\alpha}$ is the neutron current density in the state $\psi_{\alpha'\alpha}$. In calculating $\mathbf{J}'_{\alpha'\alpha}$ one must include the contributions from both the incident and scattered waves in Eq. 2.6.20. One then finds that

$$\sigma_a = \frac{4\pi}{k}\sum_{\alpha} P_\alpha \operatorname{Im}[\langle \mathbf{k}\alpha|\, F\, |\mathbf{k}\alpha\rangle] - \sum_{\alpha'\alpha} P_\alpha \frac{k'}{k}\int_{4\pi} |\langle \mathbf{k}'\alpha'|\, F\, |\mathbf{k}\alpha\rangle|^2\, d\Omega \tag{2.6.37}$$

The first term in this expression is due to the interference between the incident and the scattered waves, while the second term is due to the scattered wave alone.

The total collision cross section can now be obtained with the help of Eq. 2.6.35 and we see that

$$\sigma_t = \sigma_s + \sigma_a = \frac{4\pi}{k}\sum_{\alpha} P_\alpha \operatorname{Im}[\langle \mathbf{k}\alpha|\, F\, |\mathbf{k}\alpha\rangle] \tag{2.6.38}$$

which is the generalized form of the optical theorem (Eq. 2.4.34).

Finally, it is worth noting that the expressions (2.6.34) for the double differential scattering cross section and (2.6.38) for the total collision cross section are formally exact within the framework of nonrelativistic quantum mechanics. Thus, these expressions include implicitly through the scattering amplitude operator F all effects of neutron absorption and multiple scattering within the system and, through the states $|\alpha\rangle$ and energies E_α, all effects of the interatomic forces on the motion of the atoms in the system.

2.7. Neutron scattering by a system of bound atoms

In the preceding section we considered neutron scattering by a system of atoms that is constrained by external forces to remain within some fixed volume of space. The system could, in fact, be an ideal gas. In the present section we shall consider the effect of the internal interatomic forces that were ignored in Sec. 2.6. In other words, in the preceding section it was the system as a whole that was bound, but in the present section we consider atoms that may themselves be bound together by interatomic forces.

To calculate the scattering amplitude operator F, or the equivalent transition operator T (see Eq. 2.6.21), we first need to know the one-body transition operator T_i that was introduced in Sec. 2.1. This latter quantity cannot be dealt with exactly, so that it is necessary at this point to make some suitable approximation.

Impulse approximation

The approximation we shall make was originally introduced by Fermi (1936) and was later called the impulse approximation (Chew, 1950). Fermi introduced this approximation entirely on the basis of physical intuition. The T-matrix theory described in Sec. 2.1 was later developed by Lippmann and Schwinger (1950) primarily to put the Fermi theory on a more rigorous footing.

The one-body transition operator T_i satisfies Eq. 2.1.16, the formal solution of which is

$$T = V(1 - GV)^{-1} \tag{2.7.1}$$

in which we now omit the subscript i. Here $V = V(\mathbf{r})$ is the complex optical potential that represents the interaction between the neutron and the nucleus, and $\mathbf{r}$ is the position of the neutron relative to the nucleus. In general, $V(\mathbf{r})$ can be thought of as a potential well with a depth of about 40 MeV and a radius that is typically 7 fm. In contrast, the interatomic potential that binds an atom in a molecule or a crystal has a magnitude of about 1 eV and a range of about 1 Å, i.e., 10^5 fm.

It is evident from Eq. 2.7.1 that T is likewise appreciably different from zero only if $r \lesssim 7$ fm. During the instant of collision, when $T \neq 0$, the nuclear force that the neutron exerts on the scattering nucleus is therefore many orders of magnitude larger than the interatomic forces. As a result we can, to a good approximation, neglect the interatomic forces during the instant of collision by putting T equal to the transition operator for a free atom. This is the impulse approximation (Lippmann and Schwinger, 1950; Chew, 1950; Goldberger and Watson, 1964).

The transition operator for a fixed atom is given by Eq. 2.3.42, so that for a free atom

$$T = \frac{2\pi\hbar^2}{m_0} a\, \delta(\mathbf{r}) \tag{2.7.2}$$

where m_0 is the reduced mass (Eq. 2.5.2). Alternatively,

$$T = \frac{2\pi\hbar^2}{m} b\, \delta(\mathbf{r}) \tag{2.7.3}$$

where (see Eq. 2.5.31),

$$b = \frac{m}{m_0} a = \left(\frac{A+1}{A}\right) a \tag{2.7.4}$$

in which $A = M/m$ is the atom/neutron mass ratio as before. The impulse approximation is therefore implicit in the use of the Fermi pseudopotential (Eq. 2.7.3) to describe the scattering of thermal neutrons by chemically bound atoms. It is clear that the Fermi pseudopotential is valid, not because the atoms are assumed to be bound, but in spite of the fact that they may be bound.

Numerous authors have investigated the accuracy of the impulse approximation and have calculated corrections to it (Breit, 1947; Breit and Zilsel, 1947; Zilsel et al., 1947; Lippmann, 1950; Plummer and Summerfield, 1963; Summerfield, 1964; Trammell and Chalk, 1966; Dietze and Nowak, 1981; Nowak, 1982a; Adli and Summerfield, 1984). They find quite generally that the error in the scattering cross section due to the impulse approximation is always $\lesssim 0.3$ percent, which is negligible for most practical purposes.

It will be noted that in the impulse approximation the effect of the interatomic forces is neglected only in T and F but not in the energy eigenstates $|\alpha\rangle$ and eigenvalues E_α. Thus, for example, the double differential scattering cross section Eq. 2.6.34 is not independent of the interatomic forces in the impulse approximation. In the limit of large momentum transfer the effect of the interatomic forces on the final states $|\alpha'\rangle$ in Eq. 2.6.34 is, however, negligible. The approximation of neglecting these final-state interactions is often included in the term *impulse approximation* (Hohenberg and Platzman, 1966; Sears, 1969, 1984b). However, in the present book we shall always use the term impulse approximation in the former, less restricted sense.

Neutron scattering by a chemically bound atom

As a simple application of the impulse approximation, let us consider the scattering of a neutron by a single atom that is bound by a potential similar to the interatomic potential in a molecule or a crystal. If the atom is initially in the ground state $|0\rangle$ then $P_\alpha = \delta_{\alpha 0}$ and Eq. 2.6.34 becomes

$$\frac{d^2\sigma}{d\Omega\, d\varepsilon'} = \frac{k'}{k} \sum_{\alpha'} |\langle \mathbf{k}'\alpha'| F |\mathbf{k}0\rangle|^2 \,\delta(\omega - \omega_{\alpha'0}) \tag{2.7.5}$$

If the incident neutron energy is less than the energy separation of the ground and first excited states, the preceding expression reduces to

$$\frac{d^2\sigma}{d\Omega\, d\varepsilon'} = \frac{d\sigma}{d\Omega}\,\delta(\omega) \tag{2.7.6}$$

where

$$\frac{d\sigma}{d\Omega} = |\langle \mathbf{k}'0| F |\mathbf{k}0\rangle|^2 \tag{2.7.7}$$

and we have used the fact that $k' = k$, since the scattering is elastic.

In the impulse approximation,

$$T = \frac{2\pi\hbar^2}{m} b\,\delta(\mathbf{r} - \mathbf{R}) \tag{2.7.8}$$

where **r** now denotes the position of the neutron in the laboratory system and **R** the position of the atom. Hence,

$$\begin{aligned}\langle \mathbf{k}'|\,T\,|\mathbf{k}\rangle &= \int \phi_{\mathbf{k}'}(\mathbf{r})^* T(\mathbf{r})\phi_{\mathbf{k}}(\mathbf{r})\,d\mathbf{r}\\ &= \frac{2\pi\hbar^2}{m} b\,\phi_{\mathbf{k}'}(\mathbf{R})^*\phi_{\mathbf{k}}(\mathbf{R}) \qquad (2.7.9)\\ &= \frac{\hbar^2}{4\pi^2 m} b\exp(i\mathbf{q}\cdot\mathbf{R})\end{aligned}$$

where $\mathbf{q}=\mathbf{k}-\mathbf{k}'$ as before (see Eq. 2.5.5). It then follows from Eq. 2.6.21 that

$$\langle \mathbf{k}'|\,F\,|\mathbf{k}\rangle = -b\exp(i\mathbf{q}\cdot\mathbf{R}) \qquad (2.7.10)$$

and, hence, that

$$\langle \mathbf{k}'0|\,F\,|\mathbf{k}0\rangle = -bg(\mathbf{q}) \qquad (2.7.11)$$

where $g(\mathbf{q})$ is the *ground-state form factor,*

$$g(\mathbf{q}) = \langle 0|\exp(i\mathbf{q}\cdot\mathbf{R})\,|0\rangle \qquad (2.7.12)$$

The differential scattering cross section is, therefore,

$$\frac{d\sigma}{d\Omega} = |b|^2\,|g(\mathbf{q})|^2 \qquad (2.7.13)$$

and the total scattering cross section is

$$\sigma_s = \int d\sigma = |b|^2 \int_{4\pi} |g(\mathbf{q})|^2\,d\Omega \qquad (2.7.14)$$

Since $k' = k$ we get

$$q = 2k\sin\frac{\theta}{2} \qquad (2.7.15)$$

where θ is the angle of scattering (Fig. 2.1). Hence,

$$q\,dq = k^2\sin\theta\,d\theta \qquad (2.7.16)$$

Also, $d\Omega = \sin\theta\,d\theta\,d\phi$, so that if $g(\mathbf{q})$ is isotropic,

$$\sigma_s = \frac{2\pi}{k^2}|b|^2\int_0^{2k} |g(q)|^2\,q\,dq \qquad (2.7.17)$$

Strong versus rigid binding

In general, the ground-state form factor $g(q)$ decreases monotonically from the value 1 at $q=0$ to the value 0 at $q=\infty$. Let $g(R_0^{-1})=\frac{1}{2}$, where R_0 is a characteristic length that can be called the *radius of the bound state*. The atom is said to be *strongly bound* (Fermi, 1936) if, for the q

values of interest,

$$7\text{ fm} \ll R_0 \ll q^{-1} \tag{2.7.18}$$

The lower limit ensures that the impulse approximation is valid, while the upper limit ensures that $g(q) = 1$ for the q values of interest. For such a strongly bound atom, therefore,

$$\frac{d\sigma}{d\Omega} = |b|^2 \tag{2.7.19}$$

and

$$\sigma_s = 4\pi\,|b|^2 \tag{2.7.20}$$

For a free atom that is initially at rest, σ_s is given according to Eq. 2.5.42 by

$$\sigma_s = 4\pi\,|a|^2 \tag{2.7.21}$$

Hence, it follows from Eq. 2.7.4 that (Fermi, 1936)

$$\sigma_s(\text{bound}) = \left(\frac{A+1}{A}\right)^2 \sigma_s(\text{free}) \tag{2.7.22}$$

The scattering cross section of a strongly bound atom is, therefore, larger than that of a free atom by a factor $[(A+1)/A]^2$. In the case of hydrogen, where $A \simeq 1$, this factor is 4, which is a very large effect.

The quantity b can be interpreted as the scattering length of a strongly bound atom. It is not, however, the scattering length of a rigidly bound atom. If the atom is rigidly bound to some fixed position in space, then the binding forces are not negligible during the instant of collision and the impulse approximation is not valid.

To calculate the scattering length b_r of a rigidly bound atom we note first that, for a given optical potential $V(\mathbf{r})$, the free-atom scattering length a is some definite function of the reduced mass m_0, so that we can write

$$a = a(m_0) \tag{2.7.23}$$

A rigidly bound atom is one for which the atomic mass M is effectively infinite. Thus,

$$b_r = \lim_{M\to\infty} a(m_0) = a(m) \tag{2.7.24}$$

The explicit value of b_r depends on $V(\mathbf{r})$. For example, if $V(\mathbf{r})$ is represented by a hard-sphere potential then, according to Eq. 2.3.37, $a(m_0) = r_0$, which is independent of m_0, so that $b_r = a$. In this case, the scattering length of a rigidly bound atom is the same as when it is free. However, for a realistic model of $V(\mathbf{r})$ there is, in general, no simple analytic expression for $a(m_0)$ and, hence, no simple relation between b_r and a.

Neutron absorption by a chemically bound atom

With $P_\alpha = \delta_{\alpha 0}$ as before, the optical theorem Eq. 2.6.38 gives

$$\sigma_t = \frac{4\pi}{k} \operatorname{Im}[\langle \mathbf{k}0 | F | \mathbf{k}0 \rangle] \tag{2.7.25}$$

With $b = b' - ib''$, it then follows from Eq. 2.7.11 that

$$\langle \mathbf{k}0 | F | \mathbf{k}0 \rangle = -b = -b' + ib'' \tag{2.7.26}$$

so that

$$\sigma_t = \frac{4\pi}{k} b'' \tag{2.7.27}$$

At the end of Sec. 2.4 it was pointed out that when σ_t is calculated from the optical theorem the scattering contribution comes from a higher order term that is neglected in the Fermi pseudopotential. Thus, Eq. 2.7.27 is, in fact, simply the absorption cross section,

$$\sigma_a = \frac{4\pi}{k} b'' \tag{2.7.28}$$

and this is the same as the expression (2.5.48) for a free atom. Thus, as first pointed out by Lamb (1937),

$$\sigma_a(\text{bound}) = \sigma_a(\text{free}) \tag{2.7.29}$$

There is an important difference between Eqs. 2.7.22 and 2.7.29 quite apart from the factor $[(A+1)/A]^2$. The relation (2.7.22) is valid only if the bound atom is strongly bound and the free atom is initially at rest. In contrast, Eq. 2.7.29 is true quite generally for any chemically bound atom and any free atom. Both relations are, however, restricted by the assumption at the end of Sec. 2.4 that the incident-neutron energy is not near an (n, γ) resonance. The effect of Doppler and chemical binding effects on neutron resonances has been discussed by a number of authors (Bethe and Placzek, 1937; Lamb, 1939; Trammell, 1962; Word and Trammell, 1981; Word and Werner, 1982) but will not concern us here.

Spin incoherence

Up to this point we have ignored the fact that the neutron has spin $\mathbf{s}$ and the nucleus has spin $\mathbf{I}$. If $I \neq 0$, the optical potential $V(\mathbf{r})$, and hence the bound scattering length b will in general be spin dependent (Schwinger and Teller, 1937). Since $s = \frac{1}{2}$, the most general rotationally invariant expression for b is

$$b = b_c + 2[I(I+1)]^{-1/2} b_i \mathbf{s} \cdot \mathbf{I} \tag{2.7.30}$$

in which the coefficients b_c and b_i are called the *bound coherent and incoherent scattering lengths* for reasons that will become clear later.

The total spin $\mathbf{J} = \mathbf{I} + \mathbf{s}$ is a constant of the motion with eigenvalues

$J = I \pm \frac{1}{2}$. Hence, the bound scattering lengths for given values of J are

$$\begin{aligned} b_+ &= b_c + [I/(I+1)]^{1/2} b_i \\ b_- &= b_c - [(I+1)/I]^{1/2} b_i \end{aligned} \tag{2.7.31}$$

Conversely,

$$\begin{aligned} b_c &= g_+ b_+ + g_- b_- \\ b_i &= (g_+ g_-)^{1/2}(b_+ - b_-) \end{aligned} \tag{2.7.32}$$

in which $g_\pm$ are the statistical weight factors,

$$g_+ = \frac{I+1}{2I+1}, \qquad g_- = \frac{I}{2I+1} \tag{2.7.33}$$

with the normalization property

$$g_+ + g_- = 1 \tag{2.7.34}$$

Let us again consider the scattering of a neutron by a single chemically bound atom. When the spin dependence of the scattering length is taken into account, the relation corresponding to Eq. 2.7.7 is

$$\frac{d\sigma}{d\Omega} = \sum_{M'm'Mm} P_{Mm} \, |\langle \mathbf{k}'0M'm'| \, F \, |\mathbf{k}0Mm\rangle|^2 \tag{2.7.35}$$

where M and m now denote the eigenvalues of I_z and s_z, respectively. In the impulse approximation this becomes

$$\frac{d\sigma}{d\Omega} = |g(\mathbf{q})|^2 \sum_{M'm'Mm} P_{Mm} \, |\langle M'm'| \, b \, |Mm\rangle|^2 \tag{2.7.36}$$

where $g(\mathbf{q})$ is the ground-state form factor as before. It then follows with the help of the completeness relation

$$\sum_{M'm'} |M'm'\rangle\langle M'm'| = 1 \tag{2.7.37}$$

that

$$\frac{d\sigma}{d\Omega} = |g(\mathbf{q})|^2 \, \langle |b|^2 \rangle \tag{2.7.38}$$

where the brackets denote an average over the neutron and nuclear spin states,

$$\langle \cdots \rangle = \sum_{Mm} P_{Mm} \langle Mm| \cdots |Mm\rangle \tag{2.7.39}$$

For a strongly bound atom,

$$\frac{d\sigma}{d\Omega} = \langle |b|^2 \rangle \tag{2.7.40}$$

and

$$\sigma_s = 4\pi \, \langle |b|^2 \rangle \tag{2.7.41}$$

Similarly, it follows from the optical theorem (Eq. 2.6.38) that

$$\sigma_a = \frac{4\pi}{k} \langle b'' \rangle \tag{2.7.42}$$

If the nucleus is unpolarized,

$$\langle I_\mu \rangle = 0, \qquad \langle I_\mu I_\nu \rangle = \tfrac{1}{3} I(I+1)\delta_{\mu\nu} \tag{2.7.43}$$

where μ, $\nu = x, y, z$. If the neutron is unpolarized,

$$\langle s_\mu \rangle = 0, \qquad \langle s_\mu s_\nu \rangle = \tfrac{1}{4}\delta_{\mu\nu} \tag{2.7.44}$$

In either case it follows that

$$\langle b \rangle = b_c \tag{2.7.45}$$

and

$$\sigma_s = \sigma_c + \sigma_i \tag{2.7.46}$$

where σ_c is the *bound coherent scattering cross section,*

$$\sigma_c = 4\pi\, |b_c|^2 = 4\pi\, |g_+ b_+ + g_- b_-|^2 \tag{2.7.47}$$

and σ_i is the *bound incoherent scattering cross section,*

$$\sigma_i = 4\pi\, |b_i|^2 = 4\pi g_+ g_-\, |b_+ - b_-|^2 \tag{2.7.48}$$

Alternatively,

$$\sigma_s = g_+ \sigma_{s+} + g_- \sigma_{s-} \tag{2.7.49}$$

where

$$\sigma_{s\pm} = 4\pi\, |b_\pm|^2 \tag{2.7.50}$$

The absorption cross section is given by

$$\sigma_a = \frac{4\pi}{k} b_c'' = g_+ \sigma_{a+} + g_- \sigma_{a-} \tag{2.7.51}$$

where

$$\sigma_{a\pm} = \frac{4\pi}{k} b_\pm'' \tag{2.7.52}$$

We saw in Sec. 2.4 that the imaginary part of the scattering length is of the same order of magnitude as the real part only if the absorption is large ($\sigma_a \simeq 10^4 \sigma_s$). When this happens, the absorption is usually large for only one spin state of the compound nucleus. For nuclides such as ^{113}Cd, ^{155}Gd, or ^{157}Gd, where the absorption is large only in the $J = I + \frac{1}{2}$ state, we have $b_-'' \simeq 0$, so that from Eq. 2.7.32,

$$b_c'' = g_+ b_+'', \qquad b_i'' = (g_+ g_-)^{1/2} b_+'' \tag{2.7.53}$$

and, hence,

$$b_c'' = [(I+1)/I]^{1/2} b_i'' = \frac{k}{4\pi} \sigma_a \tag{2.7.54}$$

In this case the measured value of σ_a determines both b_c'' and b_i''.

On the other hand, for nuclides such as ^{3}He or ^{10}B, for which the absorption is large only in the $J = I - \frac{1}{2}$ state, we have $b''_+ \simeq 0$ so that from Eq. 2.7.32,

$$b''_c = g_- b''_-, \qquad b''_i = -(g_+ g_-)^{1/2} b''_- \tag{2.7.55}$$

and, hence,

$$b''_c = -[I/(I+1)]^{1/2} b''_i = \frac{k}{4\pi}\sigma_a \tag{2.7.56}$$

Again, the measured value of σ_a determines both b''_c and b''_i. Note that in this case b''_i is negative, whereas in the previous case it was positive. On the other hand, b''_c is always positive.

For those few nuclides, such as ^{6}Li, where the absorption is large for both the $J = I \pm \frac{1}{2}$ states, we get

$$\begin{aligned} b''_c &= \frac{k}{4\pi}\sigma_a = \frac{k}{4\pi}(g_+\sigma_{a+} + g_-\sigma_{a-}) \\ b''_i &= \frac{k}{4\pi}(g_+ g_-)^{1/2}(\sigma_{a+} - \sigma_{a-}) \end{aligned} \tag{2.7.57}$$

In this case the measured value of σ_a determines only b''_c, and σ_{a+} and σ_{a-} must be measured separately in order to obtain b''_i. Here b''_i is positive or negative depending on the relative values of σ_{a+} and σ_{a-}.

Isotope incoherence

The coefficients b_c and b_i in Eq. 2.7.30 depend on the particular isotope under consideration. We shall see later that, as first pointed out by Wick (1937), the isotope dependence of the scattering length provides an additional source of incoherence in the scattering of a neutron by a mixture of isotopes.

If the brackets $\langle \cdots \rangle$ are now taken to denote an average over both the spin and isotope distributions then the expressions (2.7.41) for σ_s, (2.7.42) for σ_a, and (2.7.45) for b_c also apply to a mixture of isotopes. Thus, if C_l is the fraction of isotopes of type l, so that

$$\sum_l C_l = 1 \tag{2.7.58}$$

then

$$\sigma_s = \sum_l C_l \sigma_{sl} \qquad \sigma_a = \sum_l C_l \sigma_{al} \qquad b_c = \sum_l C_l b_{cl} \tag{2.7.59}$$

If we define

$$\sigma_c = 4\pi\,|b_c|^2 \tag{2.7.60}$$

as before, then we get

$$\sigma_s = \sigma_c + \sigma_i \tag{2.7.61}$$

where

$$\sigma_i = \sigma_i(\text{spin}) + \sigma_i(\text{isotope}) \tag{2.7.62}$$

in which the contribution from spin incoherence is given by

$$\sigma_i(\text{spin}) = \sum_l C_l \sigma_{il} = 4\pi \sum_l C_l \, |b_{il}|^2 \tag{2.7.63}$$

and that from isotope incoherence is given by

$$\sigma_i(\text{isotope}) = 4\pi \sum_{l<l'} C_l C_{l'} \, |b_{cl} - b_{cl'}|^2 \tag{2.7.64}$$

For more information on thermal neutron scattering lengths and cross sections, and for a tabulation of measured values of these quantities for the elements and their isotopes, we refer the reader to the "barn book" (Mughabghab et al., 1981, Mughabghab, 1984) and to the recent review article by Sears (1986b).

2.8. Coherent scattering

Coherence

The concept of coherence is so ubiquitous in scattering theory that it is useful to have terms that distinguish the many different kinds of coherent effects that can occur. The most important distinction is between what Lax (1951) calls *absolute* and *relative* coherence.

The scattering is said to be *coherent in the absolute sense* if the scattered waves interfere with the incident wave. Since it is this kind of coherence that is responsible for all optical phenomena, we shall be concerned almost entirely with absolute coherence. In general, waves can interfere only if they have the same frequency, so that the absolute coherent scattering is necessarily elastic. We shall follow the usual terminology in optics and refer to all scattering that is not coherent in the absolute sense as *diffuse scattering*.

The scattering is said to be *coherent in the relative sense* if the waves scattered by different atoms interfere with each other, though not necessarily with the incident wave. This kind of coherent scattering is therefore not necessarily elastic. The coherent elastic scattering is, in fact, coherent in the absolute sense but the coherent inelastic scattering is not. In books on the application of thermal neutron scattering to condensed-matter physics, such as those by Squires (1978) or Lovesey (1984), the word *coherent* is always used in the relative sense.

In addition to absolute and relative coherence, one can also distinguish between *spatial* and *temporal* coherence, *longitudinal* and *transverse* coherence, *second-order* and *fourth-order* coherence, etc. (Mandel and Wolf, 1965). These latter types of coherence will be discussed later in Chap. 7 but can be ignored for the time being.

Fluctuations

In general, any dynamical variable X can be expressed as

$$X = \langle X \rangle + \delta X \tag{2.8.1}$$

where $\langle X \rangle$ is the equilibrium value of X,

$$\langle X \rangle = \sum_{\alpha} P_\alpha \langle \alpha | X | \alpha \rangle \tag{2.8.2}$$

and $\delta X \equiv X - \langle X \rangle$ is the fluctuation of X about its equilibrium value and has the property

$$\langle \delta X \rangle = 0 \tag{2.8.3}$$

It then follows that

$$\langle XY \rangle = \langle X \rangle \langle Y \rangle + \langle \delta X\, \delta Y \rangle \tag{2.8.4}$$

In particular, with $Y = X^*$,

$$(\Delta X)^2 \equiv \langle |\delta X|^2 \rangle = \langle |X|^2 \rangle - |\langle X \rangle|^2 \tag{2.8.5}$$

where ΔX is the root-mean-square fluctuation in X.

In what follows, we shall often make use of these results. For example, the bound coherent and incoherent scattering lengths, which were introduced in the last section, can be expressed as

$$b_c = \langle b \rangle, \qquad b_i = \Delta b \tag{2.8.6}$$

Coherent versus diffuse scattering

We now return to where we left off at the end of Sec. 2.6 and begin by expressing the scattering amplitude operator in the form

$$F = f + \delta F \tag{2.8.7}$$

where $f = \langle F \rangle$. Then,

$$\langle \mathbf{k}'\alpha' | F | \mathbf{k}\alpha \rangle = \langle \mathbf{k}' | f | \mathbf{k} \rangle\, \delta_{\alpha'\alpha} + \langle \mathbf{k}'\alpha' | \delta F | \mathbf{k}\alpha \rangle \tag{2.8.8}$$

It now follows with the help of Eq. 2.6.6 that the double differential scattering cross section (Eq. 2.6.34) becomes the sum of two terms,

$$\frac{d^2\sigma}{d\Omega\, d\varepsilon'} = \frac{d^2\sigma_c}{d\Omega\, d\varepsilon'} + \frac{d^2\sigma_d}{d\Omega\, d\varepsilon'} \tag{2.8.9}$$

where

$$\frac{d^2\sigma_c}{d\Omega\, d\varepsilon'} = \frac{d\sigma_c}{d\Omega}\, \delta(\omega) \tag{2.8.10}$$

in which

$$\frac{d\sigma_c}{d\Omega} = |\langle \mathbf{k}' | f | \mathbf{k} \rangle|^2 \tag{2.8.11}$$

and

$$\frac{d^2\sigma_d}{d\Omega\, d\varepsilon'} = \frac{k'}{k} \sum_{\alpha'\alpha} P_\alpha \,|\langle \mathbf{k}'\alpha'|\, \delta F\, |\mathbf{k}\alpha\rangle|^2\, \delta(\omega - \omega_{\alpha'\alpha}) \tag{2.8.12}$$

Here $d^2\sigma_c$ is determined by the equilibrium value of the scattering amplitude operator and represents scattering that is elastic in the laboratory system. On the other hand, $d^2\sigma_d$ is determined by the equilibrium fluctuations in the scattering amplitude operator and represents scattering that is distributed in an essentially continuous way over all ω.

We shall show later in this chapter that $d^2\sigma_c$ represents the absolute coherent scattering, and hence describes all the neutron optical phenomena, while $d^2\sigma_d$ represents the diffuse scattering. We shall also show in later chapters that the coherent scattering is usually confined to a small number of well-defined directions in space, while the diffuse scattering is, as its name suggests, distributed continuously over all directions.

According to the optical theorem (Eq. 2.6.38),

$$\sigma_t = \frac{4\pi}{k} \,\mathrm{Im}[\langle \mathbf{k}|\, f\, |\mathbf{k}\rangle] \tag{2.8.13}$$

Thus, the total collision cross section is determined by the equilibrium value of the scattering amplitude operator even though it contains contributions from both coherent and diffuse scattering.

Coherent wave

Our next task is to show that the coherent scattering is determined by a one-body Schrödinger equation. We begin by putting

$$T = t + \delta T \tag{2.8.14}$$

where $t = \langle T \rangle$. Then

$$\langle \alpha'|\, T\, |\alpha\rangle = t\delta_{\alpha'\alpha} + \langle \alpha'|\, \delta T\, |\alpha\rangle \tag{2.8.15}$$

and Eq. 2.6.16 becomes

$$\psi_{\alpha'\alpha} = \psi\delta_{\alpha'\alpha} + \delta\psi_{\alpha'\alpha} \tag{2.8.16}$$

where ψ is called the *coherent state* and is given by

$$\psi = (1 + gt)\, |\mathbf{k}\rangle \tag{2.8.17}$$

in which g is the one-body Green function,

$$g = (E_\mathbf{k} + i\varepsilon - h)^{-1} \tag{2.8.18}$$

and

$$\delta\psi_{\alpha'\alpha} = g'\langle \alpha'|\, \delta T\, |\alpha\rangle\, |\mathbf{k}\rangle \tag{2.8.19}$$

in which g' is given by Eq. 2.6.14 as before. One can easily verify that

$$\psi = \sum_\alpha P_\alpha \psi_{\alpha\alpha} \tag{2.8.20}$$

The *coherent wave* is defined as the wave function of the coherent state in the coordinate representation,

$$\psi(\mathbf{r}) = \langle \mathbf{r} \mid \psi \rangle \tag{2.8.21}$$

Since Eq. 2.8.17 is formally identical to Eq. 2.2.19, it follows that $\psi(\mathbf{r})$ also has the asymptotic property of Eq. 2.2.26. Thus, as $r \to \infty$,

$$\psi(\mathbf{r}) \to (2\pi)^{-3/2}\left[\exp(i\mathbf{k}\cdot\mathbf{r}) + f(\theta)\frac{\exp(ikr)}{r}\right] \tag{2.8.22}$$

where

$$f(\theta) = -\frac{4\pi^2 m}{\hbar^2}\langle \mathbf{k}' | \, t \, | \mathbf{k} \rangle \tag{2.8.23}$$

Since $t = \langle T \rangle$, $f = \langle F \rangle$, and T and F are related by Eq. 2.6.21, we then get

$$f(\theta) = \langle \mathbf{k}' | f \, | \mathbf{k} \rangle \tag{2.8.24}$$

If now follows from Eq. 2.4.21 that the differential scattering cross section for the coherent wave is given by

$$\frac{d\sigma}{d\Omega} = |\langle \mathbf{k}' | f \, | \mathbf{k} \rangle|^2 \tag{2.8.25}$$

which is identical to Eq. 2.8.11. Thus, the coherent scattering is completely described by the coherent wave.

Optical potential

Let us now define a quantity v such that

$$v = (1 + tg)^{-1} t \tag{2.8.26}$$

Then v satisfies an equation analogous to Eq. 2.1.11,

$$t = v + tgv \tag{2.8.27}$$

When such an equation is used to define v in terms of t, instead of the other way around, the quantity v is called an *optical potential.*

Iterating Eq. 2.8.27, we get

$$\begin{aligned} t &= v + vgv + vgvgv + \cdots \\ &= v + vgt \\ &= v(1 + gt) \end{aligned} \tag{2.8.28}$$

It then follows from Eq. 2.8.17 that

$$v\psi = v(1 + gt)\,|\mathbf{k}\rangle = t\,|\mathbf{k}\rangle \tag{2.8.29}$$

and, hence, that ψ satisfies the Lippmann–Schwinger equation (2.1.5),

$$\psi = |\mathbf{k}\rangle + gv\psi \tag{2.8.30}$$

Operating on both sides of this equation with g^{-1}, we see that in the limit $\varepsilon \to 0^+$,

$$(h + v)\psi = E_{\mathbf{k}}\psi \tag{2.8.31}$$

Omitting the subscript $\mathbf{k}$ we obtain, finally, that in the coordinate representation,

$$\left\{-\frac{\hbar^2}{2m}\Delta + v(\mathbf{r})\right\}\psi(\mathbf{r}) = E\psi(\mathbf{r}) \tag{2.8.32}$$

Thus, the coherent wave and, hence, the coherent scattering are given rigorously by the solution of a one-body Schrödinger equation.

Neutron optics

Since Eq. 2.8.32 is the fundamental equation on which all of neutron optics is based, it is appropriate to comment on it in some detail. We note, first of all, that the expression (2.8.17) shows that ψ is a coherent superposition of the incident state $|\mathbf{k}\rangle$ and the scattered state $gt\,|\mathbf{k}\rangle$. Thus, as asserted earlier in this section, the scattering that is described by $\psi(\mathbf{r})$ is coherent in the absolute sense. In contrast, the diffuse scattering, which is described by $\delta\psi_{\alpha'\alpha}$, does not interfere with the incident wave (as is evident from Eq. 2.8.19) and, hence, is incoherent in the absolute sense.

There are two kinds of elastic scattering. In particular, scattering can be said to be *elastic in the strong sense* if $\alpha' = \alpha$ and, in the *weak sense*, if $E_{\alpha'} = E_\alpha$. It is evident from Eq. 2.8.20 that the coherent wave describes scattering that is elastic in the strong sense. It is the fact that the coherent scattering is not accompanied by any change in the microscopic state of the system $|\alpha\rangle$ that allows $\psi(\mathbf{r})$ to be calculated from a one-body Schrödinger equation with an effective potential $v(\mathbf{r})$. The collision of a thermal neutron with a single nucleus is also elastic in the strong sense as far as the internal degrees of freedom of the nucleus are concerned. It is this fact that allows us, without approximation, to describe the collision in terms of a complex nuclear optical potential $V(\mathbf{r})$.

According to Eq. 2.8.20, the coherent state ψ is the thermodynamic average of $\psi_{\alpha\alpha}$. The quantities $\psi(\mathbf{r})$ and $v(\mathbf{r})$ are therefore macroscopic thermodynamic quantities; that is, they are independent of the instantaneous positions of the atoms of which the system is composed and depend only on the thermodynamic variables (temperature, pressure, etc.) that define the macroscopic state of the system. In other words, the system behaves like a continuous macroscopic medium as far as the coherent scattering is concerned. We shall show later that in a gas, liquid, or amorphous solid, $v(\mathbf{r})$ is a constant potential, independent of $\mathbf{r}$, while, in a crystal, $v(\mathbf{r})$ is a smoothly varying function of $\mathbf{r}$ with the same periodicity as the lattice. Outside the system $v(\mathbf{r}) = 0$. Thus, Eq. 2.8.32

describes neutron optics in terms of the collision of the neutron with a potential barrier $v(\mathbf{r})$.

In elementary treatments of neutron optics (Hughes, 1954; Cowley, 1975; Squires, 1978; Klein and Werner, 1983), the authors simply assume without proof that a macroscopic system of atoms behaves like a continuous medium as far as neutron-optical phenomena are concerned, and that these phenomena are governed by a one-body Schrödinger equation of the type of Eq. 2.8.32. The reader might get the impression that, while physically plausible, such an approach is valid only in some crude, approximate sense. Such an impression would be reinforced by the fact that in the elementary treatments one must resort to heuristic arguments to determine a model for $v(\mathbf{r})$.

However, such an impression would be quite incorrect, since Eq. 2.8.32 is, in fact, an exact result. In going from the microscopic many-body Schrödinger equation (2.1.7) to the macroscopic one-body Schrödinger equation (2.8.32), no approximations have been made. It is, of course, necessary to make approximations (such as the impulse approximation) in order to calculate $v(\mathbf{r})$ for the particular system of interest. Nevertheless, the existence of $v(\mathbf{r})$ per se, and the fact that $\psi(\mathbf{r})$ satisfies the one-body Schrödinger equation (2.8.32) are formally exact results.

The remaining theory of neutron optics divides naturally into two parts: (1) the theory of dispersion and (2) the theory of reflection, refraction, and diffraction. Part 1 is concerned with the explicit calculation of $v(\mathbf{r})$ for the particular material of interest, and part 2 with the solution of Eq. 2.8.32 subject to boundary conditions appropriate to the experimental arrangement of interest.

Parts 1 and 2 both involve the solution of a multiple-scattering problem. In the theory of dispersion one is faced with the microscopic multiple-scattering problem (or local-field problem) that is implicit in the Ewald equations (2.1.24) and (2.1.25). This problem will be addressed in the rigorous theory of dispersion given in Chap. 4, where $v(\mathbf{r})$ is calculated on the basis of a systematic self-consistent solution of the Ewald equations. In the elementary theory of dispersion given later we will show that, if one neglects the local-field effects, one obtains the familiar expression for $v(\mathbf{r})$ that is given in elementary treatments of neutron optics such as those mentioned.

In the theory of reflection, refraction, and diffraction one is faced with the macroscopic multiple-scattering problem that is implicit in a rigorous solution of the one-body Schrödinger equation (2.8.32). The various aspects of this multiple-scattering problem are addressed in Chaps. 3, 6, 7, and 8. In the case of Bragg reflection in crystals, these macroscopic multiple-scattering effects are referred to as *dynamical diffraction effects* and include, for example, extinction and anomalous absorption. If one neglects the macroscopic multiple-scattering effects by solving Eq. 2.8.32 in the Born approximation, one obtains the *kinematical theory* of neutron diffraction that is given in Chap. 5. This is equivalent to the theory of coherent elastic scattering that is given in books such as those by Squires (1978) or Lovesey (1984).

Elementary theory of dispersion

The macroscopic optical potential $v(\mathbf{r})$, unlike the nuclear optical potential, is a weak potential and hence can be calculated, to a good approximation, using the inverse Born approximation,

$$v \simeq t = \langle T \rangle \tag{2.8.33}$$

In the elementary theory of dispersion we neglect local-field effects and use the kinematical expression (Eq. 2.1.26) for T,

$$T = \sum_j T_j \tag{2.8.34}$$

Finally, we use the impulse approximation (Eq. 2.7.3) for the one-body T-matrix,

$$T_j = \frac{2\pi\hbar^2}{m} b_j\, \delta(\mathbf{r} - \mathbf{r}_j) \tag{2.8.35}$$

where b_j is the bound scattering length of the jth atom and $\mathbf{r}_j$ is its position. Combining the previous results we get,

$$v(\mathbf{r}) = \left\langle \sum_j \frac{2\pi\hbar^2}{m} b_j\, \delta(\mathbf{r} - \mathbf{r}_j) \right\rangle \tag{2.8.36}$$

which is simply the equilibrium value of the total Fermi pseudopotential.

Alternatively,

$$v(\mathbf{r}) = \frac{2\pi\hbar^2}{m} \langle b(\mathbf{r}) \rangle \tag{2.8.37}$$

where $b(\mathbf{r})$ is the *bound scattering length density*,

$$b(\mathbf{r}) = \sum_j b_j\, \delta(\mathbf{r} - \mathbf{r}_j) \tag{2.8.38}$$

Except for a few special cases, the internal degrees of freedom of the atoms associated with the nuclear spin and isotopic composition are statistically independent of the atomic positions. Thus, for a monatonic system,

$$\langle b(\mathbf{r}) \rangle = b_c \langle \rho(\mathbf{r}) \rangle \tag{2.8.39}$$

where $b_c \equiv \langle b_j \rangle$ is the bound coherent scattering length (see Eq. 2.8.6) and $\rho(\mathbf{r})$ is the *number density*,

$$\rho(\mathbf{r}) = \sum_j \delta(\mathbf{r} - \mathbf{r}_j) \tag{2.8.40}$$

If the atoms are confined by the external constraints to a volume of space V, then for a homogeneous system such as a gas, liquid, or amorphous solid,

$$\langle \rho(\mathbf{r}) \rangle = \begin{cases} \rho & \text{inside } V \\ 0 & \text{outside } V \end{cases} \tag{2.8.41}$$

where $\rho = N/V$ is the average number density, N being the total number of atoms. Hence,

$$v(\mathbf{r}) = \begin{cases} v_0 & \text{inside } V \\ 0 & \text{outside } V \end{cases} \tag{2.8.42}$$

where (Fermi, 1934),

$$v_0 = \frac{2\pi\hbar^2}{m} \rho b_c \tag{2.8.43}$$

We shall see later in Chap. 6 that in a crystal $v(\mathbf{r})$ is a smooth periodic function with the same periodicity as the lattice, and that the average value of $v(\mathbf{r})$ within a single unit cell is given by Eq. 2.8.43. The difference between $v(\mathbf{r})$ and v_0 is important only when the incident neutron wave vector satisfies the condition for Bragg reflection. When this condition is not satisfied, the expression (2.8.42) can also be used for crystals.

Kinematical theory of coherent scattering

Using the kinematical approximation (Eq. 2.8.34) and the impulse approximation (Eq. 2.8.35) we get

$$T = \sum_j \frac{2\pi\hbar^2}{m} b_j\, \delta(\mathbf{r} - \mathbf{r}_j) = \frac{2\pi\hbar^2}{m} b(\mathbf{r}) \tag{2.8.44}$$

in which $b(\mathbf{r})$ is again the bound scattering length density (Eq. 2.8.38). The scattering amplitude operator (Eq. 2.6.21) then becomes

$$F = -(2\pi)^3 b(\mathbf{r}) \tag{2.8.45}$$

so that

$$\langle \mathbf{k}' |\, F \,| \mathbf{k} \rangle = -b_{\mathbf{q}} \tag{2.8.46}$$

where

$$\begin{aligned} b_{\mathbf{q}} &= \int b(\mathbf{r}) \exp(i\mathbf{q} \cdot \mathbf{r})\, d\mathbf{r} \\ &= \sum_j b_j \exp(i\mathbf{q} \cdot \mathbf{r}_j) \end{aligned} \tag{2.8.47}$$

and $\mathbf{q} = \mathbf{k} - \mathbf{k}'$ as before. Also,

$$f = \langle F \rangle = -(2\pi)^3 \langle b(\mathbf{r}) \rangle \tag{2.8.48}$$

so that

$$\langle \mathbf{k}' | f | \mathbf{k} \rangle = -\langle b_{\mathbf{q}} \rangle \tag{2.8.49}$$

and the expression (2.8.11) for the differential cross section for coherent scattering becomes

$$\frac{d\sigma_c}{d\Omega} = |\langle b_{\mathbf{q}} \rangle|^2 \tag{2.8.50}$$

This is the basic relation that is used for the determination of crystal structures in neutron crystallography. We shall evaluate this quantity for various kinds of materials in Chap. 5. The important point to note here is that Eq. 2.8.50, unlike Eq. 2.8.11, is not an exact result. For example, it is based on the kinematical approximation for T and, hence, neglects the effect of multiple scattering (e.g., extinction). Whether this is a good approximation depends, as we shall show later, on both the size of the crystal and its texture. Suffice it to say that in neutron crystallography experiments the samples are prepared such that Eq. 2.8.50 is valid to a good approximation.

However, in most neutron-optical experiments (Eq. 2.8.50) is not a valid approximation, since it is equivalent to solving the macroscopic one-body Schrödinger equation (2.8.32) in the Born approximation. We shall see in Chap. 3, for example, that it is only when Eq. 2.8.32 is solved exactly that the phenomena of reflection and refraction at the boundary of the system occur. These characteristic optical phenomena are absent in the kinematical expression (Eq. 2.8.50).

2.9. Diffuse scattering

Since all the neutron-optical phenomena are contained in the coherent scattering, the reader might well wonder why we need consider the diffuse scattering at all. There are two reasons why this is necessary. First, one cannot measure the coherent scattering alone but only the total scattering. There are situations, particularly in neutron crystallography, in which one must understand the nature of diffuse scattering in order to be able to subtract it correctly from the observed total scattering.

However, there is a second and more important reason why we cannot simply ignore diffuse scattering. Once a neutron has been scattered diffusely, it no longer belongs to the coherent wave. As a result, the diffuse scattering leads to an attenuation of the coherent wave in the medium in addition to the attenuation due to neutron absorption by the nuclei. The main purpose of the present section is to derive an expression for the total diffuse scattering cross section that will enable us to verify that the rigorous theory of dispersion given in Chap. 4 does, in fact, yield an expression for the optical potential $v(\mathbf{r})$ whose imaginary part describes correctly the attenuation of the coherent wave by both absorption and diffuse scattering.

Kinematical theory of diffuse scattering

The conditions for the validity of the kinematical theory are not as stringent for diffuse scattering as they are for coherent scattering. As long as the linear dimensions of the system are small in comparison with the neutron mean free path, which is of the order of 10 cm for most materials, the diffuse scattering can be calculated within the kinematical theory (Sears, 1975).

It follows from Eq. 2.8.46 that in the kinematical theory,

$$\langle \mathbf{k}'|\, \delta F\, |\mathbf{k}\rangle = -\delta b_{\mathbf{q}} \tag{2.9.1}$$

Hence, the expression (2.8.12) for the double differential cross section for diffuse scattering becomes

$$\frac{d^2\sigma_d}{d\Omega\, d\varepsilon'} = N\frac{k'}{k} S(\mathbf{q}, \omega) \tag{2.9.2}$$

where $S(\mathbf{q}, \omega)$ is called the scattering function and is given by

$$S(\mathbf{q}, \omega) = \frac{1}{N} \sum_{\alpha'\alpha} P_\alpha\, |\langle \alpha'|\, \delta b_{\mathbf{q}}\, |\alpha\rangle|^2\, \delta(\omega - \omega_{\alpha'\alpha}) \tag{2.9.3}$$

Here N is an arbitrary normalization factor, which in a monatomic system is usually taken to be the total number of atoms.

The Fourier integral representation of the delta function,

$$\delta(\omega) = \frac{1}{2\pi} \int_{-\infty}^{\infty} \exp(-i\omega t)\, dt \tag{2.9.4}$$

can be used to express $S(\mathbf{q}, \omega)$ in the form

$$S(\mathbf{q}, \omega) = \frac{1}{2\pi} \int_{-\infty}^{\infty} C(\mathbf{q}, t) \exp(-i\omega t)\, dt \tag{2.9.5}$$

where

$$C(\mathbf{q}, t) = \frac{1}{N} \sum_{\alpha'\alpha} P_\alpha\, |\langle \alpha'|\, \delta b_{\mathbf{q}}\, |\alpha\rangle|^2 \exp(i\omega_{\alpha'\alpha} t) \tag{2.9.6}$$

It then follows with the help of the completeness relation (Eq. 2.6.5) that

$$C(\mathbf{q}, t) = \frac{1}{N} \langle \delta b_{\mathbf{q}}(0)^\dagger\, \delta b_{\mathbf{q}}(t)\rangle \tag{2.9.7}$$

where $\delta b_{\mathbf{q}}(t)$ denotes the operator $\delta b_{\mathbf{q}}$ in the Heisenberg picture,

$$\delta b_{\mathbf{q}}(t) = \exp(iH_s t/\hbar)\, \delta b_{\mathbf{q}} \exp(-iH_s t/\hbar) \tag{2.9.8}$$

and the brackets $\langle \cdots \rangle$ represent the thermodynamic average (Eq. 2.8.2) as before. The representation of $S(\mathbf{q}, \omega)$ in terms of the time-dependent correlation function $C(\mathbf{q}, t)$ is due essentially to Van Hove (1954).

Relative coherence

In what follows we shall, for simplicity, confine our attention to a monatomic system. It then follows from Eq. 2.8.38 that

$$\delta b(\mathbf{r}) = b_c\,\delta\rho(\mathbf{r}) + \sum_j \delta b_j\,\delta(\mathbf{r}-\mathbf{r}_j) \qquad (2.9.9)$$

where $b_c = \langle b_j \rangle$ is the bound coherent scattering length as before. Thus, the fluctuations in the bound scattering length density are due partly to fluctuations in the number density and partly to the spin and isotope disorder that is responsible for the statistical fluctuations δb_j.

With

$$\rho_{\mathbf{q}} = \int \rho(\mathbf{r}) \exp(i\mathbf{q}\cdot\mathbf{r})\,d\mathbf{r} = \sum_j \exp(i\mathbf{q}\cdot\mathbf{r}_j) \qquad (2.9.10)$$

we get

$$\delta b_{\mathbf{q}} = b_c\,\delta\rho_{\mathbf{q}} + \sum_j \delta b_j \exp(i\mathbf{q}\cdot\mathbf{r}_j) \qquad (2.9.11)$$

In most cases the quantities δb_j are statistically independent, both of each other and of the atomic positions. Thus,

$$\langle \delta b_j^\dagger\,\delta b_{j'} \rangle = |b_i|^2\,\delta_{jj'} \qquad (2.9.12)$$

where $b_i = \Delta b_j$ is the bound incoherent scattering length as before (see Eq. 2.8.6). It then follows from Eq. 2.9.7 that

$$C(\mathbf{q}, t) = |b_c|^2\,C_c(\mathbf{q}, t) + |b_i|^2\,C_i(\mathbf{q}, t) \qquad (2.9.13)$$

where

$$C_c(\mathbf{q}, t) = \frac{1}{N}\langle \delta\rho_{\mathbf{q}}(0)^\dagger\,\delta\rho_{\mathbf{q}}(t)\rangle \qquad (2.9.14)$$

and

$$\begin{aligned} C_i(\mathbf{q}, t) &= \frac{1}{N}\sum_j \langle \exp(-i\mathbf{q}\cdot\mathbf{r}_j(0)) \exp(i\mathbf{q}\cdot\mathbf{r}_j(t))\rangle \\ &= \langle \exp(-i\mathbf{q}\cdot\mathbf{r}_1(0)) \exp(i\mathbf{q}\cdot\mathbf{r}_1(t))\rangle \end{aligned} \qquad (2.9.15)$$

Hence, from Eq. 2.9.5,

$$S(\mathbf{q}, \omega) = |b_c|^2\,S_c(\mathbf{q}, \omega) + |b_i|^2\,S_i(\mathbf{q}, \omega) \qquad (2.9.16)$$

where $S_c(\mathbf{q}, \omega)$ is called the *coherent scattering function* or the *dynamic structure factor*, and $S_i(\mathbf{q}, \omega)$ the *incoherent scattering function*. The word *coherent* is now used in the relative sense as defined at the beginning of Sec. 2.8.

It is evident that $S_c(\mathbf{q}, \omega)$ characterizes the collective motion of the atoms in the system, such as the lattice vibrations in a crystal, while $S_i(\mathbf{q}, \omega)$ characterizes the single-particle motion such as self-diffusion.

The calculation of $S_c(\mathbf{q}, \omega)$ and $S_i(\mathbf{q},\omega)$ for solids, liquids, and gases is described in books such as those by Squires (1978) or Lovesey (1984) but will not concern us here. It will be noted, however, that for an ideal gas of Boltzmann particles $S_c(\mathbf{q}, \omega)$ and $S_i(\mathbf{q}, \omega)$ are both given by Eq. 2.5.37.

Since the two factors in Eq. 2.9.14 are statistically independent in the limit $t \to \infty$ and $\langle \delta\rho_{\mathbf{q}} \rangle = 0$, it follows that $C_c(\mathbf{q}, \infty) = 0$ and, hence, that $S_c(\mathbf{q}, \omega)$ has no delta function component at $\omega = 0$. Thus, $S_c(\mathbf{q}, \omega)$ represents the coherent inelastic scattering. The coherent elastic scattering (i.e., the absolute coherent scattering) was discussed in Sec. 2.8. In the kinematical theory, the coherent elastic and inelastic scattering can both be included in $S_c(\mathbf{q}, \omega)$ by replacing $\delta\rho_{\mathbf{q}}$ by $\rho_{\mathbf{q}}$ in Eq. 2.9.14. However, we do not do this because, in neutron optics, the important distinction is between the absolute coherent and the diffuse scattering, and not between the relative coherent and incoherent scattering.

Static structure factor

Inverting the Fourier transform Eq. 2.9.5 we get

$$C(\mathbf{q}, t) = \int_{-\infty}^{\infty} S(\mathbf{q}, \omega) \exp(i\omega t)\, d\omega \tag{2.9.17}$$

The integrated scattering function is therefore given by

$$\begin{aligned} S(\mathbf{q}) &= \int_{-\infty}^{\infty} S(\mathbf{q}, \omega)\, d\omega \\ &= C(\mathbf{q}, 0) \\ &= |b_c|^2\, C_c(\mathbf{q}, 0) + |b_i|^2\, C_i(\mathbf{q}, 0) \end{aligned} \tag{2.9.18}$$

Similarly, the integrated coherent scattering function, or *static structure factor,* is given by

$$\begin{aligned} S_c(\mathbf{q}) &= \int_{-\infty}^{\infty} S_c(\mathbf{q}, \omega)\, d\omega \\ &= C_c(\mathbf{q}, 0) \\ &= \frac{1}{N} \langle |\delta\rho_{\mathbf{q}}|^2 \rangle \end{aligned} \tag{2.9.19}$$

and the integrated incoherent scattering function is given by

$$\begin{aligned} S_i(\mathbf{q}) &= \int_{-\infty}^{\infty} S_i(\mathbf{q}, \omega)\, d\omega \\ &= C_i(\mathbf{q}, 0) \\ &= 1 \end{aligned} \tag{2.9.20}$$

Hence,

$$S(\mathbf{q}) = |b_c|^2\, S_c(\mathbf{q}) + |b_i|^2 \tag{2.9.21}$$

It follows from Eq. 2.9.10 that

$$\delta\rho_{\mathbf{q}} = \int_V \delta\rho(\mathbf{r}) \exp(i\mathbf{q}\cdot\mathbf{r})\, d\mathbf{r} \tag{2.9.22}$$

in which the integration can be restricted to V since $\rho(\mathbf{r}) = 0$ outside V. Hence,

$$S_c(\mathbf{q}) = \frac{1}{N}\int_V d\mathbf{r}\int_V d\mathbf{r}'\, \langle \delta\rho(\mathbf{r})\, \delta\rho(\mathbf{r}')\rangle \exp[i\mathbf{q}\cdot(\mathbf{r}'-\mathbf{r})] \tag{2.9.23}$$

In general, according to Eq. 2.8.4,

$$\langle \delta\rho(\mathbf{r})\, \delta\rho(\mathbf{r}')\rangle = \langle \rho(\mathbf{r})\, \rho(\mathbf{r}')\rangle - \langle \rho(\mathbf{r})\rangle\langle \rho(\mathbf{r}')\rangle \tag{2.9.24}$$

For a system such as a gas, liquid, or amorphous solid that is macroscopically homogeneous and isotropic, $\langle \rho(\mathbf{r})\rangle = \rho$ according to Eq. 2.8.41 and

$$\langle \rho(\mathbf{r})\rho(\mathbf{r}')\rangle = \rho\delta(\mathbf{r}'-\mathbf{r}) + \rho^2 g(\mathbf{r},\mathbf{r}') \tag{2.9.25}$$

where $g(\mathbf{r},\mathbf{r}')$ is the *pair correlation function* that is defined such that

$$\rho^2 g(\mathbf{r},\mathbf{r}') = \left\langle \sum_{i\neq j} \delta(\mathbf{r}-\mathbf{r}_i)\, \delta(\mathbf{r}'-\mathbf{r}_j)\right\rangle \tag{2.9.26}$$

Except at points $\mathbf{r}$ and $\mathbf{r}'$ in the immediate neighborhood of the surface of V,

$$g(\mathbf{r},\mathbf{r}') = g(|\mathbf{r}'-\mathbf{r}|) \tag{2.9.27}$$

Hence,

$$\langle \delta\rho(\mathbf{r})\, \delta\rho(\mathbf{r}')\rangle = \rho\delta(\mathbf{r}'-\mathbf{r}) + \rho^2[g(|\mathbf{r}'-\mathbf{r}|) - 1] \tag{2.9.28}$$

and

$$S_c(\mathbf{q}) = 1 + \rho\int [g(r) - 1] \exp(i\mathbf{q}\cdot\mathbf{r})\, d\mathbf{r} \tag{2.9.29}$$

Static approximation

In the rigorous theory of dispersion that is given in Chap. 4, we idealize the medium by considering the scattering of a neutron by a static array of strongly bound atoms. In this "static approximation,"

$$\begin{aligned} C_c(\mathbf{q}, t) &= C_c(\mathbf{q}, 0) = S_c(\mathbf{q}) \\ C_i(\mathbf{q}, t) &= C_i(\mathbf{q}, 0) = 1 \end{aligned} \tag{2.9.30}$$

so that

$$\begin{aligned} S_c(\mathbf{q}, \omega) &= S_c(\mathbf{q})\, \delta(\omega) \\ S_i(\mathbf{q}, \omega) &= \delta(\omega) \end{aligned} \tag{2.9.31}$$

and

$$S(\mathbf{q}, \omega) = [|b_c|^2 S_c(\mathbf{q}) + |b_i|^2]\, \delta(\omega) \tag{2.9.32}$$

In this case the diffuse scattering is elastic so that Eq. 2.9.2 becomes

$$\frac{d^2\sigma_d}{d\Omega\, d\varepsilon'} = \frac{d\sigma_d}{d\Omega}\,\delta(\omega) \tag{2.9.33}$$

where

$$\frac{d\sigma_d}{d\Omega} = N[|b_c|^2\, S_c(\mathbf{q}) + |b_i|^2] \tag{2.9.34}$$

The total cross section for diffuse scattering is therefore given by

$$\sigma_d = \int d\sigma_d = N\left[\frac{\sigma_c}{4\pi}\int_{4\pi} S_c(\mathbf{q})\, d\Omega + \sigma_i\right] \tag{2.9.35}$$

in which $\sigma_c = 4\pi\,|b_c|^2$ and $\sigma_i = 4\pi\,|b_i|^2$ are, as in Sec. 2.7, the bound coherent and incoherent scattering cross sections. If $S_c(\mathbf{q})$ is isotropic, as in Eq. 2.9.29, then σ_d can be expressed with the help of Eq. 2.7.16 as

$$\sigma_d = N\left[\frac{\sigma_c}{2k^2}\int_0^{2k} S_c(q)q\, dq + \sigma_i\right] \tag{2.9.36}$$

Finally, it will be noted that experiments in which one measures $d\sigma_d/d\Omega$ so as to determine $S_c(q)$ and, hence, $g(r)$ are conventionally called *diffraction experiments*. However, strictly speaking, this terminology is incorrect. It is only the absolute coherent scattering $d\sigma_c/d\Omega$ that represents diffraction in the optical sense.

3

GEOMETRICAL OPTICS

Geometrical optics is concerned with experimental arrangements in which diffraction is negligible so that reflection and refraction at the boundaries between different media are the only significant optical phenomena that occur. In practice these boundaries are usually plane, spherical or cylindrical because such surfaces are easier to grind and polish than other shapes. Since the rays within a homogeneous medium are straight lines (apart from the effect of gravity) optics problems then involve only straight lines and circles and, hence, can be solved by the methods of Euclidean geometry. This is the origin of the term *geometrical optics*.

Many of the optical instruments employing mirrors, lenses, prisms, etc., with which the reader will be familiar in the context of ordinary optics have been adapted to neutron optics. Among the many important uses of such instruments the most notable are for the accurate measurement of neutron scattering lengths and for the production and analysis of polarized neutron beams. Before discussing these subjects we must first consider in detail the way in which neutrons are reflected and refracted at surfaces.

3.1. Reflection and refraction

In Chap. 2 we showed that the coherent wave $\psi(\mathbf{r})$ that describes all neutron optical phenomena satisfies the one-body Schrödinger equation (2.8.32). The optical potential $v(\mathbf{r})$, which represents the effective interaction of the neutron with the medium, is given for a homogeneous medium by Eq. 2.8.42. The general solution of Eq. 2.8.32 can then be expressed as a superposition of plane waves,

$$\psi(\mathbf{r}) = \begin{cases} \sum A \exp(i\mathbf{K} \cdot \mathbf{r}) & \text{inside } V \\ \sum a \exp(i\mathbf{k} \cdot \mathbf{r}) & \text{outside } V \end{cases} \tag{3.1.1}$$

where the magnitudes of the wave vectors are determined by the incident neutron energy,

$$E = \frac{(\hbar K)^2}{2m} + v_0 = \frac{(\hbar k)^2}{2m} \tag{3.1.2}$$

The directions of the various wave vectors in Eq. 3.1.1 and the corresponding amplitudes are then determined by requiring that $\psi(\mathbf{r})$ and $\nabla\psi(\mathbf{r})$ be continuous at the boundary of V.

Neutron refractive index

The *index of refraction* is defined as

$$n = \frac{K}{k} \tag{3.1.3}$$

so that from Eq. 3.1.2

$$n^2 = 1 - \xi \tag{3.1.4}$$

where

$$\xi = \frac{v_0}{E} \tag{3.1.5}$$

In the elementary theory of dispersion v_0 is given by Eq. 2.8.43,

$$v_0 = \frac{2\pi\hbar^2}{m}\rho b \tag{3.1.6}$$

where we now omit the subscript c in the scattering length since we shall be dealing exclusively with coherent scattering. Hence,

$$\xi = \frac{4\pi}{k^2}\rho b = \frac{\lambda^2}{\pi}\rho b \tag{3.1.7}$$

where $\lambda = 2\pi/k$ is the incident-neutron wavelength.

In general, ρb is the average bound coherent scattering length density. Thus, for a monatomic system ρ is the average number of atoms per unit volume and b the bound coherent scattering length of an atom. For a polyatomic system,

$$\rho b = \sum_l \rho_l b_l \tag{3.1.8}$$

where l labels the contributions of the various atomic species. For a system of identical molecules, on the other hand, ρ can be taken to be the average number of molecules per unit volume, in which case b is the bound coherent scattering length of a single molecule,

$$b = \sum_l N_l b_l \tag{3.1.9}$$

where N_l is the number of atoms of type l per molecule.

The quantity ξ can also be expressed as

$$\xi = \left(\frac{\lambda}{\lambda_c}\right)^2 \tag{3.1.10}$$

Table 3.1 Values of quantities that characterize the neutron refractive index for various liquids

Liquid	ρ ($Å^{-3}$)	b (fm)	λ_c (Å)	ξ^* (10^{-5})	ϕ_c^* (degree)
Na	0.0243	3.63	1887	0.449	0.121
Al	0.0529	3.449	1312	0.929	0.175
S	0.0338	2.847	1807	0.490	0.127
Ar	0.0213	1.909	2780	0.207	0.083
Ga	0.0527	7.288	904	1.956	0.253
Sn	0.0354	6.226	1194	1.123	0.192
Hg	0.0407	12.69	780	2.630	0.294
Pb	0.0311	9.402	1037	1.489	0.221
H_2O	0.0333	−1.679		−0.285	
D_2O	0.0332	19.145	703	3.237	0.326
CCl_4	0.00625	44.95	1057	1.431	0.217
C_8H_{10}	0.00490	15.781	2016	0.394	0.114

* For $\lambda = 4.0$ Å.

where λ_c is called the *critical wavelength* and is given by

$$\lambda_c = (\pi/\rho b)^{1/2} \tag{3.1.11}$$

If b is real and positive, which it usually is to a good approximation, then the index of refraction is real if $\lambda < \lambda_c$ and imaginary if $\lambda > \lambda_c$. The values of λ_c for a number of liquids are shown in Table 3.1, where it is seen that λ_c is typically 1400 Å. For thermal neutrons with $\lambda = 4.0$ Å (say), this corresponds to $\xi = 0.82 \times 10^{-5}$.

At the boundary between two media A and B the *relative index of refraction* is defined as

$$n = \frac{K_A}{K_B} = \frac{n_A}{n_B} \tag{3.1.12}$$

Thus,

$$n^2 = \frac{1 - \xi_A}{1 - \xi_B} \equiv 1 - \xi \tag{3.1.13}$$

where, to first order,

$$\begin{aligned} \xi &= \xi_A - \xi_B \\ &= \frac{1}{E}(v_{0A} - v_{0B}) \\ &= \frac{\lambda^2}{\pi}(\rho_A b_A - \rho_B b_B) \end{aligned} \tag{3.1.14}$$

Reflection and refraction at a plane boundary

Suppose a neutron with wave vector **k** is incident on a medium with a plane boundary. We choose a system of Cartesian coordinate axes such

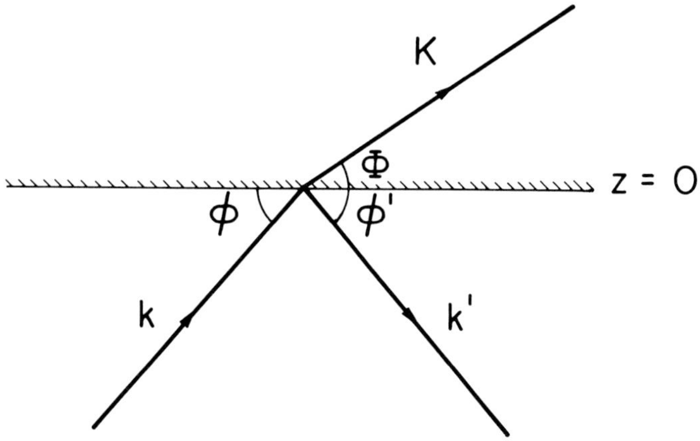

Fig. 3.1 Reflection and refraction at a plane boundary.

that the boundary is the $z = 0$ plane. In particular, the region $z < 0$ is taken to be the vacuum and the region $z > 0$ to be the medium. We look for a solution of the Schrödinger equation in which the incident wave is partially reflected and partially refracted at the boundary, as illustrated in Fig. 3.1. Then Eq. 3.1.1 becomes

$$\psi(\mathbf{r}) = \begin{cases} A \exp(i\mathbf{K} \cdot \mathbf{r}), & z > 0 \\ a \exp(i\mathbf{k} \cdot \mathbf{r}) + a' \exp(i\mathbf{k}' \cdot \mathbf{r}), & z < 0 \end{cases} \tag{3.1.15}$$

Suppose the coordinate axes are chosen such that **k** is in the $y = 0$ plane,

$$\mathbf{k} = (k \cos \phi,\ 0,\ k \sin \phi) \tag{3.1.16}$$

Since $\psi(\mathbf{r})$ and $\nabla\psi(\mathbf{r})$ must be continuous at $z = 0$ it then follows that

$$\begin{aligned} \mathbf{k}' &= (k \cos \phi,\ 0,\ -k \sin \phi) \\ \mathbf{K} &= (k \cos \phi,\ 0,\ n_z k \sin \phi) \end{aligned} \tag{3.1.17}$$

and

$$\begin{aligned} a' &= \left(\frac{1 - n_z}{1 + n_z}\right) a \\ A &= \left(\frac{2}{1 + n_z}\right) a \end{aligned} \tag{3.1.18}$$

in which

$$n_z \equiv \frac{K_z}{k_z} = (1 - \xi \operatorname{cosec}^2 \phi)^{1/2} \tag{3.1.19}$$

We note that, given k, ϕ, and a, all the other quantities are determined. In other words, the reflected and refracted waves are uniquely determined by the incident wave.

In general b, and hence ξ and n_z, are complex so that

$$n_z = n_z' + in_z'' \tag{3.1.20}$$

Then

$$\mathbf{K} = \mathbf{K}' + i\mathbf{K}'' \tag{3.1.21}$$

where

$$\begin{aligned} \mathbf{K}' &= (k \cos \phi, 0, n_z' k \sin \phi) \\ \mathbf{K}'' &= (0, 0, n_z'' k \sin \phi) \end{aligned} \tag{3.1.22}$$

If b is complex the coherent wave in the medium is therefore an inhomogeneous plane wave. That is, the planes of constant phase ($\mathbf{K}' \cdot \mathbf{r} = \text{const}$) are not parallel to the planes of constant amplitude ($\mathbf{K}'' \cdot \mathbf{r} = \text{const}$).

In most cases the refracted wave vector $\mathbf{K}$ can, to a good approximation, be taken to be real, in which case

$$K_z = n_z k \sin \phi = K \sin \Phi, \tag{3.1.23}$$

where Φ is the angle of refraction in Fig. 3.1. Hence,

$$n_z \sin \phi = n \sin \Phi \tag{3.1.24}$$

or, equivalently,

$$\cos \phi = n \cos \Phi \tag{3.1.25}$$

This is the familiar *law of refraction* (Snell's law). The *law of reflection,*

$$\phi = \phi' \tag{3.1.26}$$

is implicit in Eq. 3.1.17.

Fresnel's formulas

It follows from Eq. 2.4.12 that the neutron current densities in the incident, reflected, and refracted waves are given, respectively, by

$$\begin{aligned} \mathbf{j} &= \left(\frac{\hbar \mathbf{k}}{m}\right) |a|^2 \\ \mathbf{j}' &= \left(\frac{\hbar \mathbf{k}'}{m}\right) |a'|^2 \\ \mathbf{J} &= \left(\frac{\hbar \mathbf{K}'}{m}\right) |A|^2 \exp(-2\mathbf{K}'' \cdot \mathbf{r}) \end{aligned} \tag{3.1.27}$$

The *reflectivity* R is defined as the fraction of incident neutrons that are reflected at the boundary. Thus,

$$R = -\frac{j_z'}{j_z} = \left|\frac{a'}{a}\right|^2 \tag{3.1.28}$$

The *transmissivity* T is similarly defined as the fraction of incident neutrons that are transmitted at the boundary,

$$T = \frac{J_z}{j_z} = n_z' \left|\frac{A}{a}\right|^2 \tag{3.1.29}$$

Hence, it follows from Eq. 3.1.18 that

$$R = \left|\frac{1 - n_z}{1 + n_z}\right|^2, \qquad T = \frac{4n_z'}{|1 + n_z|^2} \tag{3.1.30}$$

which are the neutron analogs of Fresnel's formulas in ordinary optics (Born and Wolf, 1975). We note that

$$R + T = 1 \tag{3.1.31}$$

as required for neutron conservation.

Since $|\xi| \simeq 10^{-5}$ for thermal neutrons, it follows from Eq. 3.1.19 that for a general angle of incidence,

$$R = \frac{|\xi|^2}{16 \sin^4 \phi} \tag{3.1.32}$$

Thus, $R \simeq 10^{-10}$ which is negligibly small. Reflection is appreciable only for glancing incidence where $\phi \to 0$. In this case,

$$R = \left|\frac{1 - [1 - (\phi_c/\phi)^2]^{1/2}}{1 + [1 - (\phi_c/\phi)^2]^{1/2}}\right|^2 \tag{3.1.33}$$

where

$$\phi_c = \sqrt{\xi} = \frac{\lambda}{\lambda_c} = \lambda\left(\frac{\rho b}{\pi}\right)^{1/2} \tag{3.1.34}$$

The reflectivity is illustrated in Fig. 3.2 for the case in which b is real, which is usually a good approximation. Then ϕ_c is real if $b > 0$ and purely imaginary if $b < 0$. In the former case $R = 1$ if $\phi < \phi_c$ so that ϕ_c is the *critical angle for total reflection.* Since $|\xi| \simeq 10^{-5}$ for thermal neutrons, we see that $\phi_c \simeq 3 \times 10^{-3}$ radians $\simeq$0.2 degree. The width of the reflected beam at the critical angle is given by

$$W = L \sin \phi_c \tag{3.1.35}$$

where L is the length of the mirror. Thus, for example, $W = 1$ mm if $L = 33$ cm. If, on the other hand, $b < 0$ then $R < 1$ for all $\phi > 0$.

In general, b is complex,

$$b = b' - ib'' \tag{3.1.36}$$

and for strongly absorbing materials b'' is nonnegligible. If $b' > 0$ the primary effect of absorption is to round off the reflectivity curve at the critical angle. Let

$$\phi = \lambda\left(\frac{\rho b'}{\pi}\right)^{1/2} \tag{3.1.37}$$

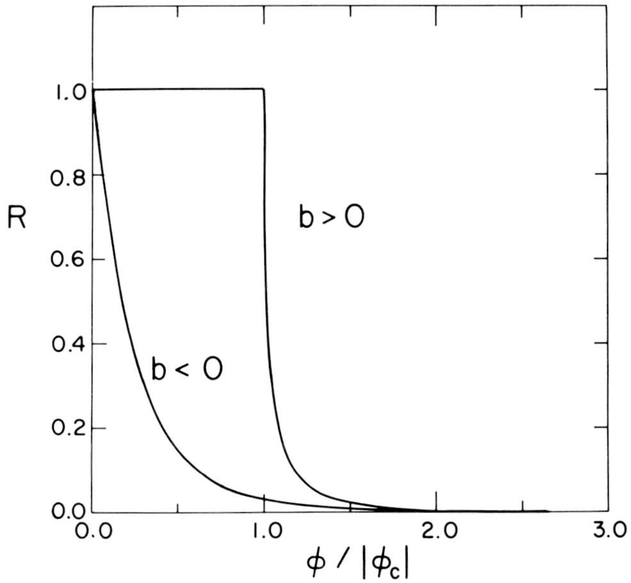

Fig. 3.2 The reflectivity as a function of the reduced angle of incidence for mirror reflection at a plane boundary. The bound coherent scattering length b is assumed to be real.

so that ϕ is what would be the critical angle if $b''=0$. At this angle of incidence,

$$R = \frac{b' + b'' - (2b'b'')^{1/2}}{b' + b'' + (2b'b'')^{1/2}} \tag{3.1.38}$$

For an element that is not strongly absorbing ($\sigma_a \lesssim \sigma_s$, say) we have $b''/b' \lesssim 10^{-4}$ so that $R \gtrsim 0.97$. In this case absorption reduces the reflectivity by at most a few percent. The values of R for a number of strongly absorbing elements are listed in Table 3.2 in order of increasing σ_a. It is seen that even for gadolinium, for which $\sigma_a \simeq 49{,}000$ barns, approximately 18 percent of the incident neutrons are reflected at the critical angle.

Since ϕ_c^2 is proportional to b it is evident from Eq. 3.1.33 that a measurement of the reflectivity can be used to determine both the magnitude and sign of the bound coherent scattering length. This important application of neutron optics will be discussed later in Sec. 3.4. In the remainder of the present section we shall describe briefly some of the other practical applications of the reflection and refraction of neutrons.

Neutron guide tubes

The most important practical application of neutron reflection is in the construction of guide tubes for the transport of thermal neutron beams

Table 3.2 Values of the absorption cross section, σ_a, the bound coherent scattering length, b, and the reflectivity at the critical angle, R, for a selection of strongly absorbing elements. In column 2 the unit "b" is the barn where $1\ \mathrm{b} = 100\ \mathrm{fm}^2$

Element	σ_a (b)	b (fm)	R
Cl	33.5	$9.5770 - 0.0093i$	0.916
Co	37.18	$2.50 - 0.0103i$	0.834
Nd	50.5	$7.69 - 0.0140i$	0.886
Ag	63.3	$5.922 - 0.0176i$	0.857
In	193.8	$4.065 - 0.0539i$	0.723
Hg	372.3	$12.69 - 0.104i$	0.775
B	767	$5.30 - 0.213i$	0.572
Dy	940	$16.9 - 0.261i$	0.705
^{3}He	5333	$5.74 - 1.48i$	0.273
Sm	5670	$4.2 - 1.58i$	0.227
Gd	48,890	$9.5 - 13.59i$	0.179

(Christ and Springer, 1962; Maier-Leibnitz and Springer, 1963). These neutron guides are analogous to light pipes in ordinary optics and the basic principle is illustrated schematically in Fig. 3.3. A neutron entering the guide tube with an angle of incidence at the wall that is less than the critical angle for its particular wavelength will be transported along the tube by multiple total reflection. If a slight curvature is incorporated in the tube, only those neutrons with wavelengths greater than some minimum value (typically $\simeq 1$ Å) are transported down the tube. The shorter-wavelength fast neutrons, as well as any γ rays in the incident beam, pass through the wall of the tube where they can be absorbed in a suitable beam stop. In this way a very clean and highly collimated beam

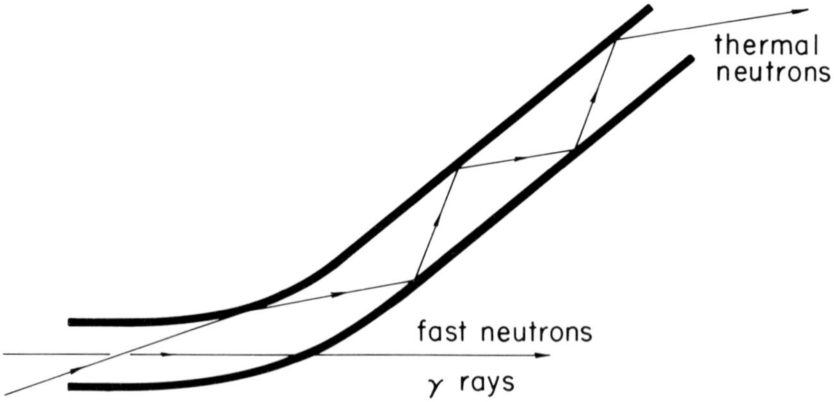

Fig. 3.3 Schematic illustration of a neutron guide tube.

of thermal neutrons can be transported to a low-background location far from the primary neutron source where it is then available for thermal neutron scattering experiments. In practice the length of such guide tubes ranges from 10 to 100 m and the transmitted intensity is typically 80 percent (Maier-Leibnitz, 1972).

Ultracold neutrons

The total reflection of thermal neutrons below the critical angle is conventionally referred to in neutron optics as *mirror reflection*. Strictly speaking this is incorrect since, in the usual sense of the word, a mirror gives total reflection for all angles of incidence. This happens when n is purely imaginary, that is, when $\lambda > \lambda_c$ and $\xi > 1$. We saw in Table 3.1 that λ_c is typically 1400 Å. This corresponds to a neutron energy of 4×10^{-8} eV or to a temperature of 0.5 mK. It is only for such *ultracold neutrons* that true mirror reflection can occur.

It was long recognized (Zeldovich, 1959; Foldy, 1966) that the mirror reflection of ultracold neutrons could be used to confine neutrons in material cavities, or "bottles." Ultracold neutrons were first produced and detected by Lushchikov et al. (1969) and Steyerl (1969), and subsequently confined for a period of about 13 seconds by Groshev et al. (1971). It was later found that smooth clean surfaces were needed for good storage and it is now possible to confine neutrons for periods of the order of 100 seconds. The storage is ultimately limited by the β-decay of the free neutron, for which the mean lifetime is 937 seconds.

The practical importance of confining neutrons in bottles lies in the fact that there are a number of fundamental measurements, such as the free neutron lifetime or the neutron electric dipole moment, for which the accuracy is limited mainly by the length of time the neutron can be kept within the region of observation. For such measurements, ultracold neutrons have a great advantage over thermal neutrons in spite of the fact that ultracold neutron fluxes are many orders of magnitude smaller than thermal neutron fluxes.

Since ultracold neutrons are totally reflected by most materials, even at normal incidence, a spherical mirror can be used to focus a collimated ultracold neutron beam. The focal length of a spherical mirror is given by

$$f = \frac{r}{2}, \tag{3.1.39}$$

where r is the radius of the mirror. Since f is independent of λ, the mirror is achromatic. Thus, a good quality image can be obtained for paraxial rays without having to use monoenergetic neutrons, an important consideration in view of the low fluxes of available ultracold neutron beams. A recent application of such spherical mirrors is in the construction and demonstration of an ultracold neutron microscope (Herrmann et al., 1985; Steyerl and Malik, 1986).

For more information on the production, confinement, and applications of ultracold neutrons, we refer the reader to the review articles by Lushchikov (1977), Steyerl (1977), and Golub and Pendlebury (1979).

Prisms

The refraction of neutrons by a prism is illustrated in Fig. 3.4. By applying Snell's law (Eq. 3.1.25) to each refraction, one can easily show that the total angle of deviation δ is given to first order in $n - 1$ by

$$\begin{aligned} \delta &= (1 - n)[\tan \theta + \tan (\alpha - \theta)] \\ &= \tfrac{1}{2}\xi[\tan \theta + \tan (\alpha - \theta)] \end{aligned} \tag{3.1.40}$$

If the incident beam is parallel to the base of the prism then $\theta = \alpha/2$ and

$$\delta = \xi \tan \frac{\alpha}{2} \tag{3.1.41}$$

In Fig. 3.4 it is tacitly assumed that $b > 0$, and hence that $n < 1$, which is the usual case for neutrons. Consequently, δ has the opposite sign to the sign it has in ordinary optics, where $n > 1$.

Since δ is of order ξ it is very much smaller than the critical angle for total reflection (Eq. 3.1.34) which is equal to $\sqrt{\xi}$. The quantity ξ is proportional to $\lambda^2 b$. Hence, an accurate measurement of δ with monoenergetic neutrons can be used to determine both the magnitude and sign of b, as will be discussed later in Sec. 3.4.

Since δ is proportional to λ^2, a prism can also be used as a monochromator for neutrons just as it can for light. This is illustrated in Fig. 3.5, where a quartz prism was used to produce a monoenergetic beam of 20 Å neutrons in a neutron-optical experiment that established a new upper limit of $10^{-20}e$ for the electric charge of the neutron (Gähler et al., 1980, 1982). Here $\alpha = 120$ degrees, $\xi = 4.6 \times 10^{-4}$ for 20 Å neutrons in quartz, and $\delta = 0.11$ degree for the value of θ that was used.

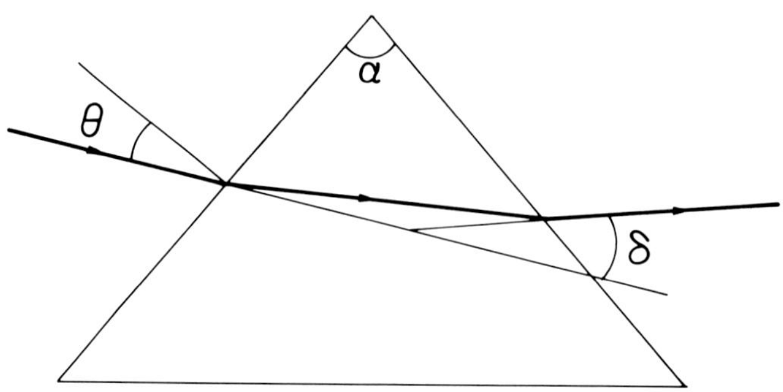

Fig. 3.4 Refraction by a prism.

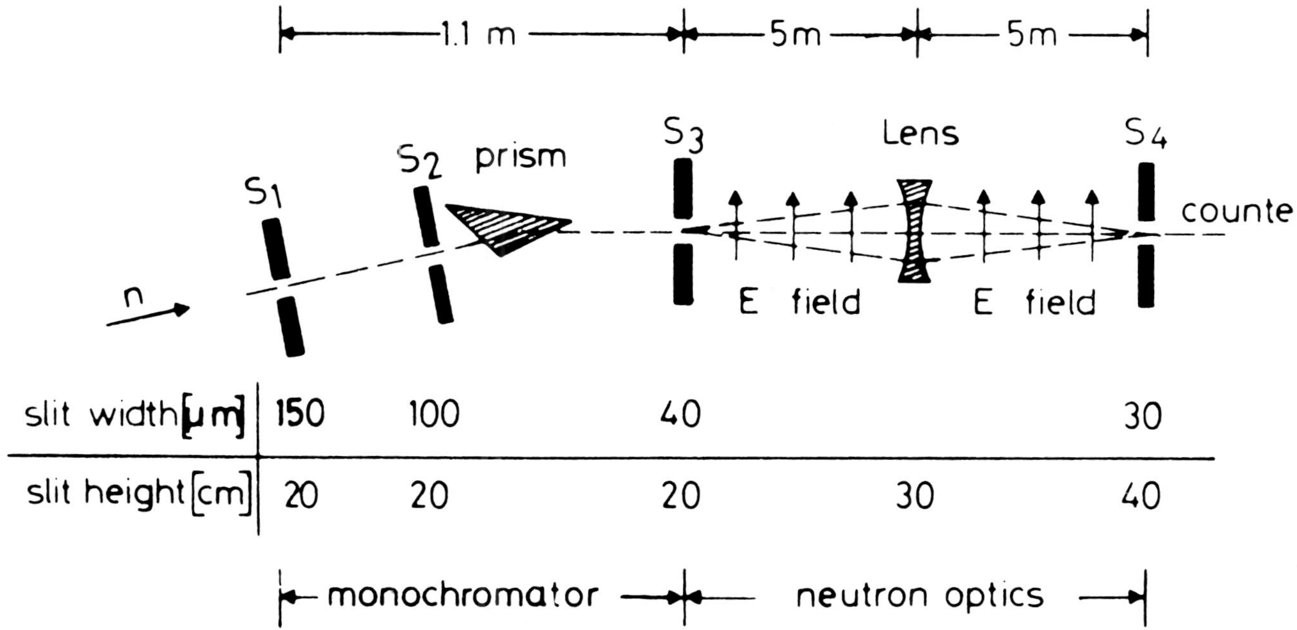

Fig. 3.5 Illustration of the use of a prism monochromator and a convergent lens in a neutron-optical experiment to measure the neutron electric charge (from Gähler et al., 1982).

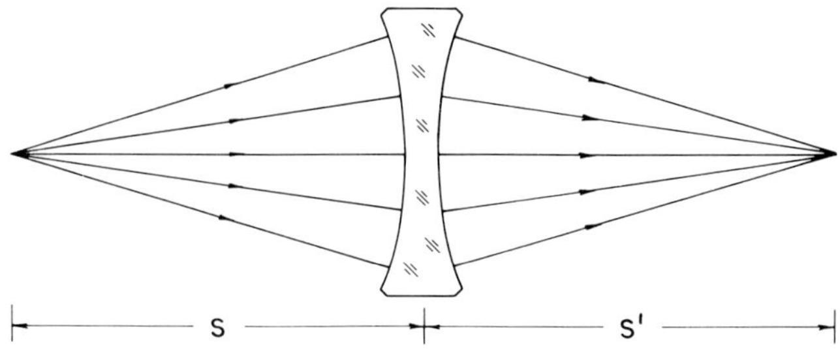

Fig. 3.6 Focal property of a concave lens.

Lenses

Figure 3.6 illustrates the focal property of a lens. For a thin lens the distances s and s' from the center of the lens to the object and image points obey the fundamental relation of Gaussian optics (Born and Wolf, 1975),

$$\frac{1}{s}+\frac{1}{s'}=\frac{1}{f} \tag{3.1.42}$$

where f is the focal length ($s'=f$ when $s=\infty$). For the equiconcave lens shown in Fig. 3.6,

$$f=\frac{r}{2(1-n)}=\frac{r}{\xi} \tag{3.1.43}$$

where r is the radius of curvature of the two faces. Again, it is tacitly assumed that $b>0$, and hence that $n<1$, which is the usual case for neutrons. Here a concave lens is convergent (producing a real image), while a convex lens is divergent (producing a virtual image). This is the opposite to the situation in ordinary optics, where $n>1$.

A practical example of the use of a lens in neutron optics is shown in Fig. 3.5. For the cylindrical quartz lens that is illustrated there, $r=2.4$ mm and $\xi=4.6\times10^{-4}$ for 20 Å neutrons so that, according to Eq. 3.1.43, $f=5.2$ m. However, Fig. 3.5 is merely schematic. The actual lens in question is a compound lens formed by two identical equiconcave cylindrical lenses. The effective focal length of this doublet is a factor of 2 smaller than the preceding value, i.e., 2.6 m.

The fact that $f\propto\lambda^{-2}$ has two important consequences. First, very long-wavelength neutrons are needed to obtain reasonably small values of f. If thermal neutrons had been used in the preceding example, with $\lambda=2$ Å (say), the focal length would have been 260 m. Second, a lens is highly nonachromatic, so that monoenergetic neutrons must be used to obtain a sharp image of the entrance slit S_3 at the position of the exit slit S_4 in Fig. 3.5. This is the reason for the presence of the prism monochromator.

3.2. Transmission

Transmission through a plane slab

We now consider the transmission of neutrons through a plane slab of thickness d, as illustrated in Fig. 3.7. The angle of incidence ϕ is assumed to be sufficiently large that reflection at the boundaries can be neglected. Then the coherent wave is given by

$$\psi(\mathbf{r}) = \begin{cases} a \exp(i\mathbf{k} \cdot \mathbf{r}), & z < 0 \\ A \exp(i\mathbf{K} \cdot \mathbf{r}), & 0 < z < d \\ a' \exp(i\mathbf{k}' \cdot \mathbf{r}), & d < z \end{cases} \tag{3.2.1}$$

Since reflection at the boundaries is neglected, we must, for consistency, also ignore the requirement that $\nabla\psi(\mathbf{r})$ be continuous at the boundaries. With $\mathbf{k}$ in the $y = 0$ plane, the continuity of $\psi(\mathbf{r})$ then requires that

$$\begin{aligned} \mathbf{k}' = \mathbf{k} &= (k \cos \phi,\ 0,\ k \sin \phi) \\ \mathbf{K} &= (k \cos \phi,\ 0,\ n_z k \sin \phi) \end{aligned} \tag{3.2.2}$$

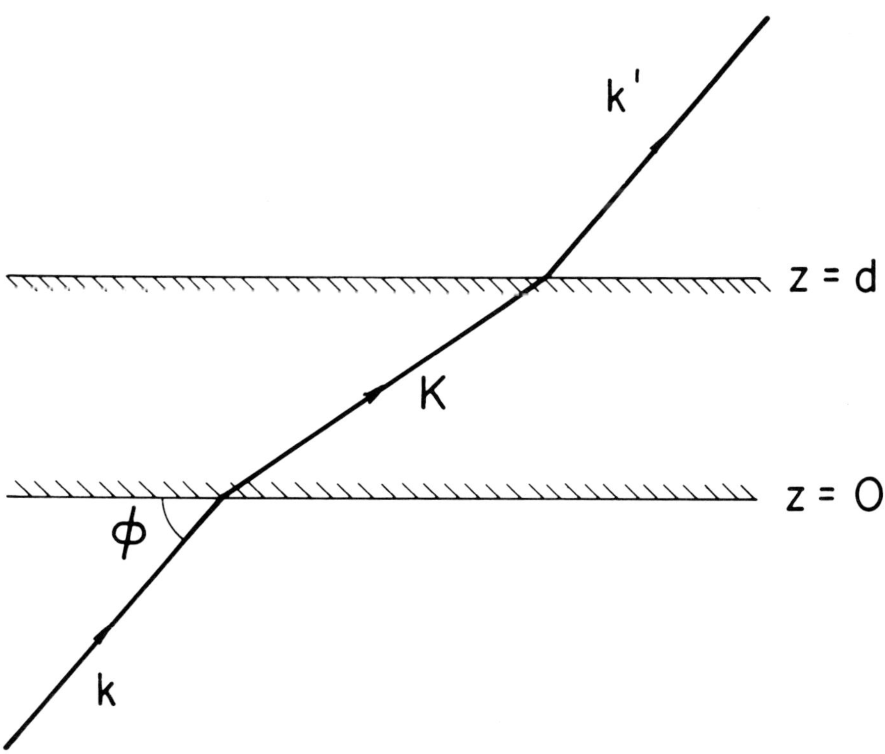

Fig. 3.7 Transmission through a plane slab.

and

$$
\begin{aligned}
a' &= a \exp(i\chi) \\
A &= a
\end{aligned}
\tag{3.2.3}
$$

Here χ is the phase shift produced by the slab and is given by

$$
\chi = (n_z - 1)kd \sin \phi \tag{3.2.4}
$$

where n_z is given by Eq. 3.1.19 as before.

Since $|\xi| \simeq 10^{-5}$ for thermal neutrons, we need only retain terms of lowest order in ξ. Then

$$
\begin{aligned}
n &= 1 - \tfrac{1}{2}\xi \\
n_z &= 1 - \tfrac{1}{2}\xi \operatorname{cosec}^2 \phi
\end{aligned}
\tag{3.2.5}
$$

and, hence,

$$
\chi = (n-1)kD = -\tfrac{1}{2}\xi kD \tag{3.2.6}
$$

where D is the thickness of the slab in the direction of the incident beam,

$$
D = d \operatorname{cosec} \phi \tag{3.2.7}
$$

Note that for normal incidence, where $\phi = \pi/2$, $D = d$, and $n_z = n$, the relation $\chi = (n-1)kD$ is exact.

Lambert's law

The transmissivity T is defined as the fraction of incident neutrons that are transmitted by the slab. Hence, as in Eq. 3.1.29,

$$
T = \frac{j_z'}{j_z} = \left|\frac{a'}{a}\right|^2 = |\exp(i\chi)|^2 \tag{3.2.8}
$$

With $\chi = \chi' + i\chi''$ we then get

$$
T = \exp(-2\chi'') \tag{3.2.9}
$$

so that the transmissivity is determined by the imaginary part of the phase shift.

Alternatively, it follows from Eq. 3.2.6 with $n = n' + in''$ that

$$
T = \exp(-\mu D) \tag{3.2.10}
$$

where

$$
\mu = 2kn'' \tag{3.2.11}
$$

Thus, one can state equivalently that the transmissivity is determined by the imaginary part of the index of refraction. The expression (3.2.10) is the neutron analog of Lambert's law of absorption in ordinary optics and μ is called the *absorption coefficient.* It follows with the help of Eq. 3.2.5

and Eq. 3.1.7 that

$$\mu = \frac{4\pi}{k}\rho b'' \tag{3.2.12}$$

or

$$\mu = \rho\sigma_a \tag{3.2.13}$$

where σ_a is the absorption cross section (Eq. 2.7.28).

The elementary theory of dispersion on which the preceding expression for μ is based only includes the attenuation of the coherent wave in the medium due to absorption by the nuclei. In reality, the coherent wave is also attenuated by diffuse scattering. The latter effect is included in the rigorous theory of dispersion given in Chap. 4, where it is shown that

$$\mu = \rho\sigma_t \tag{3.2.14}$$

where $\sigma_t = \sigma_a + \sigma_d$ is the total collision cross section, including contributions from both absorption and diffuse scattering.

Lambert's law can also be expressed as

$$T = \exp(-D/L) \tag{3.2.15}$$

where L is the *mean free path* between collisions,

$$L = \mu^{-1} = (\rho\sigma_t)^{-1} \tag{3.2.16}$$

If $D \ll L$ then $T \simeq 1$ and the medium is *transparent* to neutrons of the particular energy under consideration. On the other hand, if $D \gg L$ then $T \simeq 0$ and the medium is *opaque*. In general, ρ is typically of the order of 0.03 atoms/Å^3 and, for a weakly absorbing material, $\sigma_t \simeq 5$ b/atom so that $\mu \simeq 0.15$ cm^{-1} and $L \simeq 6.7$ cm. If $D = 1.0$ cm (say), then we get $T = 0.86$.

Total cross section measurement

Lambert's law (Eq. 3.2.10) provides the basis for an absolute measurement of the total collision cross section σ_t (Koester, 1977). The results of such measurements for the elements and their isotopes are summarized in the "barn book" (Mughabghab et al., 1981; Mughabghab, 1984).

The accuracy with which σ_t can be determined is limited not only by experimental uncertainties in ρ and D, but also by statistical fluctuations in the neutron counts that furnish the measurement of T. One can appreciate intuitively that the uncertainty in σ_t due to counting statistics will be very large if μD is either much greater than or much less than unity. There must, therefore, be some value of D for which the uncertainty in σ_t is a minimum. Since the fluctuations in the neutron counts obey Poisson statistics, one can show (Sears, 1983a) that the error in σ_t due to counting statistics is a minimum when $T = 0.109$. The optimum value of T is reduced to 0.078 if one optimizes not only the sample thickness but also the relative time spent accumulating counts in the presence and absence of the sample (Sears, 1985c).

Other applications

Lambert's law also has a number of important applications in neutron technology. For example, it is used in the design of shielding, filters, and detectors. In addition, the changes in transmissivity resulting from variations in composition in heterogeneous objects form the basis of neutron radiography (Hawkesworth and Walker, 1969; Berger, 1971; Von Der Hardt and Röttger, 1981). One reason for the importance of neutron radiography lies in the fact that neighboring elements in the periodic table, or even different isotopes of the same element, often have very different values of σ_t. As a result, good contrast can be obtained in situations where X-ray radiography would be quite ineffective. Another advantage of neutron over X-ray radiography is in the location of regions with high hydrogen concentration, which is made possible by the anomalously large scattering cross section of hydrogen (82 b/atom).

3.3. Eikonal equation

In geometrical optics it is conventional to draw ray diagrams to indicate how a beam of radiation is reflected and refracted at surfaces (Figs. 3.1 and 3.3), deflected by prisms (Fig. 3.4) and parallel plates (Fig. 3.7), or focused by lenses (Fig. 3.6). The eikonal equation, which we derive later, shows that in neutron optics these "rays" are simply the classical trajectories of the neutrons. More generally, it shows that classical mechanics is the geometrical optics limit of quantum mechanics.

In neutron optics experiments the neutron velocities are often sufficiently small, and the path lengths sufficiently long, that the effect of gravity is not negligible as it is in ordinary optics. This is especially true for ultracold neutrons. Thus, the neutron rays are in general parabolas rather than straight lines. In the presence of an inhomogeneous magnetic field, more complex trajectories may also occur. The total potential energy is given in general by

$$v = v_{\text{op}} - m\mathbf{g} \cdot \mathbf{r} - \boldsymbol{\mu} \cdot \mathbf{B} \tag{3.3.1}$$

where the optical potential that represents the effective interaction of the neutron with the medium is now denoted by v_{op}. The second term is the gravitational potential in which $\mathbf{g}$ denotes the gravitational acceleration. Finally, the third term is the energy of interaction of the neutron magnetic dipole moment $\boldsymbol{\mu}$ with the magnetic field $\mathbf{B}$.

The quantity v is discontinuous at the boundaries between different media as a result of the discontinuities in v_{op} and $\mathbf{B}$. However, within a homogeneous medium or in free space, v is a continuous, slowly varying function of **r** and is virtually constant over distances of the order of the neutron wavelength. Our first task, therefore, is to investigate the nature of the solution of the Schrödinger equation for such a slowly varying potential.

General state

Since the coherent wave $\Psi(\mathbf{r}, t)$ is in general complex, it can be expressed as

$$\Psi = A \exp(iS/\hbar) \tag{3.3.2}$$

where A and S are real scalar fields that represent the amplitude and phase of the coherent wave. In particular, S is called the *eikonal.* Substituting the preceding expression for Ψ into the time-dependent Schrödinger equation (2.4.2) with the complex optical potential $v = v' - iv''$, we find that the real and imaginary parts of the equation give rise to two coupled equations for determining A and S (Messiah, 1964). In particular, the amplitude equation is

$$\frac{\partial A}{\partial t} + \frac{1}{m}(\nabla A \cdot \nabla S + \tfrac{1}{2}A\,\Delta S) + \frac{1}{\hbar}v''A = 0 \tag{3.3.3}$$

and the eikonal equation is

$$\frac{\partial S}{\partial t} + \frac{1}{2m}(\nabla S)^2 + v' = \frac{\hbar^2}{2m}\frac{\Delta A}{A} \tag{3.3.4}$$

The neutron number density and current density (Eq. 2.4.3) can be expressed as

$$\rho = A^2, \qquad \mathbf{J} = \frac{A^2}{m}\nabla S \tag{3.3.5}$$

Multiplying the amplitude equation (3.3.3) by $2A$ we then find that this equation reduces to the equation of continuity (2.4.5). In other words, the amplitude equation is simply an expression of neutron balance. The implications of the equation of continuity were discussed in detail in Sec. 2.4, so that in what follows we shall confine our attention to the eikonal equation.

The neutron current density can be expressed as

$$\mathbf{J} = \rho\mathbf{v} \tag{3.3.6}$$

where the velocity field is

$$\mathbf{v} = \frac{1}{m}\nabla S \tag{3.3.7}$$

and the corresponding momentum field is

$$\mathbf{p} = m\mathbf{v} = \nabla S \tag{3.3.8}$$

The eikonal equation (3.3.4) then becomes

$$\frac{\partial S}{\partial t} + \frac{1}{2m}\mathbf{p}^2 + v_{\text{eff}} = 0 \tag{3.3.9}$$

in which the effective potential energy is given by

$$v_{\text{eff}} = v' - \frac{\hbar^2}{2m}\frac{\Delta A}{A} \tag{3.3.10}$$

The eikonal equation (3.3.9) is formally the same as the Hamilton–Jacobi equation in classical mechanics (Goldstein, 1950),

$$\frac{\partial S}{\partial t} + H = 0 \tag{3.3.11}$$

in which S is Hamilton's principal function and H is the Hamiltonian. In fact, Schrödinger originally obtained his celebrated equation by a synthesis of the Hamilton–Jacobi formulation of classical mechanics with the de Broglie relation between the momentum and the wavelength of a particle.

Taking the gradient of the eikonal equation (3.3.9) we get

$$\frac{d\mathbf{p}}{dt} = \mathbf{F} \tag{3.3.12}$$

in which

$$\frac{d}{dt} = \frac{\partial}{\partial t} + \mathbf{v} \cdot \nabla \tag{3.3.13}$$

is the material derivative and

$$\mathbf{F} = -\nabla v_{\text{eff}} \tag{3.3.14}$$

is the effective force. The relation (3.3.12) between the vector fields $\mathbf{p}$ and $\mathbf{F}$ is formally the same as Newton's law of motion.

Stationary state

For a stationary state,

$$\Psi(\mathbf{r}, t) = \psi(\mathbf{r}) \exp(-iEt/\hbar) \tag{3.3.15}$$

so that the eikonal is of the form

$$S(\mathbf{r}, t) = S(\mathbf{r}, 0) - Et \tag{3.3.16}$$

When one speaks of a *wave front* in optics one means a surface of constant phase, $S = \text{const}$. Equation 3.3.16 shows how the wave fronts propagate in time when the system is in a stationary state. Thus, for example, if $S(\mathbf{r}, 0) = c$ denotes some particular wave front at $t = 0$, then the same wave front is given at a later time t by $S(\mathbf{r}, t) = c - Et$. It is evident that as the wave front propagates through space it will generally change its shape.

When one speaks of a *ray* in optics one means a line that is everywhere perpendicular to the wave fronts. Thus, the rays are given by the momentum field $\mathbf{p} = \nabla S$ and, hence, are governed by an equation (Eq. 3.3.12) that is formally the same as Newton's law of motion.

According to Eq. 3.3.16,

$$E = -\frac{\partial S}{\partial t} \tag{3.3.17}$$

so that for a stationary state the eikonal equation (3.3.9) becomes

$$E = \frac{1}{2m}\mathbf{p}^2 + v_{\text{eff}} \tag{3.3.18}$$

which is a statement of energy conservation.

Geometrical optics and classical mechanics

When a thermal neutron is moving in the slowly varying potential of Eq. 3.3.1 the amplitude A is also slowly varying and we can neglect the $\hbar^2$ term in (Eq. 3.3.10). The equation (Eq. 3.3.12) that determines the neutron rays is then precisely the classical equation of motion. In this sense, therefore, classical mechanics is the geometrical optics limit of quantum mechanics.

In regions of space where v is a rapidly varying function of $\mathbf{r}$, such as at the boundary between different media, A will also be rapidly varying and the $\hbar^2$ term in Eq. 3.3.10 will no longer be negligible. In this case reflection and refraction or, in some situations, diffraction will become important and the motion of the neutron can no longer be treated classically. If v does not vary too rapidly, quantum corrections can be calculated by expanding A and S in powers of $\hbar$, a procedure known as the WKB method (Messiah, 1964).

In the geometrical optics limit, where $v_{\text{eff}} = v$ in the absence of absorption, the eikonal equation (Eq. 3.3.18) becomes

$$E = \frac{1}{2m}\mathbf{p}^2 + v = \frac{1}{2m}p_0^2 \tag{3.3.19}$$

where p_0 is the magnitude of the momentum of a neutron with the same energy in free space. Hence, it follows from Eq. 3.3.8 that

$$(\nabla S)^2 = n^2 p_0^2 \tag{3.3.20}$$

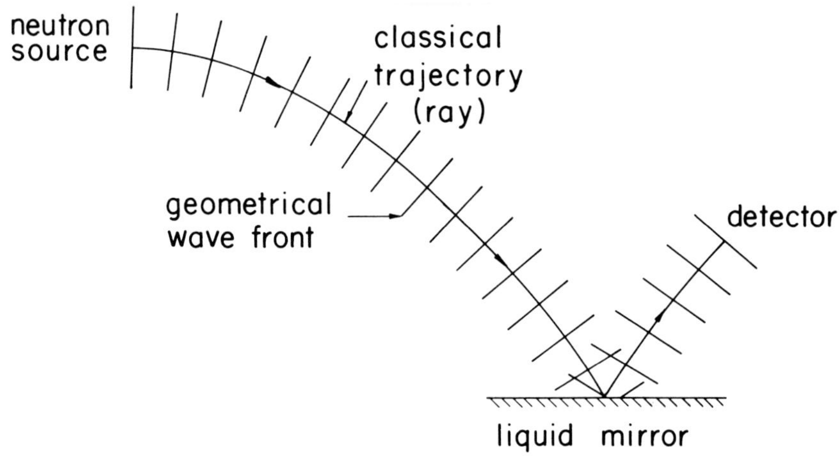

Fig. 3.8 Schematic representation of the neutron gravity refractometer as an illustration of the relationship between geometrical optics and classical mechanics.

where $n = p/p_0$ is the index of refraction and

$$n^2 = 1 - \frac{v}{E} \tag{3.3.21}$$

Note that in this equation v, and hence n, are not necessarily constant, as was assumed earlier in Eq. 3.1.4, but may in general be functions of $\mathbf{r}$. Equation 3.3.20 is the conventional form of the eikonal equation that determines the geometrical wave fronts in ordinary optics (Born and Wolf, 1975).

Finally, an example of the above results is illustrated in Fig. 3.8, which represents schematically the neutron gravity refractometer, which will be discussed in detail in the next section. Here an initially horizontal collimated beam of neutrons falls in the gravitational field onto the surface of a liquid mirror, from which it is reflected into a detector. The free fall of the neutron can be treated classically as described, since the gravitational potential is a slowly varying function of $\mathbf{r}$. However, the reflection of the neutron must be treated quantum-mechanically, since the optical potential of the mirror is a rapidly varying function of $\mathbf{r}$ in the neighborhood of its surface.

3.4. Neutron refractometry

The measurement of the index of refraction is referred to in general as *refractometry*. It is evident from Eq. 3.1.7 that neutron refractometry can be used to determine b when ρ is known or, conversely, ρ when b is known. The former application is the more common one, and values of the bound coherent scattering lengths have been determined with increasing accuracy during the past 40 years by the methods discussed later for all the elements in the periodic table and for most of the individual isotopes as well. A summary of the results of these various measurements was given some years ago by Koester et al. (1981) and a compilation of the best current values has been given by Sears (1986b).

In the present section we describe the methods of neutron refractometry that depend on the geometrical optics phenomena of reflection, refraction, and transmission. Those that depend on the physical optics phenomena of diffraction and interference will be discussed later in Chaps. 5, 6, and 7. Our treatment will be brief, focusing mainly on the principles involved. For more technical details, we refer the reader to the review articles of Koester (1977), Klein and Werner (1983), and Werner and Klein (1986).

Mirror reflection

The mirror reflection of neutrons was first demonstrated by Fermi and Zinn in 1944 (Segrè, 1962). It was initially used simply to determine the sign of the bound coherent scattering length of a number of elements (Fermi and Zinn, 1946; Fermi and Marshall, 1946). Later it was found that accurate values of the magnitude of the scattering length could also be obtained from mirror reflection measurements (Fermi and Marshall, 1947).

The conventional mirror reflection method has been discussed in detail by Dickinson et al. (1962). According to Eq. 3.1.33 one could, in principle, determine b by measuring the reflectivity R as a function of the angle of incidence ϕ using a collimated monoenergetic incident neutron beam. In practice, however, it is necessary to use a polyenergetic beryllium-filtered beam. The reason is simply that ϕ_c is so small ($\simeq 0.2$ degree) that the incident beam must be narrow and very finely collimated. Consequently, one cannot afford the additional loss in reflected intensity that would result from using an incident-beam monochromator.

The finite angular divergence of the incident beam, together with the continuous distribution of wavelengths, broadens the measured reflectivity curve and, hence, complicates the analysis of the data. As a result, the absolute measurement of b by the conventional mirror reflection technique is limited to accuracies of about 1 percent. However, higher accuracies of up to 0.1 percent can be obtained in measurements of the relative scattering lengths with different mirror materials in the same experimental setup.

In general, the bound coherent scattering length is complex, $b = b' - ib''$, and b'' is determined by the measured absorption cross section according to Eq. 2.7.51. Thus, the quantity that is actually obtained from the reflection measurements is the real part of the scattering length, b'.

The imaginary part of the scattering length, b'', has only a relatively small effect on the reflectivity, as we showed earlier in Sec. 3.1. Hence, b' can be accurated determined from reflectivity measurements even for strongly absorbing materials such as ^{3}He (Kitchens et al., 1974).

Gravity refractometry

An improvement in the mirror reflection method was made by Maier-Leibnitz (1962) that has led to an increase of two orders of magnitude in the accuracy with which neutron scattering lengths can be determined. We refer here to the neutron gravity refractometer that was mentioned at the end of Sec. 3.3 and is illustrated schematically in Fig. 3.8. This instrument is based on the fact that when a neutron with an initially horizontal velocity v falls through a vertical distance h, the angle at which it strikes a horizontal plane is given to first order by

$$\phi = \frac{1}{v}\sqrt{2gh} \tag{3.4.1}$$

where g is the gravitational acceleration. Since $mv = 2\pi\hbar/\lambda$, the reflectivity (Eq. 3.1.33) can be expressed as

$$R = \left|\frac{1-(1-h_c/h)^{1/2}}{1+(1-h_c/h)^{1/2}}\right|^2 \tag{3.4.2}$$

where h_c is the critical height for total reflection,

$$h_c = \left(\frac{2\pi\hbar^2}{m^2 g}\right)\rho b \tag{3.4.3}$$

It will be noted that, as one would expect,

$$mgh_c = v_0 \tag{3.4.4}$$

where v_0 is the optical potential in the mirror (Eq. 3.1.6).

The important point to note here is that h_c is independent of λ. Thus, the reflectivity curve (R versus h) is independent of the wavelength spread in the incident beam. Using the values of ρ and b for liquid lead in Table 3.1 we find, for example, that $h_c = 74$ cm.

The horizontal distance that a neutron travels while falling through a vertical distance h is given by

$$L = v\left(\frac{2h}{g}\right)^{1/2} \tag{3.4.5}$$

and this depends on the initial horizontal velocity v and, hence, on the neutron wavelength λ. For $L = 100$ m and $h = 74$ cm we get $v = 257$ m s^{-1} which corresponds to $\lambda = 15.4$ Å. To obtain sufficient reflected intensity for such long-wavelength neutrons, the length of the reflecting surface must be of the order of 1 m. Consequently, the method is applied most conveniently to liquid mirrors.

A neutron gravity refractometer of the type just described was constructed at the FRM reactor in Munich by Koester (1965) who, together with his students and colleagues, has measured the coherent scattering lengths of many elements with accuracies that, in the most favorable cases, are of the order of 0.01 percent. A typical reflectivity curve is that shown in Fig. 3.9 for a liquid lead mirror at 662.6 K (Koester et al., 1986). To obtain such high accuracy a number of effects that are usually negligible in other situations in neutron optics must be taken into account. For example, the centrifugal and Coriolis forces due to the rotation of the Earth give a significant correction of the order of 0.2 percent in gravity refractometry measurements (Koester and Nistler, 1971). For more details on the accurate measurement of scattering lengths by gravity refractometry, we refer the reader to the article by Koester and Nistler (1975).

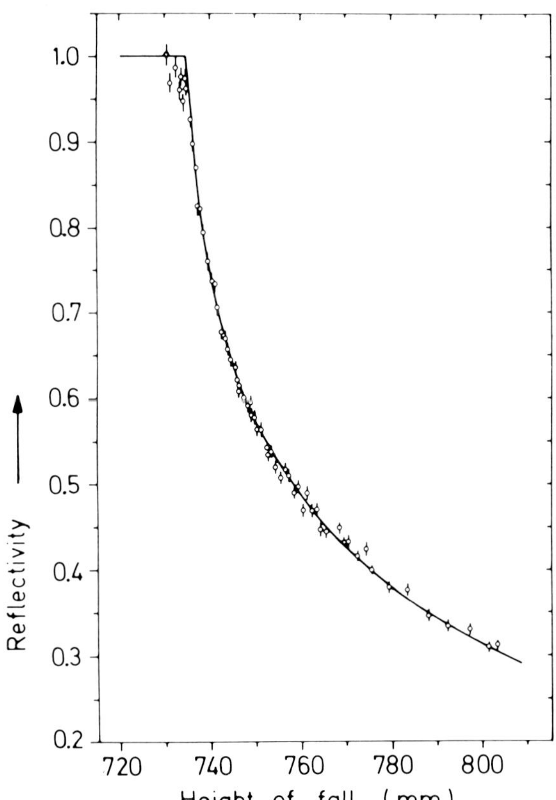

Fig. 3.9 Reflectivity of liquid lead as measured by neutron gravity refractometry (from Koester et al., 1986).

The measured free fall of a thermal neutron beam has also been used to obtain the gravitational acceleration g (McReynolds, 1951; Dabbs et al., 1965). Also, a comparison of values of neutron scattering lengths measured by neutron gravity refractometry with the corresponding values obtained by methods in which gravity plays a negligible role has been used to verify Galileo's principle of the universality of free fall with an accuracy of 3×10^{-4} (Koester, 1976; Sears, 1982b).

Prism refraction

It is evident from Eq. 3.1.40 that the refractive bending of a monoenergetic neutron beam by a prism can be used to determine ξ and, hence, the coherent scattering length b. Although the angle of deviation δ is very small, typically a few seconds of arc for thermal neutrons, it can nevertheless be measured very accurately using a double perfect-crystal spectrometer (Schneider and Shull, 1971). Values of b have been determined for a few elements in this way with an accuracy of about 0.03 percent (Schneider and Shull, 1971; Schneider, 1973, 1976).

According to Eq. 3.1.43, it is also possible, at least in principle, to determine the scattering length by measuring the focal length of a lens with monoenergetic neutrons. However, to the best of our knowledge, this has not yet been done.

Christiansen filter

The Christiansen filter technique employs a heterogeneous sample consisting of a finitely divided powder immersed in a liquid. The relative index of refraction is given according to Eq. 3.1.14 by

$$n = 1 - \frac{\lambda^2}{2\pi}(\rho_p b_{\mathrm{p}} - \rho_l b_l) \tag{3.4.6}$$

where p labels the powder and l the liquid. If $\rho_p b_p \neq \rho_l b_l$ then the angular divergence of the incident neutron beam will be broadened by small-angle scattering (caused by multiple refraction or diffraction, depending on the size of the powder particles [Weiss, 1951]). On the other hand, if $\rho_p b_p = \rho_l b_l$ then the sample is optically homogeneous and no broadening will occur. This technique was originally invented by Christiansen in 1884 in connection with the ordinary optical index of refraction. Later Weiss (1951) suggested that it might be used to determine neutron scattering lengths. This was eventually done with considerable success by Koester and Ungerer (1969).

The method consists in measuring the small-angle scattering as a function of $\rho_l b_l$. The curve is then a minimum when

$$\rho_p b_p = \rho_l b_l \tag{3.4.7}$$

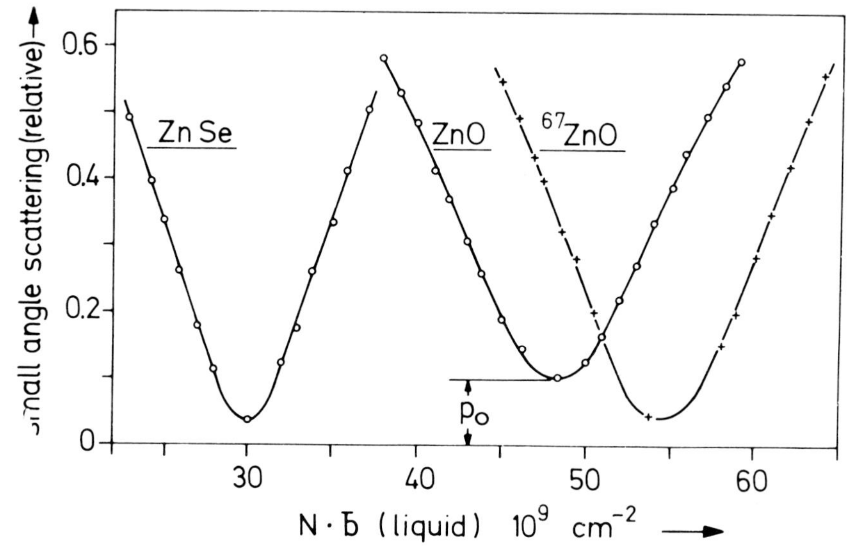

Fig. 3.10 The Christiansen filter method as applied in the determination of the bound coherent scattering length of zinc (from Koester et al., 1985). The abscissa is $\rho_l b_l$ in our notation.

By determining precisely the position of this minimum, one can then obtain b_p from the measured values of ρ_p and ρ_l and the known value of b_l. The latter quantity can be obtained by an absolute measurement such as gravity refractometry. The values of b_p obtained in this way have a typical accuracy of about 0.1 percent. The value of $\rho_l b_l$ is most conveniently varied by using a two-component liquid and by changing the relative concentrations of the components. For example, if the powder is not soluble in water, one can use a mixture of H_2O and D_2O. Alternatively, one can use mixtures of various organic liquids. An example of the Christiansen filter method is given in Fig. 3.10, which shows part of a series of measurements by Koester et al. (1985) to determine the bound coherent scattering length of zinc and its various isotopes.

Neutron scattering lengths can also be determined by measuring the width of the small-angle scattering away from the minimum (Weiss, 1951). However, the values of b obtained in this way are not very accurate, so that this method has not been used much in practice.

Transmission

Transmission measurements provide an absolute determination of the total collision cross section σ_t, as we saw earlier in Sec. 3.2. After subtracting the contributions from absorption and incoherent scattering,

Table 3.3 Bound coherent scattering lengths (in fm) of selected elements as determined by various neutron-optical methods

Element	Gravity refractometry	Prism refraction	Dynamical diffraction	Interfer-ometry	Trans-mission	Christiansen filter	Mirror reflection	Bragg reflection
H	−3.7390(11)			−3.741(4)	−3.733(4)		−3.74(2)	−3.7(6)
C	6.6460(12)			6.6470(14)	6.648(4)			6.4(2)
O	5.803(4)	5.830(2)		5.83(5)	5.804(7)	5.805(4)	5.80(5)	5.8(2)
F	5.665(15)				5.603(11)	5.654(12)	5.74(9)	5.60(5)
Si		4.1478(16)	4.149(1)		4.145(4)	4.159(6)		
Cl	9.5770(8)					9.59(7)		9.9(2)
Ga	7.2879(16)					7.284(15)		7.21(5)
Ge		8.1929(17)	8.186(4)					8.4(2)
Sn	6.2257(19)			6.228(4)				6.1(1)
Tl	8.776(4)					8.89(2)		8.9
Pb	9.4017(20)				9.409(4)	9.39(6)		9.6(4)
Bi	8.5307(20)			8.503(12)	8.530(5)			8.9(4)
*	0.01%	0.03%	0.03%	0.1%	0.1%	0.1%	1%	1%

* Characteristic accuracy under favorable conditions.

one can then determine the coherent scattering cross section and, hence, the coherent scattering length.

The determination of b in this way is complicated by the presence of interatomic interference effects in the coherent scattering (as given, for example, in Eq. 2.9.36), and by inelastic scattering effects in both the coherent and incoherent scattering. The interatomic interference effects are negligible when the sample is a dilute gas (Fermi and Marshall, 1947). For liquids and solids the interatomic interference effects can be minimized by using epithermal neutrons with energies $\gtrsim 1$ eV (Triftshäuser, 1965). In the latter case, an additional correction for the spin–orbit (or Schwinger) interaction must, however, be taken into account (Koester et al., 1976). When the interatomic interference effects are negligible, the correction for inelastic scattering can be calculated analytically. In situations in which the various corrections are well under control, the values of b obtained from transmission measurements have a typical accuracy of about 0.1 percent.

Discussion

Values of the bound coherent scattering lengths of a selected group of elements as determined by various neutron-optical methods are listed for comparison in Table 3.3. With three exceptions, the values for any given element are all in agreement within about two standard deviations.

The traditional methods of mirror reflection and Bragg reflection in powders are seldom used today owing to their relatively low accuracy in comparison with the other, more modern methods in Table 3.3. The most accurate method is gravity refractometry. However, this requires samples in the form of large liquid mirrors with linear dimensions of the order of 1 m. Among the remaining methods, the most versatile are perfect-crystal neutron interferometry, transmission, and the Christiansen filter technique. The latter is especially useful for measuring the scattering lengths of separated isotopes of an element since it requires only small powder samples.

3.5. Neutron polarization

Refractive index for a ferromagnet

To calculate the index of refraction for a ferromagnet, the optical potential (Eq. 2.8.42) must be modified to include the energy of interaction of the neutron magnetic dipole moment $\boldsymbol{\mu}$ with the magnetic field. Then

$$v(\mathbf{r}) = \begin{cases} v_0 - \boldsymbol{\mu} \cdot \mathbf{B}, & \text{inside } V \\ -\boldsymbol{\mu} \cdot \mathbf{B}', & \text{outside } V \end{cases} \tag{3.5.1}$$

where **B** and **B**′ denote the magnetic induction inside and outside the volume V occupied by the ferromagnetic medium.

The quantity that determines the reflection and refraction of neutrons at the surface of the ferromagnet is the relative index of refraction (Eq. 3.1.12) which can, as before, be expressed as

$$n^2 = 1 - \xi \tag{3.5.2}$$

where now

$$\xi = \frac{1}{E}[v_0 - \boldsymbol{\mu} \cdot (\mathbf{B} - \mathbf{B}')] \tag{3.5.3}$$

Thus, the magnetic reflection of neutrons is due to the discontinuity in **B** at the surface of the ferromagnet (Lax, 1951).

In general,

$$\mathbf{B} = \mathbf{H} + 4\pi\mathbf{M} \tag{3.5.4}$$

where **H** is the magnetic field inside the ferromagnet and **M** is the corresponding magnetization (in gaussian units). In practice **H** is made sufficiently large that **M** has its saturation value. On the other hand,

$$\mathbf{B}' = \mathbf{H}' \tag{3.5.5}$$

since the magnetization vanishes outside V. Let us now resolve the fields into components with directions normal and tangential to the surface so that, for example, $\mathbf{B} = \mathbf{B}_n + \mathbf{B}_t$. Then,

$$\begin{aligned} \mathbf{B} - \mathbf{B}' &= (\mathbf{B}_n - \mathbf{B}'_n) + (\mathbf{B}_t - \mathbf{B}'_t) \\ &= (\mathbf{B}_n - \mathbf{B}'_n) + (\mathbf{H}_t - \mathbf{H}'_t) + 4\pi\mathbf{M}_t \end{aligned} \tag{3.5.6}$$

Since $\operatorname{div} \mathbf{B} = 0$ it follows from Gauss' theorem that the normal component of **B** is continuous at the boundary ($\mathbf{B}_n = \mathbf{B}'_n$). Furthermore, since $\operatorname{curl} \mathbf{H} = 0$ it follows from Stokes' theorem that the tangential component of **H** is continuous at the boundary ($\mathbf{H}_t = \mathbf{H}'_t$). Hence, Eq. 3.5.6 reduces to

$$\mathbf{B} - \mathbf{B}' = 4\pi\mathbf{M}_t \tag{3.5.7}$$

and Eq. 3.5.3 becomes

$$\xi = \frac{1}{E}(v_0 - 4\pi\boldsymbol{\mu} \cdot \mathbf{M}_t) \tag{3.5.8}$$

Thus, one can state equivalently that the magnetic reflection of neutrons is due to the tangential component of the magnetization.

The neutron magnetic dipole moment can be expressed as

$$\boldsymbol{\mu} = \gamma\beta_N\boldsymbol{\sigma} \tag{3.5.9}$$

where $\boldsymbol{\sigma} = 2\mathbf{s}$ is the Pauli spin operator, $\beta_N = e\hbar/2mc$ is the nuclear magneton, and $\gamma = -1.913$ is the gyromagnetic ratio. Since **M** is the average magnetic dipole moment per unit volume, we can write

$$\mathbf{M} = \rho\mu_A\hat{\mathbf{e}} \tag{3.5.10}$$

where ρ is the number of atoms per unit volume, μ_A the average magnetic dipole moment per atom, and $\hat{\mathbf{e}}$ a unit vector in the direction of $\mathbf{M}$. We can also write

$$\mu_A = 2\beta_B S \tag{3.5.11}$$

where $\beta_B = e\hbar/2m_e c$ is the Bohr magneton and S the effective spin. Combining the preceding results, we then find with the help of Eq. 3.1.6 that

$$\xi = \frac{4\pi}{k^2}\rho(b + b_M\boldsymbol{\sigma}\cdot\hat{\mathbf{e}}_t) \tag{3.5.12}$$

where b_M can be called the *magnetic scattering length* and is given by

$$b_M = -\gamma r_e S \tag{3.5.13}$$

where $r_e = e^2/m_e c^2 = 2.818$ fm is the classical electron radius.

Since it is only the component of the magnetization in the tangential direction that is effective in magnetic reflection, the mirror is always magnetized in this direction in practice. In particular, $\mathbf{M}$ is chosen to be parallel to the surface of the mirror and perpendicular to the plane of scattering in Fig. 3.1. If we now take the z direction to be the direction of $\mathbf{M}$ then Eq. 3.5.12 becomes

$$\xi = \frac{4\pi}{k^2}\rho(b + b_M\sigma_z) \tag{3.5.14}$$

Since σ_z has eigenvalues ± 1, the corresponding values of the refractive index are given by

$$n_\pm^2 = 1 - \xi_\pm \tag{3.5.15}$$

where

$$\xi_\pm = \frac{4\pi}{k^2}\rho b_\pm \tag{3.5.16}$$

in which

$$b_\pm = b \pm b_M \tag{3.5.17}$$

and the $\pm$ refers to neutrons with spin parallel (+) or antiparallel (−) to $\mathbf{M}$. The double-valued nature of the refractive index of a ferromagnet is analogous to birefringence in ordinary optics. Apart from some misprints that were later corrected by Halpern (1949a), the preceding expression for $n_\pm$ was first obtained by Halpern et al. (1941). The values of the spin-dependent scattering lengths $b_\pm$ for the 3d ferromagnets iron, cobalt, and nickel are given in Table 3.4.

Table 3.4 Values of the spin-dependent scattering lengths for 3d ferromagnets

Quantity	Fe	Co	Ni
$\mu_A/\beta_B = 2S$	2.22	1.72	0.60
b_M(fm)	5.98	4.64	1.62
b(fm)	9.54	2.50	10.3
b_+(fm)	15.52	7.14	11.9
b_-(fm)	3.56	−2.14	8.7

Reflectivity and polarization

The polarization of a neutron beam in the z direction is defined as

$$p = \langle \sigma_z \rangle \tag{3.5.18}$$

where the brackets $\langle \cdots \rangle$ denote an average over all the neutrons in the beam. In general,

$$-1 \leq p \leq 1 \tag{3.5.19}$$

The beam is said to be completely polarized if $p = \pm 1$, unpolarized if $p = 0$, and partially polarized otherwise.

If the incident neutron beam is monoenergetic its polarization is given, according to the preceding definition, by

$$p = \frac{j_+ - j_-}{j_+ + j_-} \tag{3.5.20}$$

where $j_\pm$ denotes the incident neutron flux in the $\pm$ spin states. Similarly, the polarization of the reflected beam is given by

$$p' = \frac{j'_+ - j'_-}{j'_+ + j'_-} \tag{3.5.21}$$

where $j'_\pm = R_\pm j_\pm$ is the reflected neutron flux in the $\pm$ spin states and $R_\pm$ denotes the corresponding reflectivities. Hence,

$$p' = \frac{R_+(1+p) - R_-(1-p)}{R_+(1+p) + R_-(1-p)} \tag{3.5.22}$$

Thus, we see that if the incident beam is completely polarized ($p = \pm 1$) so is the reflected beam ($p' = \pm 1$). On the other hand, if the incident beam is unpolarized ($p = 0$), then

$$p' = \frac{R_+ - R_-}{R_+ + R_-} \tag{3.5.23}$$

and the reflected beam will be at least partially polarized when the mirror is magnetized and $R_+ \neq R_-$. However, if the mirror is unmagnetized then

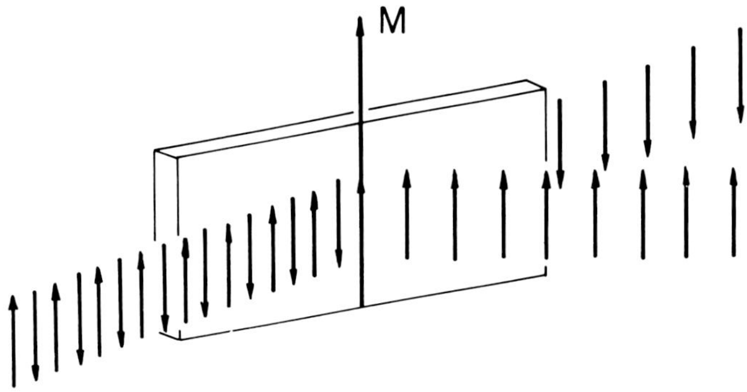

Fig. 3.11 Neutron polarization by reflection from a ferromagnetic mirror.

$R_+ = R_-$, in which case $p' = p$ and the mirror has no effect on the polarization of the reflected beam.

The critical angles for total reflection for two spin states are given according to Eq. 3.1.34 by (Halpern, 1949b)

$$\phi_{c,\pm} = \sqrt{\xi_\pm} \tag{3.5.24}$$

The magnetic scattering length b_M is always positive so that $b_+ > b_-$ and, when both critical angles are real, $\phi_{c,+} > \phi_{c,-}$. It is then evident from Fig. 3.2 that if the angle of incidence ϕ is below $\phi_{c,+}$ but still well above $\phi_{c,-}$ we shall have $R_+ = 1$ and $R_- \simeq 0$ in which case $p' \simeq +1$ as long as $p > -1$. For this angle of incidence, therefore, the reflected beam will be almost completely polarized in the direction of **M**, while the transmitted beam will be almost completely polarized in the opposite direction, as illustrated schematically in Fig. 3.11.

Production of polarized neutron beams

The production of polarized neutrons by reflection from ferromagnetic mirrors was originally proposed by Halpern (1949b) and subsequently demonstrated by Hughes and Burgy (1951) with a cobalt mirror. The advantage of cobalt is that $b_+ > 0$ while $b_- < 0$ (see Table 3.4). Consequently, total reflection occurs only for the + spin neutrons and an almost completely polarized reflected beam is obtained even with a polyenergetic incident beam. Today, polarizing mirrors are usually made from sheets of FeCo mounted on a nonmagnetic base such as brass or glass (Hamelin, 1976). The sheets are made thin ($\simeq$0.3 mm) in order that magnetic saturation can be achieved with relatively small applied fields.

The disadvantage of using conventional ferromagnetic mirrors to produce polarized neutrons is that the reflected beam is very narrow unless the mirror is very long (see Eq. 3.1.35). For thermal neutrons this disadvantage can be overcome by using, instead, the Bragg reflection of neutrons by ferromagnetic crystals, as will be discussed later in Chap. 5.

Hayter et al. (1978) have developed a magnetized curved Soller guide tube for the polarization of cold neutrons with $\lambda > 5$ Å. The main advantage of this device is that it can polarize large beam areas (5×4 cm) in a relatively short distance ($\simeq 18$ cm). Other devices for the production of polarized neutrons employing thin-film multilayers will be discussed later in Chap. 7. See also the review article by Hayter (1978).

Neutrons can also be reflected and refracted by a magnetic field in vacuo. If the magnetic field is produced within a volume V then Eq. 3.5.1 becomes

$$v(\mathbf{r}) = \begin{cases} -\boldsymbol{\mu} \cdot \mathbf{B}, & \text{inside } V \\ 0, & \text{outside } V \end{cases} \tag{3.5.25}$$

and the index of refraction is again given by Eq. 3.5.2 with

$$\xi = -\frac{\boldsymbol{\mu} \cdot \mathbf{B}}{E} \tag{3.5.26}$$

For thermal neutrons, ξ has roughly the same value as in a material medium if $B \simeq 1$ T (i.e., 10^4 gauss). The production of polarized neutrons by refraction by shaped magnetic fields in the form of prisms or lenses has been demonstrated (Just et al., 1973; Badurek et al., 1979) but has not yet found much practical application.

The force on a neutron in an inhomogeneous magnetic field is given by

$$\mathbf{F} = -\boldsymbol{\nabla} v = \boldsymbol{\nabla}(\boldsymbol{\mu} \cdot \mathbf{B}) \tag{3.5.27}$$

Since $\mathbf{F}$ has the opposite sign for the two spin states, this leads to a spatial separation of the two spin components in the incident beam. This neutron analog of the Stern–Gerlach effect is not used in practice for the production of polarized neutrons, but it can be used to advantage in the analysis of polarized neutron beams with a typical accuracy of ± 1 percent (Sherwood et al., 1954; Barkan et al., 1968).

Incoherent scattering length measurement

In deriving the macroscopic optical potential (Eq. 2.8.43) we tacitly assumed that the neutrons were unpolarized. More generally,

$$v_0 = \frac{2\pi\hbar^2}{m} \rho \langle b \rangle \tag{3.5.28}$$

where b is the spin-dependent scattering length (Eq. 2.7.30), which can also be expressed as

$$b = b_c + [I(I+1)]^{-1/2} b_i \boldsymbol{\sigma} \cdot \mathbf{I} \tag{3.5.29}$$

where $\boldsymbol{\sigma} = 2\mathbf{s}$ is again the Pauli spin operator. In the usual situation where $\boldsymbol{\sigma}$ and $\mathbf{I}$ are statistically independent,

$$\langle b \rangle = b_c + \left(\frac{I}{I+1}\right)^{1/2} b_i \mathbf{p} \cdot \mathbf{P} \tag{3.5.30}$$

where $\mathbf{p} = \langle \boldsymbol{\sigma} \rangle$ is the neutron polarization vector and $\mathbf{P} = \langle \mathbf{I} \rangle / I$ the nuclear polarization vector.

Values of the neutron polarization p approaching 100 percent can be obtained by the methods discussed earlier in this section. The nuclei can be partially polarized by applying an external magnetic field or by the spontaneous magnetic field that exists in a ferromagnet. However, the values of the nuclear polarization P obtained in this way are usually <1 percent. On the other hand, values of $P \gtrsim 50$ percent, and in favorable cases $\gtrsim 90$ percent, can be obtained by dynamic polarization techniques (Abragam and Goldman, 1978).

Since the quantity b in Eq. 3.1.7 is actually $\langle b \rangle$, the quantity that is measured using the neutron-optical methods discussed in Sec. 3.4 is likewise $\langle b \rangle$. It is only when the neutrons or the nuclei are unpolarized that $\langle b \rangle = b_c$. If the neutrons and the nuclei are both polarized, then $\langle b \rangle$ depends on both b_c and b_i. In principle, therefore, each of the methods in Sec. 3.4 is capable of determining not only b_c but also b_i if the neutrons and the nuclei are both polarized. In fact, values of b_i have been determined in this way by transmission experiments (Schermer, 1963) and Bragg scattering experiments (Shull and Ferrier, 1963; Ito and Shull, 1969; Abragam et al., 1972).

The most common and accurate method of measuring b_i today is, however, not simply an extension of one of the methods in Sec. 3.4. Rather, it is the so-called pseudomagnetic method (Baryshevskii and Podgoretskii, 1965; Abragam et al., 1973), which begins with the observation that the spin-dependent optical potential can be expressed as

$$v_0 = v_c - \boldsymbol{\mu} \cdot \mathbf{B}^* \tag{3.5.31}$$

where

$$v_c = \frac{2\pi\hbar^2}{m} \rho b_c \tag{3.5.32}$$

$\boldsymbol{\mu}$ is the neutron magnetic dipole moment (Eq. 3.5.9), and $\mathbf{B}^*$ is a pseudomagnetic field given by

$$\mathbf{B}^* = -\left(\frac{I}{I+1}\right)^{1/2} \left(\frac{2\pi\hbar^2 \rho b_i}{\gamma \beta_N m}\right) \mathbf{P} \tag{3.5.33}$$

Alternatively,

$$\mathbf{B}^* = 4\pi\rho\boldsymbol{\mu}^* \tag{3.5.34}$$

where $\boldsymbol{\mu}^*$ is the nuclear pseudomagnetic dipole moment,

$$\boldsymbol{\mu}^* = -\beta_B \left(\frac{I}{I+1}\right)^{1/2} \left(\frac{2b_i}{\gamma r_e}\right) \mathbf{P} \tag{3.5.35}$$

in which β_B denotes the Bohr magneton and r_e the classical electron radius as before. The value of $\boldsymbol{\mu}^*$ is typically three orders of magnitude larger than the actual magnetic dipole moment of the nucleus.

Although $\mathbf{B}^*$ is a purely nuclear quantity characterizing the spin dependence of the strong interaction between the neutron and the nuclei, it is nevertheless equivalent to a magnetic field. Thus, the neutron magnetic dipole moment $\boldsymbol{\mu}$ precesses about $\mathbf{B}^*$ with the Larmor frequency

$$\omega_L = \frac{2\mu B^*}{\hbar} \tag{3.5.36}$$

just as it would in a true magnetic field. The value of ω_L can be measured by standard magnetic resonance techniques whereupon the value of B^*, and hence of b_i, can be determined.

Values of b_i have now been obtained by the pseudomagnetic method for most stable nuclides with $I \neq 0$. The experimental uncertainty in these values of b_i is typically ± 0.05 fm. More information on this method can be found in the review articles by Abragam and Goldman (1978) and Glättli and Goldman (1986).

4

RIGOROUS THEORY OF DISPERSION

In Chap. 2 we showed that the coherent wave $\psi(\mathbf{r})$, which describes all neutron optical phenomena, satisfies the one-body Schrödinger equation (2.8.32) containing a complex optical potential $v(\mathbf{r})$ that represents the effective interaction of the neutron with the medium. In the elementary theory of dispersion given in that chapter we derived an approximate expression (Eq. 2.8.42) for $v(\mathbf{r})$ that formed the basis for the discussion of geometrical optics in Chap. 3. One defect in the elementary theory of dispersion, which we pointed out in Sec. 3.2, is that it includes only the attenuation of the coherent wave in the medium resulting from absorption by the nuclei and ignores the additional attenuation due to diffuse scattering that must also be present in reality.

In the present chapter we give a rigorous theory of dispersion based on a systematic self-consistent solution of the Ewald equations (2.1.24) and (2.1.25). We demonstrate that when one takes properly into account both the local-field effects and the correction to the scattering amplitude, which are neglected in the elementary theory, one obtains an expression for the optical potential that correctly describes the attenuation of the coherent wave by both absorption and diffuse scattering.

The rigorous theory of dispersion in ordinary optics was originally developed by Lorentz, Ewald, and others early in the present century and has been reviewed, by, for example, Born and Wolf (1975) and Van Kranendonk and Sipe (1977). The analogous theory for neutrons was first given by Foldy (1945), Lax (1951, 1952), and Ekstein (1951, 1953). Further extensions have been made by Lenk (1975), Blaudeck and Lenk (1976), Sears (1982a), Nowak (1982b), and Warner and Gubernatis (1985). The present treatment follows that given in Sears (1982a). A brief, nontechnical account of this subject can also be found in Sears (1988).

4.1. Coherent wave

Ewald's equations in the static approximation

The rigorous theory of dispersion requires the solution of the rather intricate multiple-scattering problem that is implicit in the Ewald equations. To avoid any unnecessary complexities in the solution of this problem, and to display the underlying physics in the simplest possible way, we shall idealize the medium by using the static approximation that was introduced in Sec. 2.9. Thus, we assume that each atom in the system is strongly bound (in the sense of Sec. 2.7) to some fixed position in space. The collision of the neutron with the system is then a one-body problem.

In the static approximation, the channel state $\Phi = |\mathbf{k}\rangle$, and the Ewald equations (2.1.24) and (2.1.25) then reduce to

$$\Psi = |\mathbf{k}\rangle + \sum_i GT_i\chi_i \tag{4.1.1}$$

and

$$\chi_i = |\mathbf{k}\rangle + \sum_{j(\neq i)} GT_j\chi_j \tag{4.1.2}$$

Here Ψ is the scattering state corresponding to the collision of a neutron, with incident wave vector $\mathbf{k}$, with a static configuration of atoms with positions $\mathbf{r}_i$.

The wave function of the state Ψ is given in the coordinate representation by

$$\begin{aligned}\Psi(\mathbf{r}) &= \langle \mathbf{r} \mid \Psi\rangle \\ &= \langle \mathbf{r} \mid \mathbf{k}\rangle + \sum_i \int d\mathbf{r}' \int d\mathbf{r}'' \langle \mathbf{r}| G |\mathbf{r}'\rangle \langle \mathbf{r}'| T_i |\mathbf{r}''\rangle \langle \mathbf{r}'' \mid \chi_i\rangle \\ &= \phi_{\mathbf{k}}(\mathbf{r}) + \sum_i \int d\mathbf{r}' \int d\mathbf{r}'' \langle \mathbf{r}| G |\mathbf{r}'\rangle \langle \mathbf{r}'| T_i |\mathbf{r}''\rangle \chi_i(\mathbf{r}'')\end{aligned} \tag{4.1.3}$$

where, as in Sec. 2.2,

$$\phi_{\mathbf{k}}(\mathbf{r}) = \langle \mathbf{r} \mid \mathbf{k}\rangle = (2\pi)^{-3/2} \exp(i\mathbf{k}\cdot\mathbf{r}) \tag{4.1.4}$$

and

$$\chi_i(\mathbf{r}) = \langle \mathbf{r} \mid \chi_i\rangle \tag{4.1.5}$$

The Green function is given by Eq. 2.2.18 and can be expressed as

$$\langle \mathbf{r}| G |\mathbf{r}'\rangle = -\frac{m}{2\pi\hbar^2} G(\mathbf{r}-\mathbf{r}') \tag{4.1.6}$$

where

$$G(\mathbf{r}) = \frac{1}{r}\exp(ikr) \tag{4.1.7}$$

The T-matrix for the ith atom is given in the coordinate representation by

$$\langle \mathbf{r}| T_i |\mathbf{r}'\rangle = T_i(\mathbf{r})\,\delta(\mathbf{r}-\mathbf{r}') \tag{4.1.8}$$

where $T_i(\mathbf{r})$ is the concrete transition operator in this representation. In the impulse approximation (Eq. 2.7.3) this operator is of the form

$$T_i(\mathbf{r}) = -\frac{2\pi\hbar^2}{m} f_i\,\delta(\mathbf{r}-\mathbf{r}_i) \tag{4.1.9}$$

where f_i is the bound scattering amplitude of the ith atom and can be expressed with the help of Eq. 2.3.38 in the form

$$f_i = -b_i + ikb_i^2 + O(k^2) \tag{4.1.10}$$

in which b_i is the bound scattering length. In the elementary theory of dispersion given in Sec. 2.8 we used the Fermi pseudopotential for $T_i(\mathbf{r})$ in which $f_i = -b_i$. However, we must now include the second term in Eq. 4.1.10, since this term is of the same order of magnitude as both the local field effects and the imaginary part of b_i.

Substituting the preceding results into Eq. 4.1.3, and neglecting the normalization factor $(2\pi)^{-3/2}$, we get

$$\Psi(\mathbf{r}) = \exp(i\mathbf{k}\cdot\mathbf{r}) + \sum_i G(\mathbf{r}-\mathbf{r}_i)f_i\chi_i \tag{4.1.11}$$

where, now,

$$\chi_i \equiv (2\pi)^{3/2}\chi_i(\mathbf{r}_i) \tag{4.1.12}$$

Similarly, Eq. 4.1.2 is given in the coordinate representation by

$$\chi_i = \exp(i\mathbf{k}\cdot\mathbf{r}_i) + \sum_{j(\neq i)} G(\mathbf{r}_i-\mathbf{r}_j)f_j\chi_j \tag{4.1.13}$$

These equations have a simple and appealing physical interpretation. Equation 4.1.11 states that, at each each point $\mathbf{r}$, the neutron wave function $\Psi(\mathbf{r})$ is the sum of the incident plane wave $\exp(i\mathbf{k}\cdot\mathbf{r})$ plus a scattered spherical wave $G(\mathbf{r}-\mathbf{r}_i)$ from each atom in the system. The local field χ_i that generates this scattered wave is, according to Eq. 4.1.13, the sum of the incident plane wave plus the scattered spherical waves from all the other atoms. Thus, it is evident that $\Psi(\mathbf{r})$ embodies all possible multiple-scattering effects in the collision of the neutron with the system of atoms.

Coherent wave and the local field

The wave function $\Psi(\mathbf{r})$ is a microscopic quantity in the sense that it depends parametrically on the positions of the atoms $\mathbf{r}_i$. To each atomic configuration there corresponds a distinct wave function $\Psi(\mathbf{r})$. In Sec. 2.8 we showed that the quantity that determines the neutron optical phenomena is the coherent wave,

$$\psi(\mathbf{r}) = \langle\Psi(\mathbf{r})\rangle \tag{4.1.14}$$

in which the brackets $\langle\cdots\rangle$ denote a thermodynamic average over all the atomic configurations. In evaluating this average we shall, for simplicity, assume that the system is macroscopically homogeneous, e.g., a liquid or amorphous solid. Again for simplicity, we shall restrict our attention to a monatomic system.

Just as in ordinary optics, where it is necessary to introduce two macroscopic vector fields **E** and **D** to describe the optical properties of dielectric media, so in neutron optics it is necessary to introduce two macroscopic scalar fields to describe the neutron-optical effects. These are the coherent wave $\psi(\mathbf{r})$ and the macroscopic local field

$$\chi(\mathbf{r}) = \frac{\left\langle \sum_i f_i \chi_i \, \delta(\mathbf{r} - \mathbf{r}_i) \right\rangle}{\left\langle \sum_i f_i \, \delta(\mathbf{r} - \mathbf{r}_i) \right\rangle} \tag{4.1.15}$$

Thus, the coherent wave $\psi(\mathbf{r})$ is the average neutron wave function at the point $\mathbf{r}$ which, according to Eq. 4.1.11, is due to the combined effects of the incident wave plus the waves scattered by all the atoms. The quantity $\chi(\mathbf{r})$ is the average local field at the point $\mathbf{r}$ occupied by an atom, which, according to Eq. 4.1.13, is due to the combined effects of the incident wave plus the waves scattered by all the other atoms.

In the usual situation in which f_i and $\mathbf{r}_i$ are statistically independent, we see that

$$\left\langle \sum_i f_i \, \delta(\mathbf{r} - \mathbf{r}_i) \right\rangle = \rho f \tag{4.1.16}$$

where ρ is the number density (Eq. 2.8.41) and $f = \langle f_i \rangle$ is the bound coherent scattering amplitude. Hence,

$$\rho f \chi(\mathbf{r}) = \left\langle \sum_i f_i \chi_i \, \delta(\mathbf{r} - \mathbf{r}_i) \right\rangle \tag{4.1.17}$$

It follows from Eq. 4.1.11 that

$$\psi(\mathbf{r}) = \exp(i\mathbf{k} \cdot \mathbf{r}) + \rho f \int_V G(\mathbf{r} - \mathbf{r}')\chi(\mathbf{r}') \, d\mathbf{r}' \tag{4.1.18}$$

so that the coherent wave $\psi(\mathbf{r})$ is determined at all points $\mathbf{r}$, inside or outside V, by the local field $\chi(\mathbf{r}')$ at points $\mathbf{r}'$ inside V. The Green function (Eq. 4.1.7) can be expressed with the help of Eq. 2.2.17 in the form

$$G(\mathbf{r}) = \frac{1}{2\pi^2} \int \frac{\exp(i\mathbf{k}' \cdot \mathbf{r})}{k'^2 - i\eta - k^2} \, d\mathbf{k}' \tag{4.1.19}$$

where $\eta \to 0^+$. Hence,

$$\begin{aligned} (\Delta + k^2)G(\mathbf{r}) &= -\frac{1}{2\pi^2} \int \exp(i\mathbf{k}' \cdot \mathbf{r}) \, d\mathbf{k}' \\ &= -4\pi \, \delta(\mathbf{r}) \end{aligned} \tag{4.1.20}$$

It then follows from Eq. 4.1.18 that

$$(\Delta + k^2)\psi(\mathbf{r}) = \begin{cases} -4\pi\rho f \chi(\mathbf{r}), & \text{inside } V \\ 0, & \text{outside } V \end{cases} \tag{4.1.21}$$

Conversely, Eq. 4.1.18 is the solution of Eq. 4.1.21 that has the required asymptotic behavior at infinity and for which $\psi(\mathbf{r})$ and $\nabla\psi(\mathbf{r})$ are continuous at the surface of V.

4.2. Phenomenological theory of dispersion

Optical potential and the index of refraction

Equation 4.1.21 plays the same role in neutron optics as do Maxwell's equations in ordinary optics. Before we can solve this equation for $\psi(\mathbf{r})$ we require an independent constitutive relation between $\psi(\mathbf{r})$ and $\chi(\mathbf{r})$ analogous to the relation $\mathbf{D} = \varepsilon\mathbf{E}$ in ordinary optics. In Sec. 4.3 it will be shown that the required constitutive relation is (Lax, 1951)

$$\chi(\mathbf{r}) = c\psi(\mathbf{r}) \tag{4.2.1}$$

where c is a material constant that depends only on the temperature, density, and chemical composition of the medium. In the remainder of the present section we shall simply regard Eq. 4.2.1 as a given phenomenological relation and examine its consequences.

Combining Eqs. 4.1.21 and 4.2.1, we note first of all that $\psi(\mathbf{r})$ satisfies the one-body Schrödinger equation (2.8.32) in which the incident-neutron energy $E = (\hbar k)^2/2m$ and the optical potential is given by

$$v(\mathbf{r}) = \begin{cases} -2\pi\hbar^2\rho fc/m, & \text{inside } V \\ 0, & \text{outside } V \end{cases} \tag{4.2.2}$$

Inside V, Eq. 4.1.21 becomes

$$(\Delta + K^2)\psi(\mathbf{r}) = 0 \tag{4.2.3}$$

in which (Lax, 1951)

$$K^2 = k^2 + 4\pi\rho fc \tag{4.2.4}$$

Hence, the index of refraction $n = K/k$ is again given by Eq. 3.1.4 but now

$$\xi = -\frac{4\pi}{k^2}\rho fc \tag{4.2.5}$$

Comparing this result with Eq. 3.1.7, we see that the elementary theory of dispersion is equivalent to putting $f = -b$ and $c = 1$.

Extinction theorem

Having established via the differential equation (4.1.21) that the coherent wave $\psi(\mathbf{r})$ represents the refracted wave in the medium, we can now regard the corresponding integral equation,

$$\psi(\mathbf{r}) = \exp(i\mathbf{k}\cdot\mathbf{r}) + \rho fc\int_V G(\mathbf{r}-\mathbf{r}')\psi(\mathbf{r}')\,d\mathbf{r}' \tag{4.2.6}$$

as a statement of the extinction theorem (Lax, 1952; Born and Wolf, 1975). When read from right to left this equation states that the scattered waves $G(\mathbf{r}-\mathbf{r}')$ from all points $\mathbf{r}'$ in V interfere with each other in such a way as to completely extinguish the incident wave $\exp(i\mathbf{k}\cdot\mathbf{r})$ and replace it with the refracted wave $\psi(\mathbf{r})$ that propagates with a different wave vector $\mathbf{K}$.

To demonstrate explicitly how the extinction of the incident wave comes about, let us introduce the abbreviations

$$G = G(\mathbf{r}-\mathbf{r}'), \qquad \psi = \psi(\mathbf{r}'), \qquad \delta = \delta(\mathbf{r}-\mathbf{r}') \tag{4.2.7}$$

so that

$$(\Delta' + k^2)G = -4\pi\delta, \qquad (\Delta' + K^2)\psi = 0 \tag{4.2.8}$$

It then follows from the identity

$$\begin{aligned}\nabla'\cdot(\psi\nabla' G - G\,\nabla'\psi) &= \psi\,\Delta' G - G\,\Delta'\psi\\ &= (K^2 - k^2)G\psi - 4\pi\,\delta\psi\\ &= 4\pi(\rho fcG\psi - \delta\psi)\end{aligned} \tag{4.2.9}$$

that the quantity

$$\begin{aligned}\Omega(\mathbf{r}) &\equiv \frac{1}{4\pi}\int_V \nabla'\cdot(\psi\nabla' G - G\,\nabla'\psi)\,d\mathbf{r}'\\ &= \rho fc\int_V G\psi\,d\mathbf{r}' - \psi(\mathbf{r})\end{aligned} \tag{4.2.10}$$

The integral equation (4.2.6) then becomes

$$\psi(\mathbf{r}) = \exp(i\mathbf{k}\cdot\mathbf{r}) + \Omega(\mathbf{r}) + \psi(\mathbf{r}) \tag{4.2.11}$$

from which we see that the incident wave is extinguished by $\Omega(\mathbf{r})$. It follows from Gauss' theorem (2.4.8) that

$$\Omega(\mathbf{r}) = \frac{1}{4\pi}\int_S (\psi\,\nabla' G - G\,\nabla'\psi)\cdot\hat{\mathbf{n}}\,dS' \tag{4.2.12}$$

where S is the surface that encloses V. This result is usually interpreted (Lax, 1952; Born and Wolf, 1975) by saying that the incident wave is extinguished by the scattered waves produced at the surface S. However, it is clear from Eq. 4.2.10 that one can also regard the incident wave as being extinguished by the scattered waves produced throughout the entire volume V. These two seemingly different physical pictures are clearly equivalent and are no more paradoxical than is Gauss' theorem itself.

Optical theorem

It follows from Eqs. 2.2.23 and 4.1.7 that as $r \to \infty$,

$$G(\mathbf{r}-\mathbf{r}') \to \frac{1}{r} \exp[i(kr - \mathbf{k}' \cdot \mathbf{r}')] \tag{4.2.13}$$

The expression (4.1.18) for the coherent wave then shows that as $r \to \infty$,

$$\psi(\mathbf{r}) \to \exp(i\mathbf{k} \cdot \mathbf{r}) + F' \frac{\exp(ikr)}{r} \tag{4.2.14}$$

where F' denotes the coherent scattering amplitude of the system as a whole and is given by

$$F' = \rho f \int_V \exp(-i\mathbf{k}' \cdot \mathbf{r}) \chi(\mathbf{r})\, d\mathbf{r} \tag{4.2.15}$$

Thus, the coherent scattering amplitude is determined by the Fourier component of the local field at the scattered neutron wave vector.

The optical theorem (Eq. 2.8.13) states that the total collision cross section is given by

$$\sigma_t = \frac{4\pi}{k} \operatorname{Im}(F) \tag{4.2.16}$$

where F now denotes the forward coherent scattering amplitude,

$$F = \rho f \int_V \exp(-i\mathbf{k} \cdot \mathbf{r}) \chi(\mathbf{r})\, d\mathbf{r} \tag{4.2.17}$$

Thus, σ_t is determined by the Fourier component of the local field at the incident neutron wave vector.

4.3. Microscopic theory of dispersion

Local field hierarchy

To complete the theory of dispersion we must derive the constitutive relation (4.2.1) and obtain an expression for the material constant c in terms of known quantities. To this end we note first that Eq. 4.1.13 can be used to generate the following hierarchy of equations,

$$\chi(\mathbf{r}) = \exp(i\mathbf{k} \cdot \mathbf{r}) + \rho f \int_V G(\mathbf{r}-\mathbf{r}') g(\mathbf{r}, \mathbf{r}') \chi(\mathbf{r}' \mid \mathbf{r})\, d\mathbf{r}' \tag{4.3.1}$$

$$\begin{aligned}\chi(\mathbf{r}' \mid \mathbf{r}) = {} & \exp(i\mathbf{k} \cdot \mathbf{r}') + fG(\mathbf{r}'-\mathbf{r}) \chi(\mathbf{r} \mid \mathbf{r}') \\ & + \rho f \int_V G(\mathbf{r}'-\mathbf{r}'') \frac{g(\mathbf{r}, \mathbf{r}', \mathbf{r}'')}{g(\mathbf{r}, \mathbf{r}')} \chi(\mathbf{r}'' \mid \mathbf{r}, \mathbf{r}')\, d\mathbf{r}''\end{aligned} \tag{4.3.2}$$

etc., where

$$(\rho f)^2 g(\mathbf{r}, \mathbf{r}')\chi(\mathbf{r}' \mid \mathbf{r}) = \left\langle \sum_{i \neq j} f_i f_j \chi_j \,\delta(\mathbf{r} - \mathbf{r}_i)\,\delta(\mathbf{r}' - \mathbf{r}_j) \right\rangle \tag{4.3.3}$$

$$(\rho f)^3 g(\mathbf{r}, \mathbf{r}', \mathbf{r}'')\chi(\mathbf{r}'' \mid \mathbf{r}, \mathbf{r}') = \left\langle \sum_{ijl \neq} f_i f_j f_l \chi_l \,\delta(\mathbf{r} - \mathbf{r}_i)\,\delta(\mathbf{r}' - \mathbf{r}_j)\,\delta(\mathbf{r}'' - \mathbf{r}_l) \right\rangle \tag{4.3.4}$$

etc. Here $g(\mathbf{r}, \mathbf{r}')$ is the pair correlation function (Eq. 2.9.26) and $g(\mathbf{r}, \mathbf{r}', \mathbf{r}'')$ is the triplet correlation function, which is defined such that

$$\rho^3 g(\mathbf{r}, \mathbf{r}', \mathbf{r}'') = \left\langle \sum_{ijl \neq} \delta(\mathbf{r} - \mathbf{r}_i)\,\delta(\mathbf{r}' - \mathbf{r}_j)\,\delta(\mathbf{r}'' - \mathbf{r}_l) \right\rangle \tag{4.3.5}$$

In the preceding equations $ijl \neq$ means i, j and l are all distinct. We see that $\chi(\mathbf{r})$ is the local field for an atom at $\mathbf{r}$, $\chi(\mathbf{r}' \mid \mathbf{r})$ is the local field for an atom at $\mathbf{r}'$ given that there is also an atom at $\mathbf{r}$, $\chi(\mathbf{r}'' \mid \mathbf{r}, \mathbf{r}')$ is the local field for an atom at $\mathbf{r}''$ given that there are also atoms at $\mathbf{r}$ and $\mathbf{r}'$, etc.

The preceding hierarchy, though formally exact, is not in a particularly useful form since, in first approximation,

$$\begin{aligned} \chi(\mathbf{r}) &= \psi(\mathbf{r}), \qquad \text{not } \exp(i\mathbf{k}\cdot\mathbf{r}) \\ \chi(\mathbf{r}' \mid \mathbf{r}) &= \chi(\mathbf{r}'), \qquad \text{not } \exp(i\mathbf{k}\cdot\mathbf{r}') \end{aligned} \tag{4.3.6}$$

etc. This situation can be remedied with the help of Eq. 4.1.18 and we find that the hierarchy can be expressed in the equivalent form,

$$\chi(\mathbf{r}) = \psi(\mathbf{r}) + \rho f \int_V G(\mathbf{r} - \mathbf{r}')\{g(\mathbf{r}, \mathbf{r}')\chi(\mathbf{r}' \mid \mathbf{r}) - \chi(\mathbf{r}')\}\, d\mathbf{r}' \tag{4.3.7}$$

$$\begin{aligned} \chi(\mathbf{r}' \mid \mathbf{r}) &= \chi(\mathbf{r}') + fG(\mathbf{r}' - \mathbf{r})\chi(\mathbf{r} \mid \mathbf{r}') \\ &\quad + \rho f \int_V G(\mathbf{r}' - \mathbf{r}'')\left\{ \frac{g(\mathbf{r}, \mathbf{r}', \mathbf{r}'')}{g(\mathbf{r}, \mathbf{r}')} \chi(\mathbf{r}'' \mid \mathbf{r}, \mathbf{r}') - g(\mathbf{r}', \mathbf{r}'')\chi(\mathbf{r}'' \mid \mathbf{r}') \right\} d\mathbf{r}'' \end{aligned} \tag{4.3.8}$$

etc.

Let us now write

$$\chi_i = \chi(\mathbf{r}_i) + \delta\chi_i \tag{4.3.9}$$

where $\chi(\mathbf{r}_i)$ is the average local field at the position of the ith atom and $\delta\chi_i$ is the fluctuation in χ_i which, according to Eq. 4.1.17, has the property

$$\left\langle \sum_i f_i \,\delta\chi_i \,\delta(\mathbf{r} - \mathbf{r}_i) \right\rangle = 0 \tag{4.3.10}$$

It follows from Eq. 4.3.3 that

$$\chi(\mathbf{r}' \mid \mathbf{r}) = \chi(\mathbf{r}') + \delta\chi(\mathbf{r}' \mid \mathbf{r}) \tag{4.3.11}$$

where $\delta\chi(\mathbf{r}' \mid \mathbf{r})$ is the fluctuation in the local field at $\mathbf{r}'$ due to the presence of the atom at $\mathbf{r}$ and is given by

$$(\rho f)^2 g(\mathbf{r}, \mathbf{r}')\, \delta\chi(\mathbf{r}' \mid \mathbf{r}) = \left\langle \sum_{i \neq j} f_i f_j\, \delta\chi_j\, \delta(\mathbf{r} - \mathbf{r}_i)\, \delta(\mathbf{r}' - \mathbf{r}_j) \right\rangle \quad (4.3.12)$$

or, from Eq. 4.3.8,

$$\delta\chi(\mathbf{r}' \mid \mathbf{r}) = fG(\mathbf{r}' - \mathbf{r})\chi(\mathbf{r} \mid \mathbf{r}') + \rho f \int_V G(\mathbf{r}' - \mathbf{r}'') \left\{ \frac{g(\mathbf{r}, \mathbf{r}', \mathbf{r}'')}{g(\mathbf{r}, \mathbf{r}')} \chi(\mathbf{r}'' \mid \mathbf{r}, \mathbf{r}') - g(\mathbf{r}', \mathbf{r}'')\chi(\mathbf{r}'' \mid \mathbf{r}') \right\} d\mathbf{r}'' \quad (4.3.13)$$

Since $\delta\chi(\mathbf{r}' \mid \mathbf{r})$ is of order f, Eq. 4.3.7 reduces to

$$\chi(\mathbf{r}) = \psi(\mathbf{r}) + \rho f \int_V G(\mathbf{r} - \mathbf{r}')\{g(\mathbf{r}, \mathbf{r}') - 1\}\chi(\mathbf{r}')\, d\mathbf{r}' + O(f^2) \quad (4.3.14)$$

which is the basic equation for the determination of the constitutive relation.

Constitutive relation

Equation 4.3.14 is an inhomogeneous integral equation that can be solved to obtain $\chi(\mathbf{r})$ in terms of $\psi(\mathbf{r})$. Since this equation does not involve the incident wave explicitly, and since the integrand is a short-range function of $|\mathbf{r}' - \mathbf{r}|$, there is now no harm in proceeding to the thermodynamic limit $N \to \infty$, $V \to \infty$, such that $\rho = N/V = \text{const}$. In this case, Eq. 4.3.14 has a plane-wave solution

$$\chi(\mathbf{r}) = c\psi(\mathbf{r}) = c \exp(i\mathbf{K} \cdot \mathbf{r}) \quad (4.3.15)$$

in which $\mathbf{K}$ is the refracted wave vector and

$$c = (1 - J)^{-1} \quad (4.3.16)$$

where

$$J = \rho f \int \exp[i\mathbf{K} \cdot (\mathbf{r}' - \mathbf{r})]G(\mathbf{r} - \mathbf{r}')\{g(\mathbf{r}, \mathbf{r}') - 1\}\, d\mathbf{r}' + O(f^2) \quad (4.3.17)$$

It is evident from Eq. 4.2.4 that $\mathbf{K} = \mathbf{k} + O(f)$. Also, $f = -b + O(b^2)$ where b is the bound coherent scattering length (see Eq. 4.3.41). Finally, $g(\mathbf{r}, \mathbf{r}') = g(|\mathbf{r}' - \mathbf{r}|)$ for the homogeneous system under consideration. Hence,

$$J = \rho b \int \exp(i\mathbf{k} \cdot \mathbf{r})G(\mathbf{r})\{1 - g(r)\}\, d\mathbf{r} + O(b^2) \quad (4.3.18)$$

It is clear from this equation that the local field correction expresses the interference between the incident wave $\exp(i\mathbf{k} \cdot \mathbf{r})$ and the scattered wave $G(\mathbf{r})$.

For an ideal gas, $g(r) = 1$ and the first term in Eq. 4.3.18 vanishes identically so that $J = O(b^2)$. Lenk (1975) investigated the effect of the local field flucutations in this case and found that

$$\begin{aligned} J &\simeq -i(K-k)b \\ &= -i(n-1)kb \\ &\simeq 2\pi i\rho b^2/k \end{aligned} \tag{4.3.19}$$

which is of order b^2 as it should be.

In general, the expansion parameter in Eq. 4.3.18 is b/r_0 where r_0 is a characteristic length which, in the case of a liquid, is of the order of magnitude of the interatomic distance. Thus, $b/r_0 \simeq 10^{-4}$, so that for all practical purposes we can neglect the b^2 term in Eq. 4.3.18. Physically, this means we can neglect the fluctuations in the local field. It will also be noted that b is generally complex, $b = b' - ib''$, and $b''/b' \lesssim 10^{-4}$ for most nuclei according to Eq. 2.4.40. In evaluating J, we can usually put $b = b'$, which we shall, in fact, do in the remainder of this chapter.

Local field correction

It follows from Eq. 4.1.7 that the expression (4.3.18) for J can be reduced to a simple radial integral,

$$J = \frac{4\pi\rho b'}{k}\int_0^\infty \exp(ikr)\sin kr\{1-g(r)\}\,dr \tag{4.3.20}$$

Thus J is complex, $J = J' + iJ''$, with

$$J' = \frac{2\pi\rho b'}{k}\int_0^\infty \sin(2kr)\{1-g(r)\}\,dr \tag{4.3.21}$$

and

$$J'' = \frac{4\pi\rho b'}{k}\int_0^\infty \sin^2(kr)\{1-g(r)\}\,dr \tag{4.3.22}$$

In the long-wavelength limit, $k \to 0$, we note that

$$J' \to J_0 = 4\pi\rho b'\int_0^\infty r\{1-g(r)\}\,dr \tag{4.3.23}$$

and $J'' \to 0$.

The quantity J can be expressed with the help of Eq. 4.1.19 in the alternative form

$$J = \frac{b'}{2\pi^2}\int \frac{1 - S_c(\mathbf{q})}{k'^2 - i\eta - k^2}\,d\mathbf{k}' \tag{4.3.24}$$

where $\mathbf{q} = \mathbf{k} - \mathbf{k}'$ and $S_c(\mathbf{q})$ is the static structure factor (Eq. 2.9.29). This is the expression for J that was obtained by Lax (1952), who called it the *quasicrystalline approximation* since, as was shown earlier, it corresponds to neglecting the fluctuations in the local field, and such fluctuations vanish identically for a rigid crystalline lattice.

Yet another expression for J can be obtained by letting $\mathbf{k}'$ have polar coordinates (k', θ, ϕ) with $\mathbf{k}$ as the polar axis. Then, as in Sec. 2.5,

$$\begin{aligned} d\mathbf{k}' &= k'^2\, dk' \sin\theta\, d\theta\, d\phi \\ &= (mk'/\hbar)\, d\Omega\, d\varepsilon' \end{aligned} \tag{4.3.25}$$

in which $d\Omega = \sin\theta\, d\theta\, d\phi$ is an element of solid angle and $\hbar\varepsilon' = (\hbar k')^2/2m$ is the "scattered-neutron" energy. Hence, with $\omega = \varepsilon - \varepsilon'$, the expression (4.3.24) becomes

$$J = \frac{b'}{4\pi^2} \int d\Omega \int_0^\infty d\varepsilon' \frac{k'[S_c(\mathbf{q}) - 1]}{\omega + i\eta} \tag{4.3.26}$$

It then follows from the familiar identity

$$\frac{1}{\omega + i\eta} = \mathscr{P}\frac{1}{\omega} - i\pi\,\delta(\omega) \tag{4.3.27}$$

in which $\mathscr{P}$ denotes the Cauchy principal part, that

$$J'' = \frac{kb'}{4\pi} \int d\Omega [1 - S_c(\mathbf{q})] \tag{4.3.28}$$

in which $k' = k$.

Dilute hard-sphere gas

There is one simple case in which the integral in Eq. 4.3.20 for J can be evaluated analytically. This is the case of a dilute hard-sphere gas with atomic diameter a for which

$$g(r) = \begin{cases} 0, & r < a \\ 1, & r > a \end{cases} \tag{4.3.29}$$

We then find that

$$J = \frac{J_0}{2(ka)^2} [1 + 2ika - \exp(2ika)] \tag{4.3.30}$$

where $J_0 = 2\pi\rho b' a^2$. Hence,

$$J' = J_0 \left(\frac{\sin ka}{ka}\right)^2 \tag{4.3.31}$$

$$J'' = \frac{J_0}{2(ka)^2} (2ka - \sin 2ka) \tag{4.3.32}$$

Note that in the long-wavelength limit, $k \to 0$,

$$J' \to J_0, \qquad J'' \to J_0(\tfrac{2}{3}ka) \tag{4.3.33}$$

Thus, the imaginary part of J is much smaller than the real part if $ka \ll 1$. This condition is usually satisfied for the cold neutrons used in neutron optical work so that, in first approximation, one can put $J = J_0$. We also note that

$$J_0 = \sqrt{2}\,\frac{b'}{l} \tag{4.3.34}$$

where l is the mean free path,

$$l = (\sqrt{2}\,\pi\rho a^2)^{-1} \tag{4.3.35}$$

Thus, the local field correction in a dilute hard-sphere gas, as well as in gases generally, depends on the ratio of the coherent scattering length to the mean free path.

Liquids

The static structure factor $S_c(\mathbf{q})$, and hence the pair correlation function $g(r)$, for a monatomic liquid can be obtained from the measured angular variation of the diffuse scattering (Eq. 2.9.34). The local field correction (Eq. 4.3.20) can then be determined by means of a numerical integration. Figure 4.1 shows the real and imaginary parts of J as a function of the incident-neutron wave vector k for liquid argon at $T = 85$ K, which were determined from the $g(r)$ curve shown in the upper part of Fig. 4.2.

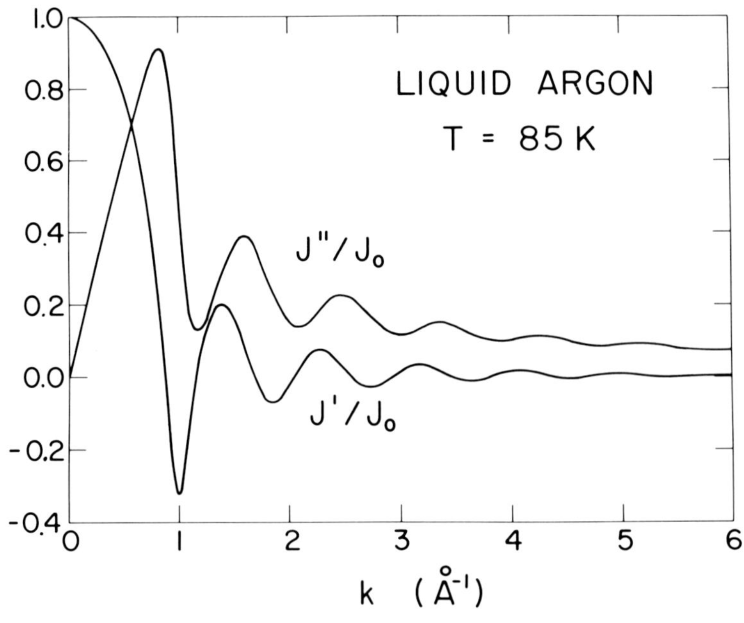

Fig. 4.1 The real and imaginary parts of the local field correction as a function of the incident neutron wave vector for liquid argon at $T = 85$ K (from Sears, 1982a).

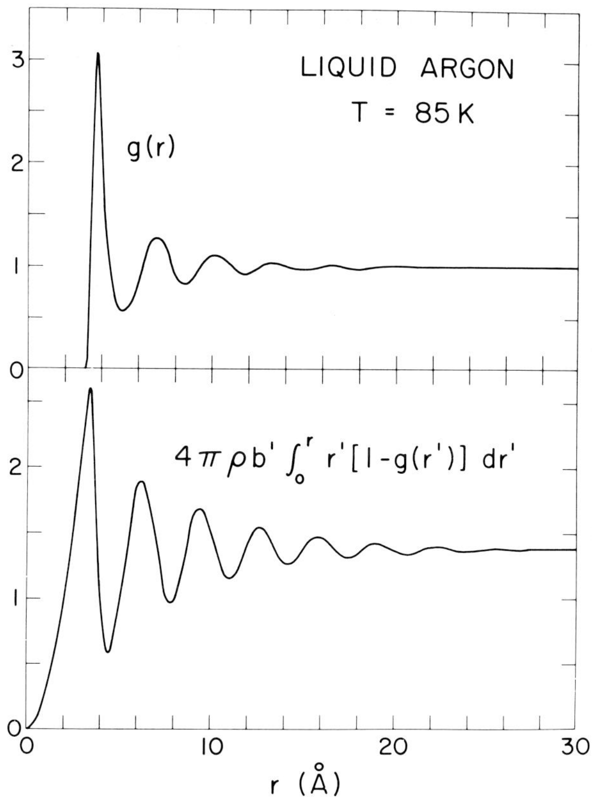

Fig. 4.2 The pair correlation function $g(r)$ of liquid argon at $T = 85$ K (upper part) and the corresponding integral (Eq. 4.3.23) for J_0 (in units of 10^{-5}) as a function of the upper limit of integration (lower part) (from Sears 1982a).

For cold neutrons ($k \ll 1\ \text{Å}^{-1}$) we can put $J = J_0$, which is then determined by the integral (4.3.23). The value of this integral is very sensitive to the oscillations in $g(r)$ at large r as can be seen in the lower part of Fig. 4.2, where we have plotted the quantity (4.3.23) as a function of the upper limit of integration. It is evident that the locus of midpoints of the upper and lower envelopes of the decay converges to J_0 much faster than does the curve itself. This fact can be used to obtain a reliable estimate of J_0 even in cases where the available $g(r)$ results include only the first few oscillations.

Table 4.1 lists the J_0 values obtained for a number of monatomic liquids, and also shows the corresponding values of the correlation length r_0, which is defined such that

$$J_0 = \frac{b'}{r_0} \tag{4.3.36}$$

and which is typically one-third the nearest-neighbor distance in the

Table 4.1 Local field correction J_0 and related quantities for various monatomic liquids (from Sears, 1982a)

Element	T	ρ ($Å^{-3}$)	b' (fm)	J_0 (10^{-5})	r_0 (Å)	$r_0\rho^{1/3}$
He	1.0 K	0.0218	3.0	2.15	1.40	0.390
Ne	35 K	0.0317	4.6	3.11	1.48	0.468
Ar	85 K	0.0213	1.8	1.38	1.30	0.362
Li	180°C	0.0437	−2.03	−1.90	1.07	0.376
Na	100°C	0.0243	3.63	3.12	1.16	0.337
K	65°C	0.0128	3.71	2.70	1.37	0.321
Rb	40°C	0.0108	7.08	3.91	1.81	0.400
Cs	30°C	0.0081	5.42	2.93	1.85	0.372
Zn	450°C	0.0606	5.7	7.20	0.79	0.311
Al	680°C	0.0529	3.449	3.50	0.99	0.370
Ga	20°C	0.0527	7.2	6.90	1.04	0.391
Tl	320°C	0.0344	8.89	9.02	0.99	0.321
Sn	250°C	0.0354	6.220	5.60	1.11	0.365
Pb	340°C	0.0311	9.4003	7.88	1.19	0.375
Bi	300°C	0.0290	8.5256	7.98	1.07	0.328

liquid. For a dilute hard-sphere gas, $r_0 = l/\sqrt{2}$ according to Eq. 4.3.34. Ekstein (1953) has shown that for a simple cubic lattice,

$$r_0\rho^{1/3} = \tfrac{1}{3} \tag{4.3.37}$$

This is in approximate agreement with the results shown in Table 4.1. Thus, the relation (4.3.37) can be used to obtain a rough estimate of r_0, and hence J_0, in liquids for which accurate $g(r)$ data are unavailable.

Local field effects in neutron refractometry

Summarizing the results obtained thus far, we note first that the index of refraction can in general be expressed in the form (Eq. 3.1.4),

$$n^2 = 1 - \xi \tag{4.3.38}$$

In the elementary theory of dispersion, ξ is given by (Eq. 3.1.7),

$$\xi = \frac{4\pi}{k^2}\rho b \tag{4.3.39}$$

where $b = \langle b_i \rangle$ is the bound coherent scattering length, b_i being the bound scattering length of one particular atom i. The rigorous theory of dispersion given in the present chapter shows that, according to Eq. 4.2.5,

$$\xi = -\frac{4\pi}{k^2}\rho f c \tag{4.3.40}$$

where $f = \langle f_i \rangle$ is the bound coherent scattering amplitude and $c = (1-J)^{-1}$, in which J is the local field correction. It follows from Eq. 4.1.10 that

$$f = -b + ik\langle b_i^2 \rangle + O(k^2) \tag{4.3.41}$$

In general, b and J are complex,

$$b = b' - ib'', \qquad J = J' + iJ'' \tag{4.3.42}$$

and considerable simplification follows from the fact that the quantities kb', b''/b', and $|J|$ are all typically $\lesssim 10^{-4}$. Keeping only the leading correction terms, we then get

$$f = -b' + i[k(b'^2 + \Delta b'^2) + b''] \tag{4.3.43}$$

where Δb is the bound incoherent scattering length (Eq. 2.8.6), and

$$c = (1 + J') + iJ'' \tag{4.3.44}$$

Hence,

$$fc = (fc)' + i(fc)'' \tag{4.3.45}$$

where,

$$\begin{aligned}(fc)' &= -b'(1 + J') \\ (fc)'' &= b'' + k(b'^2 + \Delta b'^2) - b'J''\end{aligned} \tag{4.3.46}$$

With $n = n' + in''$ and $\xi = \xi' - i\xi''$, Eq. 4.3.38 gives

$$n'^2 - n''^2 = 1 - \xi', \qquad 2n'n'' = \xi'' \tag{4.3.47}$$

Hence,

$$n' = 1 - \tfrac{1}{2}\xi' - \tfrac{1}{8}(\xi'^2 - \xi''^2) + \cdots \tag{4.3.48}$$

Since $\xi'' \simeq 10^{-4}\xi'$, we can neglect the ξ''^2 term in the preceding expansion. It then follows from Eqs. 4.3.40 and 4.3.46 that

$$n' = 1 - \frac{2\pi}{k^2}\rho b'\left(1 + J' + \frac{\pi}{k^2}\rho b'\right) \tag{4.3.49}$$

For thermal neutrons the quantity $(\pi/k^2)\rho b'$ is typically of the order of 10^{-6} and, hence, is one or two orders of magnitude smaller than the local field correction J'.

Let us write

$$(fc)' = -b'_{\text{eff}} \tag{4.3.50}$$

where

$$b'_{\text{eff}} = b'(1 + J') \tag{4.3.51}$$

Then b'_{eff} is the effective scattering length that is obtained when neutron refractometry measurements are analysed on the basis of the elementary theory of dispersion, and b' is the true scattering length. At present the

Table 4.2 Comparison of the true and effective bound coherent scattering lengths for selected elements (from Sears, 1985a)

Element	b' (fm)	b'_{eff} (fm)
H	−3.7390(11)	−3.7409(11)
C	6.6460(12)	6.6483(12)
F	5.665(15)	5.661(15)
Cl	9.5770(8)	9.5796(8)
Br	6.795(15)	6.791(15)

local field correction is larger than the experimental error only for the most accurate gravity refractometry measurements (Sears, 1985a). Table 4.2 shows a comparison of b' and b'_{eff} for selected elements. In the case of Cl, for example, the local field correction is three times larger than the experimental uncertainty in b'.

Optical theorem

It follows from Eq. 4.3.15 that the expression (4.2.17) for the forward coherent scattering amplitude reduces to

$$F = Nfc \tag{4.3.52}$$

The optical theorem (Eq. 4.2.16) then becomes

$$\sigma_t = \frac{4\pi}{k} N(fc)'' \tag{4.3.53}$$

The expression (4.3.46) for $(fc)''$ can be expressed with the help of the relation (4.3.28) for J'' in the equivalent form

$$(fc)'' = b'' + \frac{k}{4\pi}\left[b'^2 \int S_c(\mathbf{q})\, d\Omega + 4\pi\, \Delta b'^2\right] \tag{4.3.54}$$

Hence,

$$\sigma_t = \sigma_a + \sigma_d \tag{4.3.55}$$

where σ_a is the absorption cross section,

$$\sigma_a = N\frac{4\pi}{k} b'' \tag{4.3.56}$$

and σ_d is the cross section for diffuse scattering,

$$\sigma_d = N\left[\frac{\sigma_c}{4\pi}\int S_c(\mathbf{q})\, d\Omega + \sigma_i\right] \tag{4.3.57}$$

Here σ_c and σ_i are the bound coherent and incoherent scattering cross sections per atom,

$$\sigma_c = 4\pi b'^2, \qquad \sigma_i = 4\pi\,\Delta b'^2 \tag{4.3.58}$$

These expressions for σ_a and σ_d are the same as the corresponding results (Eqs. 2.7.28 and 2.9.35) that were calculated directly in Chap. 2. This demonstrates that the rigorous theory of dispersion given above provides a consistent description of the local field effects. Alternatively, one can state that the local field effects must be taken properly into account in order for σ_t to be given correctly by the optical theorem.

Absorption coefficient

The absorption coefficient μ in Lambert's law is defined by Eq. 3.2.11 and can be expressed with the help of Eqs. 4.3.40 and 4.3.47 in the form

$$\mu = \frac{4\pi}{n'k}\rho(fc)'' \tag{4.3.59}$$

Hence,

$$\mu = \rho\sigma_t/n' \tag{4.3.60}$$

where σ_t now denotes the total collision cross section per atom,

$$\sigma_t = \frac{4\pi}{k}(fc)'' \tag{4.3.61}$$

The factor n' in Eq. 4.3.60 is absent in the elementary derivations of the absorption coefficient that one finds in most textbooks. The physical significance of this factor can be seen by writing

$$\mu = \rho(\sigma_t)_{\text{eff}} \tag{4.3.62}$$

where

$$(\sigma_t)_{\text{eff}} = \frac{\sigma_t}{n'} = \frac{4\pi}{K}(fc)'' \tag{4.3.63}$$

in which $K = n'k$ is the wave vector of the refracted wave in the medium. Thus, it is the refracted wave vector K, and not the wave vector in vacuo k, that determines the attenuation of the coherent wave in the medium. While this is important in principle, it is not so in practice because $n' - 1 \simeq 10^{-5}$ for thermal neutrons.

5

KINEMATICAL THEORY OF NEUTRON DIFFRACTION

The remainder of this book is concerned with the physical optics phenomena of diffraction and interference. We begin in this chapter with a discussion of the kinematical theory of neutron diffraction. The multiple-scattering effects that are neglected in the kinematical theory will be dealt with later in Chaps. 6 and 8.

In the kinematical theory the differential cross section for coherent scattering is given according to Eq. 2.8.50 by

$$\frac{d\sigma}{d\Omega} = |\langle b_{\mathbf{q}} \rangle|^2 \tag{5.0.1}$$

where we now omit the subscript c. Here $\mathbf{q} = \mathbf{k} - \mathbf{k}'$, where $\mathbf{k}$ and $\mathbf{k}'$ are the incident and scattered neutron wave vectors, and $k' = k$ since the scattering is elastic. Also,

$$b_{\mathbf{q}} = \sum_j b_j \exp(i\mathbf{q} \cdot \mathbf{r}_j) \tag{5.0.2}$$

where $\mathbf{r}_j$ denotes the position and b_j the bound scattering length of the jth atom. Finally, the brackets $\langle \cdots \rangle$ represent a thermodynamic average. The preceding differential scattering cross section describes the diffraction of the incident-neutron beam by the system of atoms. In particular, the diffraction is of the Fraunhofer type,[1] since in the derivation of (Eq. 5.0.1) it is tacitly assumed that the source and the detector are both effectively at infinity.

5.1. Amorphous materials

Differential scattering cross section

We begin by considering the diffraction of neutrons by a body that is macroscopically homogeneous and isotropic such as a gas, liquid, or amorphous solid. As in Sec. 2.8 we put

$$b_{\mathbf{q}} = \int b(\mathbf{r}) \exp(i\mathbf{q} \cdot \mathbf{r})\, d\mathbf{r} \tag{5.1.1}$$

[1] The distinction between Fraunhofer and Fresnel diffraction is discussed in detail in Chap. 6.

where $b(\mathbf{r})$ is the bound scattering length density,

$$b(\mathbf{r}) = \sum_j b_j \, \delta(\mathbf{r} - \mathbf{r}_j) \tag{5.1.2}$$

Hence,

$$\langle b_{\mathbf{q}} \rangle = \int \langle b(\mathbf{r}) \rangle \exp(i\mathbf{q} \cdot \mathbf{r}) \, d\mathbf{r} \tag{5.1.3}$$

where $\langle b(\mathbf{r}) \rangle$ is the average bound scattering length density,

$$\langle b(\mathbf{r}) \rangle = \sum_j \langle b_j \rangle \langle \delta(\mathbf{r} - \mathbf{r}_j) \rangle \tag{5.1.4}$$

and we have assumed that, as is usually true, b_j and $\mathbf{r}_j$ are statistically independent.

If the atoms are confined by external constraints to a volume of space V, then for an amorphous material,

$$\langle \delta(\mathbf{r} - \mathbf{r}_j) \rangle = \frac{1}{V} S(\mathbf{r}) \tag{5.1.5}$$

where $S(\mathbf{r})$ is the *shape function* (Patterson, 1939; Ewald, 1940),

$$S(\mathbf{r}) = \begin{cases} 1, & \text{inside } V \\ 0, & \text{outside } V \end{cases} \tag{5.1.6}$$

Hence,

$$\langle b(\mathbf{r}) \rangle = \rho b S(\mathbf{r}) \tag{5.1.7}$$

where $\rho = N/V$ is the average number density, N being the total number of atoms, and b the average bound coherent scattering length,

$$b = \frac{1}{N} \sum_j \langle b_j \rangle \tag{5.1.8}$$

It now follows that

$$\langle b_{\mathbf{q}} \rangle = (2\pi)^3 \rho b D(\mathbf{q}) \tag{5.1.9}$$

where $D(\mathbf{q})$ is the *shape transform,*

$$D(\mathbf{q}) = \frac{1}{(2\pi)^3} \int S(\mathbf{r}) \exp(i\mathbf{q} \cdot \mathbf{r}) \, d\mathbf{r} \tag{5.1.10}$$

Hence, the differential scattering cross section (Eq. 5.0.1) is given for an amorphous body by

$$\frac{d\sigma}{d\Omega} = N(2\pi)^3 \rho \, |b|^2 \, \Delta(\mathbf{q}) \equiv A(\mathbf{q}) \tag{5.1.11}$$

where

$$\Delta(\mathbf{q}) = \frac{(2\pi)^3}{V} |D(\mathbf{q})|^2 \tag{5.1.12}$$

The quantity $\Delta(\mathbf{q})$ characterizes the shape of the Fraunhofer diffraction peak produced by the body and is accordingly called the *peak function*. The differential scattering cross section itself is denoted by $A(\mathbf{q})$ for reasons that will become clear in Sec. 5.2.

General properties of the peak function

It is easily shown that the shape transform (Eq. 5.1.10) has the properties

$$\begin{gathered} D(\mathbf{q})^* = D(-\mathbf{q}) \\ 0 \leq |D(\mathbf{q})| \leq D(0) = \frac{V}{(2\pi)^3} \\ \int D(\mathbf{q})\, d\mathbf{q} = 1 \\ \lim_{V\to\infty} D(\mathbf{q}) = \delta(\mathbf{q}) \end{gathered} \tag{5.1.13}$$

The peak function (Eq. 5.1.12) then has the properties

$$\begin{gathered} \Delta(\mathbf{q}) = \Delta(\mathbf{q})^* = \Delta(-\mathbf{q}) \\ 0 \leq \Delta(\mathbf{q}) \leq \Delta(0) = \frac{V}{(2\pi)^3} \\ \int \Delta(\mathbf{q})\, d\mathbf{q} = 1 \\ \int \mathbf{q}\, \Delta(\mathbf{q})\, d\mathbf{q} = 0 \\ \lim_{V\to\infty} \Delta(\mathbf{q}) = \delta(\mathbf{q}) \end{gathered} \tag{5.1.14}$$

It will be noted that $\Delta(\mathbf{q})$ is an even function of $\mathbf{q}$ even if the body is not centrosymmetric. Thus, for example, a right-handed body produces the same diffraction pattern as its left-handed counterpart.

According to the relations (5.1.14), $\Delta(\mathbf{q})$ has its maximum value at $\mathbf{q}=0$ and is appreciably different from zero only if $q \lesssim 2\pi/L$, where $L = V^{1/3}$ and characterizes the linear dimensions of the body. Since $q = 2k\sin(\phi/2) \simeq k\phi$, where ϕ is the angle of scattering, it follows that

$$\Delta(\mathbf{q}) \simeq 0 \qquad \text{unless } \phi \lesssim \lambda/L, \tag{5.1.15}$$

where $\lambda = 2\pi/k$ is the incident-neutron wavelength. In other words, the diffracted beam is concentrated in the forward direction with an angular divergence of the order of λ/L. Since $\lambda/L \ll 1$ this type of diffraction is usually called *small-angle scattering*.

If, for example, $\lambda = 5$ Å and $L = 5$ mm, then $\lambda/L = 10^{-7}$ radians, which is unobservably small. Thus, the small-angle scattering by a homogeneous body of macroscopic size is negligible and $\Delta(\mathbf{q}) = \delta(\mathbf{q})$ for all practical purposes. On the other hand, if $\lambda = 5$ Å and $L = 100$ Å then $\lambda/L = 0.05$ radians $\simeq 3$ degrees, and the diffraction is appreciable. This situation is realized for small-angle scattering in heterogeneous materials, which will be discussed later in Sec. 5.6.

The peak function $\Delta(\mathbf{q})$ also plays an important role in the theory of Fraunhofer diffraction in crystals, as we shall see in Sec. 5.2. In the remainder of this section we shall develop some expressions for $\Delta(\mathbf{q})$ that we shall need later.

Analytical expressions

The peak function can be calculated analytically if the body has a simple geometrical shape. Thus, for a brick with sides L_x, L_y, and L_z we get

$$\Delta(\mathbf{q}) = \Delta(q_x)\,\Delta(q_y)\,\Delta(q_z) \tag{5.1.16}$$

where

$$\Delta(q_\alpha) = \frac{L_\alpha}{2\pi}\left\{\frac{\sin(q_\alpha L_\alpha/2)}{q_\alpha L_\alpha/2}\right\}^2 \tag{5.1.17}$$

and $\alpha = x, y, z$. In the limit of an infinite plane slab where, for example, $L_x, L_y \to \infty$, the preceding expression reduces to

$$\Delta(\mathbf{q}) = \delta(q_x)\,\delta(q_y)\,\Delta(q_z) \tag{5.1.18}$$

Another case in which $\Delta(\mathbf{q})$ can easily be calculated analytically is that of a sphere of radius R, for which

$$\Delta(\mathbf{q}) = \frac{V}{(2\pi)^3}\left[\frac{3}{(qR)^3}(\sin qR - qR\cos qR)\right]^2 \tag{5.1.19}$$

where $V = (4/3)\pi R^3$. This result was first obtained by Rayleigh (1914) in connection with the diffraction of light.

Gaussian approximation

Let us now consider a body of arbitrary size and shape, and introduce the notation

$$\langle\cdots\rangle = \frac{1}{V}\int \cdots S(\mathbf{r})\,d\mathbf{r} \tag{5.1.20}$$

so that $\langle 1\rangle = 1$. Then

$$\Delta(\mathbf{q}) = \frac{V}{(2\pi)^3}\,|\langle\exp(i\mathbf{q}\cdot\mathbf{r})\rangle|^2 \tag{5.1.21}$$

We take the origin to be at the center of mass of the body, so that $\langle\mathbf{r}\rangle = 0$. Expanding the exponential in Eq. 5.1.21 and regrouping the terms, we obtain the cumulant expansion

$$\Delta(\mathbf{q}) = \frac{V}{(2\pi)^3} \exp[-\langle(\mathbf{q}\cdot\mathbf{r})^2\rangle + O(q^3)] \tag{5.1.22}$$

The Gaussian approximation is then obtained by keeping only the leading term in the exponent,

$$\Delta(\mathbf{q}) = \frac{V}{(2\pi)^3} \exp[-\mathbf{q}\cdot\langle\mathbf{rr}\rangle\cdot\mathbf{q}] \tag{5.1.23}$$

This approximation was first introduced by Guinier (1939) in the context of small-angle X-ray scattering, so that it is often called the *Guinier approximation.*

If the directions of the coordinate axes are chosen to coincide with the principal directions of the symmetric tensor $\langle\mathbf{rr}\rangle$, then

$$\langle\mathbf{rr}\rangle = \begin{vmatrix} a_x^2 & 0 & 0 \\ 0 & a_y^2 & 0 \\ 0 & 0 & a_z^2 \end{vmatrix} \tag{5.1.24}$$

and Eq. 5.1.23 becomes

$$\Delta(\mathbf{q}) = \frac{V}{(2\pi)^3} \exp[-(q_x a_x)^2 - (q_y a_y)^2 - (q_z a_z)^2] \tag{5.1.25}$$

This is in general an anisotropic Gaussian for which the locus of the half-height points is an ellipsoid.

The Gaussian approximation is asymptotically exact as $\mathbf{q}\to 0$ and usually differs significantly from the exact result only at large q where $\Delta(\mathbf{q})$ has a long oscillating tail that is absent in the Gaussian approximation. A measure of the overall accuracy of the Gaussian approximation is the amount by which the value of the integral

$$\int \Delta(\mathbf{q})\, d\mathbf{q} = \frac{V}{8\pi^{3/2} a_x a_y a_z} \tag{5.1.26}$$

differs from unity. It is evident from Table 5.1 that this amount is typically 5 percent or so.

If a_x, a_y, a_z are simply regarded as phenomenological constants they can, of course, be chosen such that $\int \Delta(\mathbf{q})\, d\mathbf{q} = 1$ identically. The

Table 5.1 Gaussian approximation for $\Delta(\mathbf{q})$

Quantity	Brick	Sphere
a_α	$L_\alpha/2\sqrt{3}$	$R/\sqrt{5}$
V	$L_x L_y L_z$	$\frac{4}{3}\pi R^3$
$\int \Delta(\mathbf{q})\, d\mathbf{q}$	$(3/\pi)^{3/2} = 0.933$	$\frac{5}{6}(5/\pi)^{1/2} = 1.051$

Gaussian approximation then becomes

$$\Delta(\mathbf{q}) = \frac{a_x a_y a_z}{\pi^{3/2}} \exp[-(q_x a_x)^2 - (q_y a_y)^2 - (q_z a_z)^2] \tag{5.1.27}$$

However, this expression is no longer asymptotically exact as $\mathbf{q} \to 0$.

5.2. Perfect crystals

Differential scattering cross section

We now consider the diffraction of neutrons by a finite but otherwise perfect crystal[2] consisting of N_0 primitive unit cells with n atoms per unit cell, so that the total number of atoms is $N = N_0 n$. Then Eq. 5.0.2 becomes

$$b_\mathbf{q} = \sum_l{}' \sum_{\nu=1}^{n} b_{l\nu} \exp(i\mathbf{q} \cdot \mathbf{r}_{l\nu}) \tag{5.2.1}$$

where $\mathbf{r}_{l\nu}$ denotes the position of the νth atom in the lth unit cell and $b_{l\nu}$ is the corresponding bound scattering length. The prime on the summation indicates that l is restricted to the N_0 lattice sites of the finite crystal.

According to Eq. B.5 in Appendix B,

$$\mathbf{r}_{l\nu} = \mathbf{R}_l + \mathbf{r}_\nu + \mathbf{u}_{l\nu} \tag{5.2.2}$$

where $\mathbf{R}_l$ is the lattice vector giving the position of the lth lattice site, $\mathbf{r}_\nu$ is the equilibrium position of the νth atom relative to this site, and $\mathbf{u}_{l\nu}$ is the displacement of this atom from its equilibrium position. Then

$$\langle b_\mathbf{q} \rangle = I(\mathbf{q}) F(\mathbf{q}) \tag{5.2.3}$$

where $I(\mathbf{q})$ is the *interference function,*

$$I(\mathbf{q}) = \sum_l{}' \exp(i\mathbf{q} \cdot \mathbf{R}_l) \tag{5.2.4}$$

and $F(\mathbf{q})$ is the *unit-cell structure factor,*

$$F(\mathbf{q}) = \sum_{\nu=1}^{n} b_\nu \exp[-W_\nu(\mathbf{q})] \exp(i\mathbf{q} \cdot \mathbf{r}_\nu) \tag{5.2.5}$$

Here b_ν is the bound coherent scattering length of the νth atom in a unit cell,

$$b_\nu = \langle b_{l\nu} \rangle \tag{5.2.6}$$

and

$$\exp[-W_\nu(\mathbf{q})] = \langle \exp(i\mathbf{q} \cdot \mathbf{u}_{l\nu}) \rangle \tag{5.2.7}$$

---

[2] The necessary background information on space lattices and crystals is summarized in Appendices A and B.

is the *Debye–Waller factor* for this atom. In deriving Eq. 5.2.3 we have made use of the fundamental property of a perfect crystal, namely, that all the unit cells are identical on average so that $\mathbf{r}_\nu$, b_ν, and $W_\nu(\mathbf{q})$ are independent of l.

For an infinite crystal the summation over l in Eq. 5.2.4 is unrestricted and the interference function becomes

$$I_\infty(\mathbf{q}) = \sum_l \exp(i\mathbf{q} \cdot \mathbf{R}_l) \tag{5.2.8}$$

According to Eqs. A.7 and A.19 in Appendix A,

$$I_\infty(\mathbf{q}) = \frac{(2\pi)^3}{V_0} \sum_h \delta(\mathbf{q} - \mathbf{K}_h) \tag{5.2.9}$$

where h labels the reciprocal lattice sites, $\mathbf{K}_h$ denotes the corresponding reciprocal lattice vectors, and V_0 is the volume of a primitive unit cell.

To obtain the interference function for a finite crystal, we note first that the restriction on the summation over l in Eq. 5.2.4 can be removed by introducing the shape function (Eq. 5.1.6). Thus,

$$I(\mathbf{q}) = \sum_l S(\mathbf{R}_l) \exp(i\mathbf{q} \cdot \mathbf{R}_l) \tag{5.2.10}$$

Since

$$S(\mathbf{R}_l) = \int D(\mathbf{q}') \exp(-i\mathbf{q}' \cdot \mathbf{R}_l)\, d\mathbf{q}' \tag{5.2.11}$$

where $D(\mathbf{q}')$ is the shape transform (Eq. 5.1.10), it follows that

$$I(\mathbf{q}) = \int I_\infty(\mathbf{q} - \mathbf{q}') D(\mathbf{q}')\, d\mathbf{q}' \tag{5.2.12}$$

and, hence, from Eq. 5.2.9 that

$$I(\mathbf{q}) = \frac{(2\pi)^3}{V_0} \sum_h D(\mathbf{q} - \mathbf{K}_h) \tag{5.2.13}$$

This is the same as for an infinite crystal except that the delta function in Eq. 5.2.9 is now replaced by the shape transform of the finite crystal.

Since $N_0 \gg 1$ the shape transforms at different reciprocal lattice sites have negligible overlap, so that in evaluating $|I(\mathbf{q})|^2$ we can neglect the cross terms. It then follows from Eq. 5.2.3 that the differential scattering cross section (Eq. 5.0.1) becomes

$$\frac{d\sigma}{d\Omega} = N_0 \frac{(2\pi)^3}{V_0} \sum_h \Delta(\mathbf{q} - \mathbf{K}_h)\, |F(\mathbf{q})|^2 \tag{5.2.14}$$

in which $\Delta(\mathbf{q})$ is the peak function (Eq. 5.1.12) and we have used the fact that the total volume of the crystal is $V = N_0 V_0$. Thus, the differential scattering cross section of a perfect crystal consists of a set of peaks in

k-space. These peaks are usually called *Bragg peaks* and their positions are given by the *Laue equation,*

$$\mathbf{q} = \mathbf{k} - \mathbf{k}' = \mathbf{K}_h \tag{5.2.15}$$

It is evident from Eq. 5.1.14 that when the Laue equation is satisfied exactly for some value of h, the cross section has a local maximum

$$\left(\frac{d\sigma}{d\Omega}\right)_{\max} = N_0^2 |F_h|^2 \tag{5.2.16}$$

where

$$F_h \equiv F(\mathbf{K}_h) = \sum_{\nu=1}^{n} b_\nu \exp(-W_{\nu h}) \exp(i\mathbf{K}_h \cdot \mathbf{r}_\nu) \tag{5.2.17}$$

and $W_{\nu h} = W_\nu(\mathbf{K}_h)$. As $\mathbf{q}$ moves away from $\mathbf{K}_h$, the cross section decreases rapidly and is appreciably different from zero only if $|\mathbf{q} - \mathbf{K}_h| \lesssim 2\pi/L$, where $L = V^{1/3}$ and characterizes the linear dimensions of the crystal.

Since $F(\mathbf{q})$ does not vary appreciably over any one peak, Eq. 5.2.14 can be expressed equivalently as

$$\frac{d\sigma}{d\Omega} = N_0 \frac{(2\pi)^3}{V_0} \sum_h |F_h|^2 \,\Delta(\mathbf{q} - \mathbf{K}_h) \tag{5.2.18}$$

We see that the positions and integrated intensities of the Bragg peaks are determined by the crystal structure and are independent of the shape of the crystal. On the other hand, the shape of the Bragg peaks is determined by the shape of the crystal and is independent of the crystal structure.

It will be noted that the expression (5.2.2) for the atomic positions, on which the preceding results are based, is not universally applicable. For example, it breaks down in the orientationally disordered phases of molecular crystals (see Appendix B). In this case the differential scattering cross section is still given by Eq. 5.2.18 but the unit-cell structure factor F_h is no longer given by Eq. 5.2.17. Expressions for F_h appropriate for some types of orientationally disordered crystals have been derived by, for example, Dolling et al. (1979). The expression (5.2.18) itself breaks down in the case of incommensurate phases (see Appendix B), for which the diffraction is characterized by the appearance of satellite peaks around the primary Bragg peaks (Axe, 1976). Apart from special cases like these, the expression (5.2.18) for the differential scattering cross section of a perfect crystal is exact within the framework of the kinematical theory of neutron diffraction.

It is instructive to compare the diffraction of neutrons by a crystal with that by an amorphous body of the same size, shape, chemical composition, and density. To this end we first note that if $h = 0$ means

$h_1h_2h_3 = 000$ (see Appendix A), then

$$\mathbf{K}_0 = 0, \qquad F_0 = \sum_{\nu=1}^{n} b_\nu = nb \tag{5.2.19}$$

where b is the average bound coherent scattering length as before (see Eq. 5.1.8). Then Eq. 5.2.18 can be expressed in the equivalent form

$$\frac{d\sigma}{d\Omega} = \sum_h |\hat{F}_h|^2 A(\mathbf{q} - \mathbf{K}_h) \tag{5.2.20}$$

where $\hat{F}_h = F_h/F_0$ is the normalized unit-cell structure factor and $A(\mathbf{q})$ is given by Eq. 5.1.11. Since $\hat{F}_0 = 1$, we see that

$$\frac{d\sigma}{d\Omega} = A(\mathbf{q}) + \sum_{h\neq 0} |\hat{F}_h|^2 A(\mathbf{q} - \mathbf{K}_h) \tag{5.2.21}$$

The first term in this expression represents the forward-diffracted beam, i.e., the small-angle scattering, and the $h \neq 0$ terms represent the so-called *Bragg scattering*. Comparing Eqs. 5.1.11 and 5.2.21, we see that the small-angle scattering is the same for the crystal as it is for the amorphous reference body. The effect of the crystal structure is that the Fraunhofer diffraction pattern of the amorphous reference body is reproduced at each reciprocal lattice site with a relative intensity $|\hat{F}_h|^2$.

Ewald construction

Equation 5.2.15 is called the *Laue equation* since it is equivalent to the interference equations that were derived by Laue in the first paper on the diffraction of X-rays by crystals (Friedrich, Knipping, and Laue, 1912). This equation is also commonly referred to as the *Bragg condition* since, as we shall show later, it is equivalent to Bragg's law (Bragg, 1913). Nevertheless, the discussion of diffraction by crystals in terms of the reciprocal lattice is actually due neither to Laue nor Bragg but to Ewald (1913, 1921).

Ewald showed that the meaning of Eq. 5.2.15 becomes especially clear when looked at from the point of view of the geometrical construction in Fig. 5.1. First, one constructs the reciprocal lattice in **k**-space. Next, one draws the incident-neutron wave vector **k** such that it ends at a reciprocal lattice site. The origin of **k** then defines the point A. Finally, one constructs the sphere with center A and radius k. Ewald called this the sphere of propagation (Ausbreitungskugel), but it is now usually called the *Ewald sphere*.

Since the scattering is elastic ($k' = k$) the scattered-neutron wave vector $\mathbf{k}'$ must also end somewhere on the Ewald sphere. The Laue equation (5.2.15) then asserts that the Bragg scattering will occur with maximum intensity whenever one (or more) additional reciprocal lattice sites lie on the surface of the Ewald sphere. For an arbitrary value of **k** it

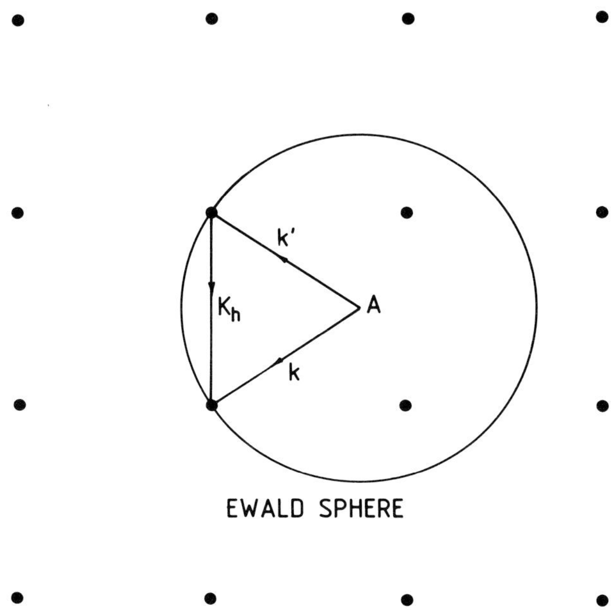

Fig. 5.1 Ewald construction.

is unlikely that any additional reciprocal lattice sites will lie on or near the surface of the Ewald sphere, in which case there will be negligible Bragg scattering. However, if **k** is varied continuously in magnitude or direction then Bragg scattering will occur whenever the surface of the Ewald sphere passes through a reciprocal lattice site. A necessary condition for Bragg scattering to occur at all is that $2k > K_c$ where K_c is the magnitude of the shortest reciprocal lattice vector. This determines the so-called *Bragg cut-off*.

The situation illustrated in Fig. 5.2a is realized when a collimated polyenergetic neutron beam is incident on the crystal. The incident-neutron wave vectors **k** are then all in essentially the same direction but have a continuous distribution of magnitudes. Those values of **k** for which the corresponding Ewald sphere intersects a reciprocal lattice site will then give rise to a collimated monoenergetic Bragg-reflected beam. This is the basis for the *Laue method* in neutron crystallography. It also illustrates how a crystal can be used as a *monochromator* for neutrons.

It is evident from Fig. 5.3 that if **k** satisfies the condition for Bragg scattering via $\mathbf{K}_h$, then $n\mathbf{k}$, where $n = 2, 3, 4, \ldots$, satisfies the condition for Bragg scattering via $n\mathbf{K}_h \equiv \mathbf{K}_{nh}$. Furthermore, the various Bragg-scattered beams are all in the same direction. This so-called *order contamination* is often a problem in monochromators but can be eliminated by the use of a suitable filter to remove the unwanted wave vectors from the incident beam, as will be described in Sec. 5.6.

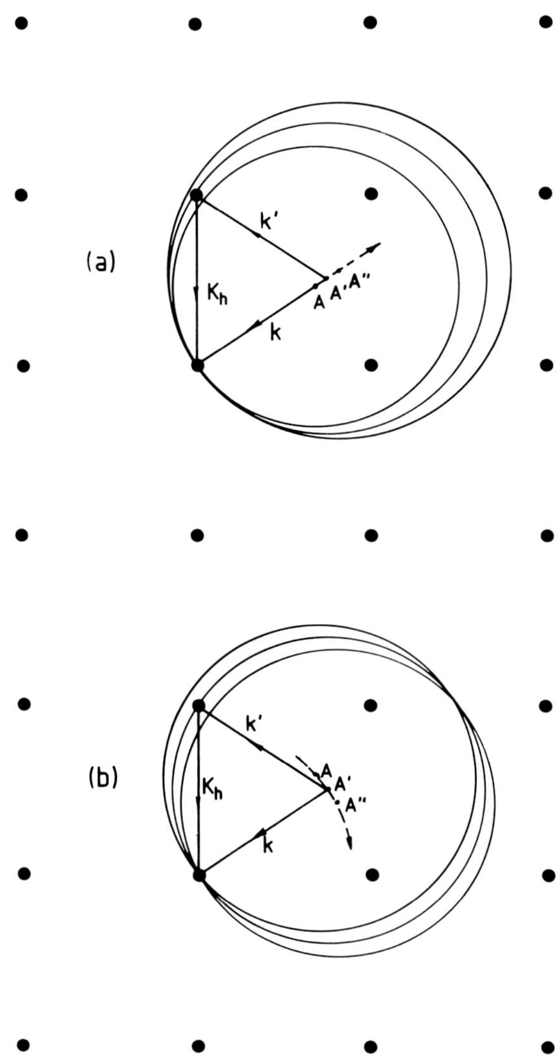

Fig. 5.2 Use of the Ewald construction to illustrate (a) the Laue method and (b) the Bragg method.

The situation illustrated in Fig. 5.2b is realized when a collimated monoenergetic neutron beam is incident on the crystal and the crystal is is rotated about an axis perpendicular to the incident beam. This produces a rotation of the Ewald sphere relative to the reciprocal lattice, and Bragg scattering occurs whenever the Ewald sphere intersects a reciprocal lattice site. This is the basis for the *Bragg method,* or *rotating-crystal method,* in neutron crystallography, which will be discussed later in the present section.

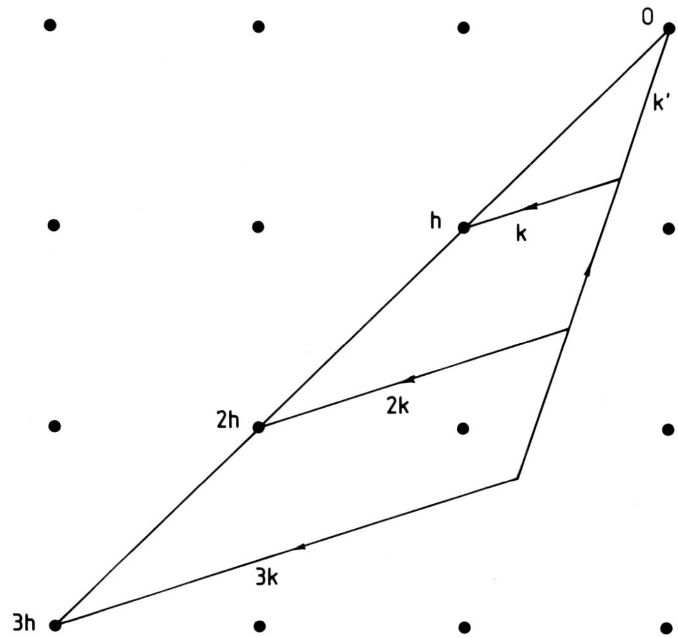

Fig. 5.3 Order contamination.

Brillouin zones

The planes in **k**-space that right-bisect the reciprocal lattice vectors $\mathbf{K}_h$ define the Brillouin zone boundaries (Brillouin, 1930). For example, the symmetric unit cell in the reciprocal lattice is the first Brillouin zone (see Appendix A). It is seen in Fig. 5.4 that when the Laue equation (5.2.15) is satisfied, the point A of Fig. 5.1 lies on the Brillouin zone boundary that right bisects $\mathbf{K}_h$ and

$$K_h = 2k \sin \theta \tag{5.2.22}$$

where θ is the angle that **k** and **k**′ make with the Brillouin zone boundary. Thus, Bragg scattering occurs whenever the incident-neutron wave vector lies on or near a Brillouin zone boundary.

Bragg's law

Since the Brillouin zone boundaries in **k**-space are perpendicular to the reciprocal lattice vectors, they are parallel to the corresponding Bragg planes in **r**-space (see Appendix A). The distance between neighboring Bragg planes is given, according to Eq. A.12, by

$$d_h = \frac{2\pi n}{K_h} \tag{5.2.23}$$

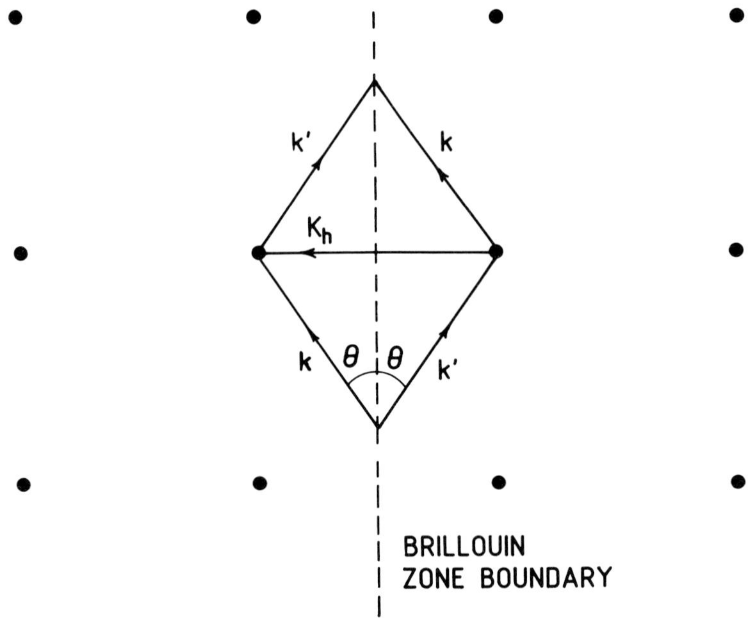

Fig. 5.4 Brillouin zones.

where n is the largest common factor of h_1, h_2, h_3. Hence, it follows from Eq. 5.2.22 that

$$n\lambda = 2d_h \sin\theta \tag{5.2.24}$$

where $\lambda = 2\pi/k$ is the incident-neutron wavelength and θ is, equivalently, the angle that $\mathbf{k}$ and $\mathbf{k}'$ make with the Bragg planes. Thus, θ is usually called the *Bragg angle*.

Equation 5.2.24 is *Bragg's law* and was originally obtained (Bragg, 1913) by means of the familiar heuristic argument that is illustrated in Fig. 5.5. One assumes that the diffraction is essentially the same as if the incident wave were partially reflected at each Bragg plane. The path difference ABC for waves reflected from neighboring planes is $2d_h \sin\theta$ and Eq. 5.2.24 is then the condition that all the reflected waves are in phase and, hence, interfere constructively. From this point of view Bragg scattering is often referred to as *Bragg reflection*.

If the reciprocal lattice sites are labeled by the conventional Miller indices *hkl* (see Appendix A), then for a cubic crystal

$$K_{hkl} = \frac{2\pi}{a}(h^2 + k^2 + l^2)^{1/2} \tag{5.2.25}$$

where a is the lattice constant. The wavelength corresponding to this reflection is then given according to Eq. 5.2.22 by

$$\lambda = 2a(h^2 + k^2 + l^2)^{-1/2} \sin\theta \tag{5.2.26}$$

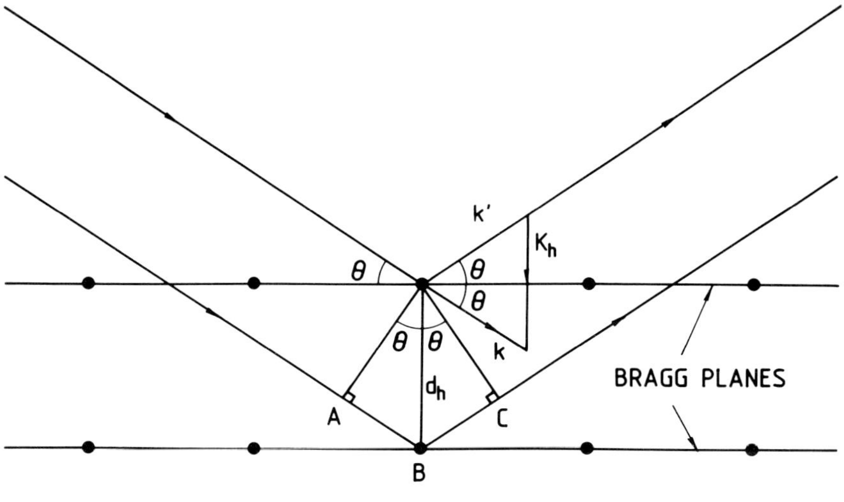

Fig. 5.5 Bragg's law.

In the Laue method, where θ has a fixed value, this means that $\lambda = \lambda_{hkl}$. In the rotating-crystal method, on the other hand, where λ has a fixed value, it means that $\theta = \theta_{hkl}$.

The relation (Eq. 5.2.26) between λ and θ is invariant under any permutation of hkl and under the reflections $h \rightarrow \bar{h}$, etc. The *multiplicity factor* M_{hkl} is defined as the total number of reflections that are symmetrically equivalent in the above sense. Table 5.2 lists the multiplicity factors for the reflections in cubic crystals with $h^2 + k^2 + l^2 \leq 16$.

Table 5.2 Multiplicity factor M_{hkl} for Bragg reflection in cubic crystals. The symbols × indicate the allowed reflections for sc, bcc, and fcc lattices.

hkl	$h^2+k^2+l^2$	M_{hkl}	sc	bcc	fcc
000	0	1	×	×	×
100	1	6	×		
110	2	12	×	×	
111	3	8	×		×
200	4	6	×	×	×
210	5	24	×		
211	6	24	×	×	
220	8	12	×	×	×
221	9	24	×		
300	9	6	×		
310	10	24	×	×	
311	11	24	×		×
222	12	8	×	×	×
320	13	24	×		
321	14	48	×	×	
400	16	6	×	×	×

Note the accidental degeneracy when $h^2+k^2+l^2=9$. Thus, $\theta_{221}=\theta_{300}$ even though these reflections are not symmetrically equivalent.

For a hexagonal crystal

$$K_{hkl}=\frac{2\pi}{a}\left[\frac{4}{3}(h^2+hk+k^2)+\frac{l^2}{\gamma^2}\right]^{1/2} \tag{5.2.27}$$

where $\gamma=c/a$, and a and c are the lattice constants. The wavelength corresponding to this reflection is then given by

$$\lambda=2a\left[\frac{4}{3}(h^2+hk+k^2)+\frac{l^2}{\gamma^2}\right]^{-1/2}\sin\theta \tag{5.2.28}$$

The above relations take a more symmetrical form when expressed in terms of the Bravais–Miller indices *hkil* (see Appendix A). Then

$$K_{hkil}=\frac{2\pi}{a}\left[\frac{2}{3}(h^2+k^2+i^2)+\frac{l^2}{\gamma^2}\right]^{1/2} \tag{5.2.29}$$

and

$$\lambda=2a\left[\frac{2}{3}(h^2+k^2+i^2)+\frac{l^2}{\gamma^2}\right]^{-1/2}\sin\theta \tag{5.2.30}$$

Note that this relation is invariant under any permutation of *hki* and under the reflections $h\to\bar{h}$, etc.

Debye–Waller factor

The Debye–Waller factor (Eq. 5.2.7) characterizes the effect of the thermal motion of the atoms on the intensities of the Bragg peaks. Omitting the subscripts we can write

$$\exp(-W)=\langle\exp(i\mathbf{q}\cdot\mathbf{u})\rangle \tag{5.2.31}$$

Expanding the exponential on the right-hand side in powers of $\mathbf{q}\cdot\mathbf{u}$, and then regrouping the various terms, we obtain the cumulant expansion

$$W=\frac{1}{2!}\langle(\mathbf{q}\cdot\mathbf{u})^2\rangle+\frac{i}{3!}\langle(\mathbf{q}\cdot\mathbf{u})^3\rangle-\frac{1}{4!}[\langle(\mathbf{q}\cdot\mathbf{u})^4\rangle-3\langle(\mathbf{q}\cdot\mathbf{u})^2\rangle^2]+\cdots \tag{5.2.32}$$

where we have used the fact that $\langle\mathbf{u}\rangle=0$ by definition (see Appendix B). If the crystal is centrosymmetric, the odd-order terms in Eq. 5.2.32 vanish identically.

An explicit calculation of W was first given by Debye (1914) using the now-famous Debye model for the lattice vibrations. A more general calculation in terms of the exact normal modes of vibration of a harmonic crystal was given later by Waller (1923). A modern account of this theory can be found in the book by Lovesey (1984) for example, where it is shown that for a harmonic crystal

$$W = \frac{1}{2!}\langle(\mathbf{q}\cdot\mathbf{u})^2\rangle \tag{5.2.33}$$

so that $\exp(-W)$ is rigorously a Gaussian function of $\mathbf{q}$.

For the Debye–Waller factor to be Gaussian it is actually not necessary that the normal modes of vibration be harmonic but only that they be statistically independent. In this case the Gaussian behavior is simply a consequence of the central limit theorem of probability theory (Sears and Khanna, 1972). Real crystals are, of course, always anharmonic to some extent and the higher-order terms in Eq. 5.2.32 are then nonvanishing. Nevertheless, the Gaussian approximation is still usually valid for most practical purposes. Even for a crystal as anharmonic as solid ^{4}He the non-Gaussian terms are $\lesssim$1 percent for most $\mathbf{q}$ values of interest (Sears and Khanna, 1972).

Let us express Eq. 5.2.33 in the form

$$2W = \mathbf{q}\cdot\langle\mathbf{uu}\rangle\cdot\mathbf{q} \tag{5.2.34}$$

If the directions of the coordinate axes are chosen to coincide with the principal directions of the symmetric tensor $\langle\mathbf{uu}\rangle$ then

$$\langle\mathbf{uu}\rangle = \begin{vmatrix} u_{0x}^2 & 0 & 0 \\ 0 & u_{0y}^2 & 0 \\ 0 & 0 & u_{0z}^2 \end{vmatrix} \tag{5.2.35}$$

and Eq. 5.2.34 becomes

$$2W = (q_x u_{0x})^2 + (q_y u_{0y})^2 + (q_z u_{0z})^2 \tag{5.2.36}$$

In this case the Debye–Waller factor is an anisotropic Gaussian for which the locus of the half-height points is an ellipsoid.

If the z axis is an axis of greater than 2-fold symmetry then Eq. 5.2.36 reduces to

$$2W = (q_x^2 + q_y^2)u_\perp^2 + q_z^2 u_\parallel^2 \tag{5.2.37}$$

where $u_\perp$ and $u_\parallel$ are the rms displacements of the atom perpendicular and parallel to the symmetry axis,

$$u_\perp = u_{0x} = u_{0y}, \qquad u_\parallel = u_{0z} \tag{5.2.38}$$

If, on the other hand, the crystal has cubic symmetry, so that

$$u_{0x} = u_{0y} = u_{0z} \equiv u_0 \tag{5.2.39}$$

then

$$2W = (q_x^2 + q_y^2 + q_z^2)u_0^2 = (qu_0)^2 \tag{5.2.40}$$

In the unit-cell structure factor (Eq. 5.2.17), the Debye–Waller factor is calculated with $\mathbf{q}$ equal to the reciprocal lattice vector. Then for a cubic crystal Eq. 5.2.40 becomes

$$2W_{hkl} = \left(\frac{2\pi u_0}{a}\right)^2 (h^2 + k^2 + l^2) \tag{5.2.41}$$

and for a hexagonal crystal Eq. 5.2.37 becomes

$$2W_{hkl} = \left(\frac{2\pi u_\perp}{a}\right)^2 \frac{4}{3}(h^2 + hk + k^2) + \left(\frac{2\pi u_\parallel}{c}\right)^2 l^2 \tag{5.2.42}$$

or, in terms of the Bravais–Miller indices,

$$2W_{hkil} = \left(\frac{2\pi u_\perp}{a}\right)^2 \frac{2}{3}(h^2 + k^2 + i^2) + \left(\frac{2\pi u_\parallel}{c}\right)^2 l^2 \tag{5.2.43}$$

In hexagonal close-packed crystals the lattice vibrations are roughly isotropic so that $u_\perp \simeq u_\parallel$. In some other hexagonal crystals, most notably graphite, the lattice vibrations are highly anisotropic and $u_\perp$ is significantly different from $u_\parallel$.

Friedel's law

The unit-cell structure factor (Eq. 5.2.17) determines the integrated intensities of the Bragg peaks. In conventional crystallographic notation (see Appendices A and B) this quantity can be expressed as

$$F_{hkl} = \sum b \exp(-W_{hkl}) \exp[2\pi i(hx + ky + lz)] \tag{5.2.44}$$

where the sum runs over all the atoms in the primitive unit cell as before.

To the extent that the Gaussian approximation for the Debye–Waller factor is valid, W is an even function of $\mathbf{q}$ and

$$W_{hkl} = W_{\bar{h}\bar{k}\bar{l}} \tag{5.2.45}$$

Except for strongly absorbing nuclei, the scattering length b can be taken to be real (see Sec. 2.4). Hence, it follows from Eq. 5.2.44 that

$$F^*_{hkl} = F_{\bar{h}\bar{k}\bar{l}} \tag{5.2.46}$$

and

$$|F_{hkl}|^2 = |F_{\bar{h}\bar{k}\bar{l}}|^2 \tag{5.2.47}$$

This result was first obtained in the context of X-ray diffraction in crystals by Friedel (1913). An immediate consequence of Friedel's law is that the differential scattering cross section (Eq. 5.2.18) is an even function of $\mathbf{q}$ even if the crystal is not centrosymmetric. Thus, for example, a right-handed crystal produces the same diffraction pattern as its left-handed counterpart.

If the crystal is centrosymmetric, so that for each atom with position

(x, y, z) there is an identical atom with position $(\bar{x}, \bar{y}, \bar{z})$, then

$$F_{hkl} = F_{\bar{h}\bar{k}\bar{l}} = F^*_{hkl} \tag{5.2.48}$$

and the unit-cell structure factor is real.

Friedel's law breaks down if one or more of the nuclei in the crystal has strong absorption at thermal-neutron energies *and* the crystal is non-centrosymmetric. This has been observed by, for example, Peterson and Smith (1961) in CdS, where the strong absorption is due to the (n, γ) resonance in the isotope ^{113}Cd.

Unit-cell structure factor

For the bcc and fcc structures there is only one atom in the primitive unit cell, so that

$$F_{hkl} = b \exp(-W_{hkl}) \tag{5.2.49}$$

All other naturally occurring structures have at least two atoms in the primitive unit cell and the complex phase factor in Eq. 5.2.44 then gives rise to interference effects. For some reflections it may happen that $F_{hkl} = 0$, in which case the reflection is said to be *forbidden*. For others, the phase factors may all equal unity, so that all the atoms in the unit cell scatter in phase, in which case hkl is called a *full-phase reflection*.

The atomic coordinates (x, y, z) for a number of common crystal structures are given in Table B.1 in Appendix B. In the diamond structure, for example, the primitive unit cell contains two identical atoms with positions $(0, 0, 0)$ and $(\frac{1}{4}, \frac{1}{4}, \frac{1}{4})$ so that

$$F_{hkl} = b \exp(-W_{hkl}) f_{hkl} \tag{5.2.50}$$

where the *phase factor* is given by

$$\begin{aligned} f_{hkl} &= 1 + \exp\left[\frac{i\pi}{2}(h + k + l)\right] \\ &= 1 + i^{h+k+l} \end{aligned} \tag{5.2.51}$$

In particular,

$$f_{hkl} = \begin{cases} 1 + 1 = 2, \\ 1 + i, \\ 1 - 1 = 0, \\ 1 - i, \end{cases} \qquad h + k + l = \begin{cases} 0, 4, 8, \ldots \\ 1, 5, 9, \ldots \\ 2, 6, 10, \ldots \\ 3, 7, 11, \ldots \end{cases} \tag{5.2.52}$$

In the diamond structure the direct lattice is fcc, so that the Miller indices h, k, and l are either all even or all odd. For the even reflections $h + k + l$ is also even and $f_{hkl} = 2$ when the two atoms in the primitive unit cell scatter in phase and $f_{hkl} = 0$ when they scatter 180 degrees out of phase. For the odd reflections $h + k + l$ is also odd so that the two atoms in the primitive unit cell scatter 90 degrees out of phase and f_{hkl} is complex.

Among the reflections for fcc lattices listed in Table 5.2 we note that for the diamond structure (200) and (222) are forbidden, (220) and (400) are full-phase reflections, and (111) and (311) are complex-phase reflections. When one takes into account the combined effect of the phase factor and the Debye–Waller factor on F_{hkl} it is evident that the strongest reflections in the diamond structure are (111) and (220).

We have discussed the diamond structure in some detail, not only because it provides a good illustration of the properties of unit-cell structure factors in general, but also because silicon has this structure and silicon is the most commonly used crystal in modern neutron-optical devices. The main advantage of silicon is that large, essentially perfect silicon crystals are available at moderate cost from the semiconductor industry. Silicon is also often used as a monochromator in crystal spectrometers. In this application the (111) reflection is sometimes employed in preference to (220) in order to minimize the effect of order contamination (i.e., (222) is forbidden while (440) is not). Note also that the second-order reflection (400) can be used as a monochromator reflection because the corresponding first-order reflection (200) is forbidden in silicon.

Reflectivity

The intrinsic intensity of Bragg reflection in crystals is usually expressed in terms of a dimensionless quantity called the reflectivity, rather than in terms of the scattering cross section. The reflectivity $R(\mathbf{k})$ is defined as the fraction of incident neutrons that are Bragg reflected when the incident-neutron wave vector has some specified value $\mathbf{k}$. Equivalently, $R(\mathbf{k})$ is the ratio of the reflected-neutron current to the incident-neutron current. The incident-neutron current (neutrons s^{-1}) is JA, where J is the incident-neutron flux (neutrons $cm^{-2}\,s^{-1}$) and A is the cross-sectional area of the crystal perpendicular to the incident beam. The reflected-neutron current is $J\int d\sigma$. Thus,

$$R(\mathbf{k}) = \frac{1}{A}\int d\sigma \tag{5.2.53}$$

Alternatively,

$$R(\mathbf{k}) = \Sigma(\mathbf{k})t \tag{5.2.54}$$

where $\Sigma(\mathbf{k})$ is the total cross section per unit volume for Bragg scattering,

$$\Sigma(\mathbf{k}) = \frac{1}{V}\int d\sigma \tag{5.2.55}$$

and $t = V/A$ is the average thickness of the crystal in the direction of the incident beam. The value of t depends on the shape of the crystal. Thus, for a sphere of radius r, we get $t = (4/3)r$. For a cylinder of radius r, with $\mathbf{k}$ perpendicular to the axis, we find that $t = (\pi/2)r$. For a plane slab of thickness d, where $\mathbf{k}$ makes an angle ϕ with the surface, we get $t = d \operatorname{cosec} \phi$.

Since the Bragg scattering is elastic, $\Sigma(\mathbf{k})$ can be expressed equivalently as

$$\Sigma(\mathbf{k}) = \frac{1}{V} \iint \frac{d\sigma}{d\Omega} \delta(\omega)\, d\Omega\, d\varepsilon' \tag{5.2.56}$$

where $\omega = \varepsilon - \varepsilon'$ is the energy transfer in units of $\hbar$ as before (see Eq. 2.5.8). Since $\mathbf{q} = \mathbf{k} - \mathbf{k}'$, we see that

$$\omega = \frac{\hbar}{2m}(k^2 - k'^2) = \frac{\hbar}{m}(\mathbf{k} \cdot \mathbf{q} - \tfrac{1}{2}q^2) \tag{5.2.57}$$

and, hence, that

$$\delta(\omega) = \frac{m}{\hbar} \delta(\mathbf{k} \cdot \mathbf{q} - \tfrac{1}{2}q^2) \tag{5.2.58}$$

Also, according to Eq. 2.5.18,

$$d\mathbf{k}' = \frac{mk'}{\hbar} d\Omega\, d\varepsilon' \tag{5.2.59}$$

and, for a fixed value of $\mathbf{k}$, it is evident that $d\mathbf{q} = d\mathbf{k}'$. Gathering the preceding results, and noting that $\delta(\omega) = 0$ unless $k' = k$, we see finally that

$$\delta(\omega)\, d\Omega\, d\varepsilon' = \frac{1}{k} \delta(\mathbf{k} \cdot \mathbf{q} - \tfrac{1}{2}q^2)\, d\mathbf{q} \tag{5.2.60}$$

and, hence, that

$$\Sigma(\mathbf{k}) = \frac{1}{Vk} \int \frac{d\sigma}{d\Omega} \delta(\mathbf{k} \cdot \mathbf{q} - \tfrac{1}{2}q^2)\, d\mathbf{q} \tag{5.2.61}$$

Since $V = N_0 V_0$ it then follows from Eq. 5.2.18 that

$$\Sigma(\mathbf{k}) = \sum_h \Sigma_h(\mathbf{k}) \tag{5.2.62}$$

where $\Sigma_h(\mathbf{k})$ is the total cross section per unit volume for Bragg scattering via $\mathbf{K}_h$,

$$\Sigma_h(\mathbf{k}) = \frac{(2\pi)^3}{V_0^2 k} |F_h|^2 \int \Delta(\mathbf{q} - \mathbf{K}_h)\, \delta(\mathbf{k} \cdot \mathbf{q} - \tfrac{1}{2}q^2)\, d\mathbf{q} \tag{5.2.63}$$

The reflectivity (Eq. 5.2.54) can similarly be expressed as

$$R(\mathbf{k}) = \sum_h R_h(\mathbf{k}) \tag{5.2.64}$$

where $R_h(\mathbf{k})$ is the reflectivity for Bragg scattering via $\mathbf{K}_h$,

$$R_h(\mathbf{k}) = \Sigma_h(\mathbf{k}) t \tag{5.2.65}$$

Rotating-crystal method

Suppose a collimated monoenergetic beam of neutrons with wave vector $\mathbf{k}$ is incident on the crystal and that the crystal is then rotated about an axis perpendicular to the plane defined by $\mathbf{k}$ and the reciprocal lattice vector $\mathbf{K}_h$ at which one wishes to measure the Bragg scattering. Then $\mathbf{k}$ will rotate relative to the reciprocal lattice in the way shown in Fig. 5.6. If θ is the angle that $\mathbf{k}$ makes with the Bragg planes, then, when the Laue equation (5.2.15) is satisfied exactly, $\theta = \theta_h$ and $\mathbf{k} = \mathbf{k}_0$. Since $K_h = 2k \sin\theta_h$, it then follows that to first order

$$\mathbf{k}\cdot\mathbf{q} - \tfrac{1}{2}q^2 = (k^2 \sin 2\theta_h)\theta_1 - (\mathbf{K}_h - \mathbf{k}_0)\cdot\mathbf{q}_1 \tag{5.2.66}$$

where

$$\theta_1 = \theta - \theta_h, \qquad \mathbf{q}_1 = \mathbf{q} - \mathbf{K}_h \tag{5.2.67}$$

Thus,

$$\delta(\mathbf{k}\cdot\mathbf{q} - \tfrac{1}{2}q^2) = \frac{1}{k^2 \sin 2\theta_h}\,\delta(\theta_1 - \Theta(\mathbf{q}_1)) \tag{5.2.68}$$

where

$$\Theta(\mathbf{q}_1) = \frac{(\mathbf{K}_h - \mathbf{k}_0)\cdot\mathbf{q}_1}{k^2 \sin 2\theta_h} \tag{5.2.69}$$

It now follows from Eq. 5.2.63 that

$$\Sigma_h(\mathbf{k}) = Q_h W(\theta_1) \tag{5.2.70}$$

where

$$Q_h = \frac{\lambda^3\,|F_h|^2}{V_0^2 \sin 2\theta_h} \tag{5.2.71}$$

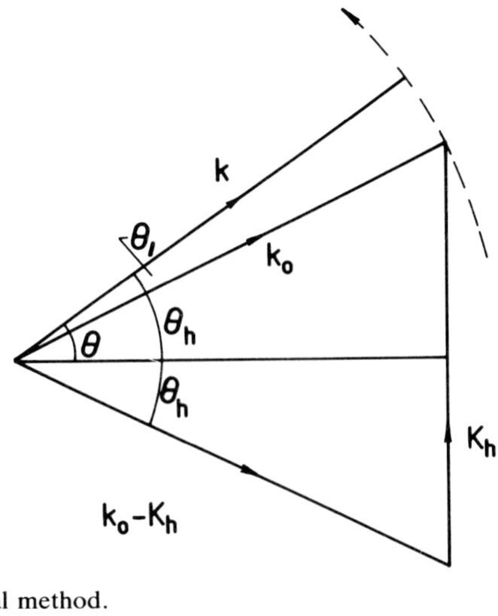

Fig. 5.6 Rotating-crystal method.

in which $\lambda = 2\pi/k$ is the incident-neutron wavelength and the factor $1/\sin 2\theta_h$ is called the *Lorentz factor*, and

$$W(\theta_1) = \int \Delta(\mathbf{q}_1)\, \delta(\theta_1 - \Theta(\mathbf{q}_1))\, d\mathbf{q}_1 \tag{5.2.72}$$

According to Eq. 5.1.14, this function has the properties

$$\begin{gathered} W(\theta_1) = W(\theta_1)^* = W(-\theta_1) \\ \int W(\theta_1)\, d\theta_1 = 1 \\ \int \theta_1 W(\theta_1)\, d\theta_1 = 0 \\ \lim_{V\to\infty} W(\theta_1) = \delta(\theta_1) \end{gathered} \tag{5.2.73}$$

The reflectivity (Eq. 5.2.65), which we now denote simply as $R_h(\theta)$, is then given by

$$R_h(\theta) = \rho_h W(\theta - \theta_h) \tag{5.2.74}$$

where ρ_h is the *integrated reflectivity,*

$$\rho_h = \int R_h(\theta)\, d\theta = Q_h t \tag{5.2.75}$$

In the rotating-crystal method, the scan portrayed in Fig. 5.6 is produced by "rocking" the crystal through the Bragg angle θ_h. Consequently, $R_h(\theta)$ is called the *rocking curve* and $W(\theta - \theta_h)$ the *normalized rocking curve.* In general, $W(\theta - \theta_h)$ consists of a sharp peak centered at $\theta = \theta_h$ with a width and shape that depends on the size and shape of the crystal. In contrast, the quantity Q_h that determines the integrated reflectivity is independent of the size and shape of the crystal and is characteristic of the crystal structure itself. Thus, in neutron crystallography studies using the rotating-crystal method, one measures the integrated reflectivity ρ_h so as to determine Q_h and, hence, $|F_h|^2$ at each reciprocal lattice site h.

The rotating-crystal method for the quantitative measurement of Bragg peak intensities in X-ray crystallography is due originally to Bragg and Bragg (1913). The use of this method was made possible by their invention of the *crystal spectrometer,* which will be discussed in Sec. 5.6. A detailed theory of the reflectivity equivalent to that just given was later developed by Bragg et al. (1921a,b).

Infinite plane slab

Let us now suppose that the crystal takes the form of a plane slab of thickness d and infinite lateral extent. If we choose a system of Cartesian coordinate axes such that the faces of the slab are the planes $z = 0$ and

$z = d$, then from Eq. 5.1.18,

$$\Delta(\mathbf{q}_1) = \delta(q_{1x})\,\delta(q_{1y})\,\Delta(q_{1z}) \tag{5.2.76}$$

where

$$\Delta(q_{1z}) = \frac{d}{2\pi}\left[\frac{\sin(q_{1z}d/2)}{q_{1z}d/2}\right]^2 \tag{5.2.77}$$

Hence, it follows from Eq. 5.2.72 that

$$W(\theta_1) = \int \Delta(q_{1z})\,\delta(\theta_1 - \Theta(\mathbf{q}_1))\,dq_{1z} \tag{5.2.78}$$

where $\mathbf{q}_1 = q_{1z}\hat{\mathbf{z}}$ and $\hat{\mathbf{z}}$ denotes a unit vector in the z-direction, i.e., the direction normal to the face of the slab.

In the discussion of scattering by an infinite plane slab, it is necessary to distinguish between reflection geometry (the *Bragg case*) and transmission geometry (the *Laue case*), as illustrated in Fig. 5.7. Then Eq.

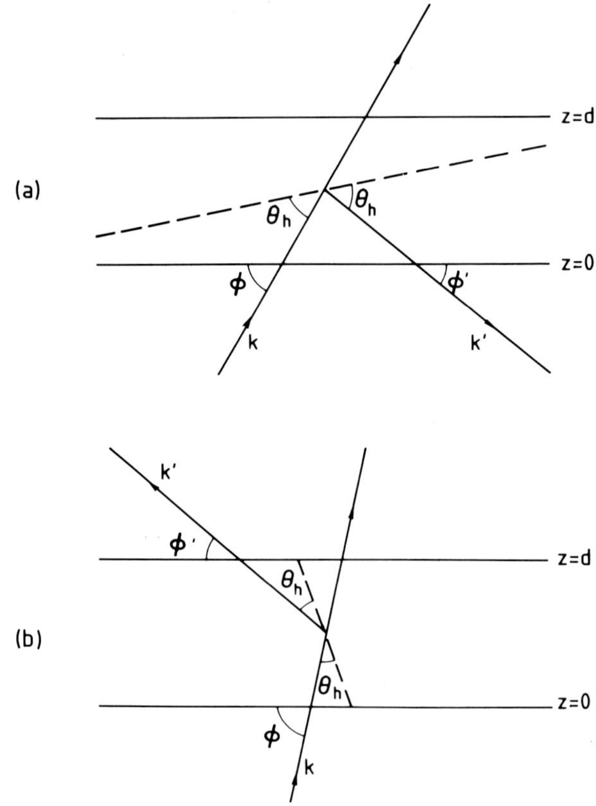

Fig. 5.7 Wave vector relations for the kinematical theory of Bragg reflection in a plane slab for (a) the Bragg case and (b) the Laue case. The dashed lines indicate the Bragg planes.

5.2.69 gives

$$\Theta(\mathbf{q}_1) = \pm\left[\frac{\sin\phi'}{k\sin 2\theta_h}\right] q_{1z} \tag{5.2.79}$$

where the plus sign is for the Bragg case and the minus sign for the Laue case, and ϕ' is the angle that the diffracted beam makes with the surface of the slab. Hence,

$$W(\theta_1) = \left|\frac{q_{1z}}{\theta_1}\right| \Delta(q_{1z}) \tag{5.2.80}$$

where now

$$q_{1z} = \pm\left[\frac{k\sin 2\theta_h}{\sin\phi'}\right]\theta_1 \tag{5.2.81}$$

We see that $W(\theta_1)$ is independent of the sign of q_{1z} and is an even function of θ_1, as it should be according to Eq. 5.2.73. The sign is, however, important in determining the scattered-neutron wave vector $\mathbf{k}'$. Thus, since $\mathbf{q}_1 = \mathbf{q} - \mathbf{K}_h$ and $\mathbf{q} = \mathbf{k} - \mathbf{k}'$, we see that

$$\mathbf{k}' = \mathbf{k} - \mathbf{K}_h - \mathbf{q}_1 \tag{5.2.82}$$

The Bragg reflection from an infinite plane slab is qualitatively different from that from a finite crystal. For an infinite plane slab the peak function $\Delta(\mathbf{q}_1)$ has the singular form of Eq. 5.2.76, as a result of which there is a single diffracted wave vector $\mathbf{k}'$ for each incident wave vector $\mathbf{k}$, determined by Eq. 5.2.82. For a finite crystal, on the other hand, $\Delta(\mathbf{q}_1)$ is nonsingular so that for each incident wave vector $\mathbf{k}$ there is a continuum of diffracted wave vectors $\mathbf{k}'$ distributed about $\mathbf{k} - \mathbf{K}_h$. Physically, this means that if the incident-neutron wave function is a plane wave with wave vector $\mathbf{k}$ then, for an infinite plane slab, the Bragg reflected wave is also a plane wave but with wave vector $\mathbf{k}'$. However, for a finite crystal, the width of the Bragg reflected wave is limited by the size of the crystal and, hence, has a finite angular divergence, so that a continuous distribution of wave vectors $\mathbf{k}'$ is required for its description.

For an infinite plane slab, the average crystal thickness in the direction of the incident beam is given by $t = d/\sin\phi$. Hence, the integrated reflectivity (Eq. 5.2.75) becomes

$$\rho_h = \frac{\lambda^3 d\,|F_h|^2}{V_0^2 \sin\phi\,\sin 2\theta_h} \tag{5.2.83}$$

The reflectivity (Eq. 5.2.74) can then be expressed with the help of Eq. 5.2.80 in the form

$$R(x, y) = \left(\frac{\sin xy}{x}\right)^2 \tag{5.2.84}$$

where

$$x = \frac{\theta - \theta_h}{\chi_h}, \qquad y = \frac{\pi d}{\Delta_h} \tag{5.2.85}$$

in which

$$\chi_h = \frac{\lambda^2 |F_h|}{\pi V_0 \sin 2\theta_h} \left(\frac{\sin \phi'}{\sin \phi}\right)^{1/2} \tag{5.2.86}$$

and

$$\Delta_h = \frac{\pi V_0}{\lambda |F_h|} (\sin \phi \sin \phi')^{1/2} \tag{5.2.87}$$

Here x is a dimensionless variable proportional to the deviation of the angle of incidence θ from the Bragg angle θ_h, and y is a dimensionless variable proportional to the thickness d of the crystal. In terms of these variables the peak reflectivity is given by

$$R(0, y) = y^2 \tag{5.2.88}$$

the integrated reflectivity by

$$\rho(y) \equiv \int R(x, y)\, dx = \pi y \tag{5.2.89}$$

and the "width" of the Bragg peak by

$$\Delta x \equiv \frac{\rho(y)}{R(0, y)} = \frac{\pi}{y} \tag{5.2.90}$$

The crystals used in neutron-optical instruments are often in the form of plane slabs. Furthermore, the crystals are usually cut so that the Bragg planes of interest are either parallel or perpendicular to the surface. This corresponds to the symmetric Bragg and Laue cases that are illustrated in Fig. 5.8. The expressions for a number of quantities of interest are listed for the symmetric Bragg and Laue cases in Table 5.3.

It will be noted that it is only in the symmetric Bragg case, where $\mathbf{q}_1$ is parallel to $\mathbf{K}_h$, that $\mathbf{k}$ and $\mathbf{k}'$ always make equal angles with the Bragg planes. In all other cases it is only at the peak of the rocking curve (where $\mathbf{q}_1 = 0$ so that the Laue equation and, hence, Bragg's law are satisfied) that the angle of reflection equals the angle of incidence. For example, it is seen in Table 5.3 that for the symmetric Laue case if the angle of incidence $\theta = \theta_h + \theta_1$ then the angle of reflection $\theta' = \theta_h - \theta_1$. This emphasizes that the elementary Bragg theory (Bragg, 1913) should not be taken literally. A single plane of atoms does not produce specular reflection of the incident wave but, rather, gives the Fraunhofer diffraction characteristic of a cross grating (Born and Wolf, 1975). There is therefore no a priori reason to expect that the angle of "reflection" must equal the angle of incidence except at the peak of the rocking curve.

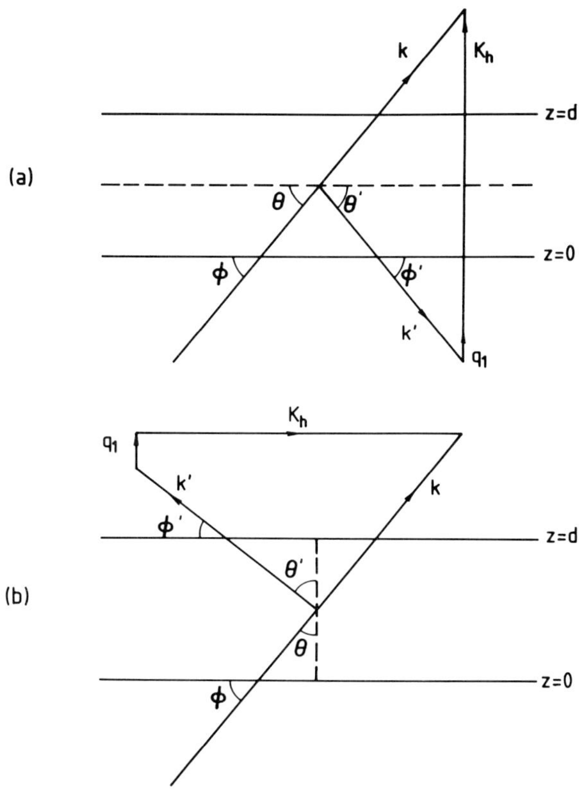

Fig. 5.8 Wave vector relations for (a) the symmetric Bragg case and (b) the symmetric Laue case.

Table 5.3 Quantities for symmetric Bragg reflection in a infinite plane slab

Quantity	Bragg case	Laue case
θ	$\theta_h + \theta_1$	$\theta_h + \theta_1$
θ'	$\theta_h + \theta_1$	$\theta_h - \theta_1$
ϕ	θ	$\dfrac{\pi}{2} - \theta$
ϕ'	θ'	$\dfrac{\pi}{2} - \theta'$
q_{1z}	$\left(\dfrac{4\pi}{\lambda}\cos\theta_h\right)\theta_1$	$-\left(\dfrac{4\pi}{\lambda}\sin\theta_h\right)\theta_1$
ρ_h	$\dfrac{\lambda^3 d\,\lvert F_h\rvert^2}{V_0^2 \sin\theta_h \sin 2\theta_h}$	$\dfrac{\lambda^3 d\,\lvert F_h\rvert^2}{V_0^2 \cos\theta_h \sin 2\theta_h}$
χ_h	$\dfrac{\lambda^2\,\lvert F_h\rvert}{\pi V_0 \sin 2\theta_h}$	$\dfrac{\lambda^2\,\lvert F_h\rvert}{\pi V_0 \sin 2\theta_h}$
Δ_h	$\dfrac{\pi V_0 \sin\theta_h}{\lambda\,\lvert F_h\rvert}$	$\dfrac{\pi V_0 \cos\theta_h}{\lambda\,\lvert F_h\rvert}$

Validity of the kinematical theory

If the incident-neutron wave vector **k** satisfies the condition for Bragg reflection via (hkl) then the scattered-neutron wave vector $\mathbf{k}'$ satisfies the condition for Bragg reflection via $(\bar{h}\bar{k}\bar{l})$. If the neutron is Bragg reflected a second time it will therefore regain its original wave vector. Consequently, each neutron will tend to be reflected back and forth between these Bragg planes until it leaves the crystal. This is illustrated in Fig. 5.9, where it is seen that the external transmitted beam contains components that have been Bragg reflected $0, 2, 4, \ldots$ times and the external Bragg reflected beam components that have been Bragg reflected $1, 3, 5, \ldots$ times.

This *multiple Bragg reflection* is precisely what is neglected in the kinematical theory. Since the reflectivity R that is calculated in the kinematical theory is the probability for a single Bragg reflection, it is evident that the kinematical theory will be valid only if $R \ll 1$. According to Eq. 5.2.84, $R \leq y^2$ for all x. Thus, the condition for the validity of the kinematical theory is that $y \ll 1$ or, equivalently, $d \ll \Delta_h$.

In the opposite limit $d \gg \Delta_h$, where $y \gg 1$, the relation (5.2.88) gives $R(0, y) \gg 1$. On the other hand, neutron conservation requires that $R(x, y) \leq 1$ for all x and y. Thus, the kinematical theory of diffraction violates neutron conservation. We shall see in Chap. 6 that when the effect of multiple Bragg reflection is taken properly into account the reflectivity is decreased in such a way that neutron conservation is

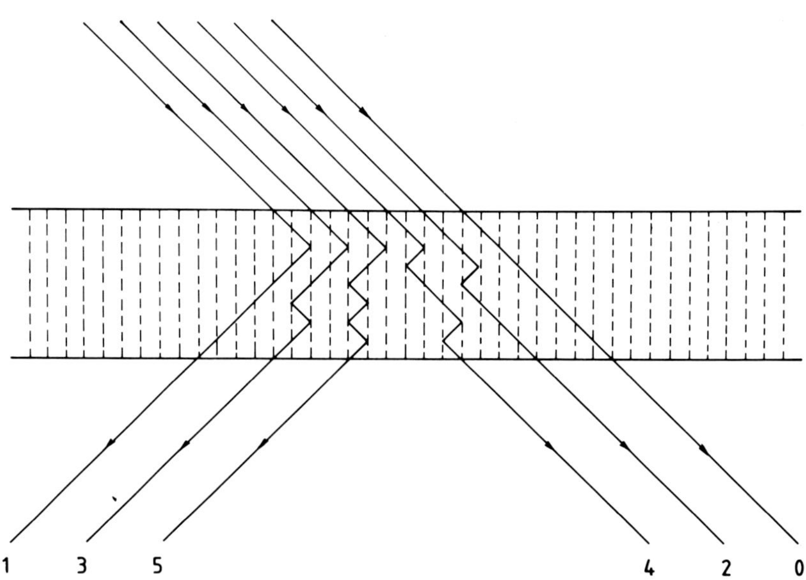

Fig. 5.9 Multiple Bragg reflection in a thick crystal.

satisfied. The correct expression for $R(x, y)$ will, however, reduce to the kinematical result (Eq. 5.2.84) in the limit $y \ll 1$.

The reduction of the Bragg peak intensities due to multiple Bragg reflection is referred to as *extinction*, and Δ_h is accordingly called the *extinction length*. Thus, it is only if the crystal thickness is much less than the extinction length that the kinematical theory is valid.

To estimate the value of Δ_h let us consider the symmetric Bragg case. Since $K_h = 2k \sin \theta_h$ it follows from Table 5.3 that

$$\Delta_h = \frac{V_0 K_h}{4 |F_h|} \tag{5.2.91}$$

For order-of-magnitude purposes we can put $V_0 \simeq a^3$, $K_h \simeq 2\pi/a$, and $F_h \simeq b$, where a is the lattice constant and b is the bound coherent scattering length. Then, apart from a constant factor of order unity,

$$\Delta_h \simeq \frac{a^2}{b} \tag{5.2.92}$$

With the typical values $a \simeq 5$ Å and $b \simeq 5$ fm we then get $\Delta_h \simeq 50$ μm = 0.05 mm.

More detailed calculations show that Δ_h generally varies from about 10 μm for very strong reflections to about 100 μm for weak reflections. Thus, it is evident that the kinematical theory of neutron diffraction is valid for a perfect crystal only if the thickness $d \lesssim 1$ μm. For a perfect crystal of macroscopic size, the reflectivity is dominated by multiple Bragg reflection and must be calculated on the basis of the dynamical theory of neutron diffraction given in Chap. 6.

Most crystalline materials are, however, not available in the form of large perfect crystals. The presence of imperfections such as dislocations or grain boundaries leads to a relaxation of the above requirement so that the kinematical theory of diffraction is, in fact, often valid to a good approximation for laboratory-size samples, as we shall see in the next two sections.

5.3. Mosaic crystals

Ideally imperfect crystals

Soon after the discovery of the diffraction of X-rays by crystals it was found that the observed Bragg reflection in single crystals differed significantly from the theoretical predictions. For example, the width of the Bragg peaks was typically two orders of magnitude larger than expected. Darwin (1914b, 1922) recognized that the problem lay in the incorrect assumption that the crystals were ideally perfect. He showed that the X-ray results can be understood if one assumes instead that a nominally single crystal is in fact composed of a large number of what are now called *mosaic blocks*. Each such block is itself assumed to be a perfect crystal, but neighboring blocks are slightly displaced and misoriented relative to one another.

Darwin introduced this mosaic model of an imperfect crystal simply because it was mathematically tractable. In fact, he thought it more likely that the imperfections in real crystals actually took the form of a "warping" of the Bragg planes. Later it was shown (Burgers, 1940; Bragg, 1940) that the theory of dislocations leads quite naturally to a mosaic structure of the Darwin type (see Appendix B), and subsequent studies have confirmed this picture (Hirsch, 1956; Van Bueren, 1961). The linear dimensions of the mosaic blocks in metallic crystals are typically of the order of 1 μm and the rms angle of misorientation of the blocks, the so-called *mosaic spread,* is typically 1 minute to 10 minutes. In exceptional cases the mosaic spread may be as large as 30 minutes but such crystals then tend to be very inhomogeneous.

A mosaic crystal will be said to be *ideally imperfect* if it scatters X-rays or neutrons in accordance with the kinematical theory of diffraction. This requires that multiple Bragg reflection is negligible. In particular, the mosaic blocks must be small enough ($\lesssim$1 μm) that multiple Bragg reflection within a single block is negligible. In addition, the crystal as a whole must be small enough ($\lesssim$1 mm, as we shall see later) that multiple Bragg reflection between different mosaic blocks is negligible.

If the individual mosaic blocks are too large, so that multiple Bragg reflection within each block is appreciable, then the Bragg peak intensities will be reduced by "primary" extinction, as discussed in Chap. 6. If, on the other hand, the crystal as a whole is too large, so that multiple Bragg reflection between different mosaic blocks is appreciable, then the Bragg peak intensities will be reduced by "secondary" extinction, as discussed in Chap. 8.

In the kinematical theory of diffraction the integrated intensities of the Bragg peaks are directly proportional to $|F_{hkl}|^2$. Since it is difficult to make adequate corrections for primary and secondary extinction when these effects are large, it is important in neutron crystallography experiments, where the object is to measure $|F_{hkl}|^2$ as accurately as possible, that the samples approximate ideally imperfect crystals as closely as possible.

On the other hand, there is no need for monochromator crystals to be ideally imperfect. In fact, since the purpose of a monochromator crystal is to produce the highest possible reflected-neutron flux, it is essential that the reflectivity be saturated by multiple Bragg reflection. Consequently, the reflecting properties of monochromator crystals cannot be understood within the kinematical theory of diffraction and we must postpone our discussion of this subject until Chap. 8.

Differential scattering cross section

We now consider the diffraction of neutrons by an ideally imperfect mosaic crystal. Then Eq. 5.2.1 is replaced by

$$b_{\mathbf{q}} = \sum_{\mu} \sum_{l}{}' \sum_{\nu=1}^{n} b_{\mu l \nu} \exp(i\mathbf{q} \cdot \mathbf{r}_{\mu l \nu}) \tag{5.3.1}$$

where $\mathbf{r}_{\mu l \nu}$ denotes the position of the νth atom in the lth unit cell of the μth mosaic block, and $b_{\mu l \nu}$ is the corresponding bound scattering length. The prime on the summation indicates that l is restricted to the N_μ lattice sites in the μth mosaic block. The total number of lattice sites is then given by

$$N_0 = \sum_{\mu} N_\mu \tag{5.3.2}$$

The position vectors are now of the form

$$\mathbf{r}_{\mu l \nu} = \mathbf{R}_\mu + \mathbf{R}_{\mu l} + \mathbf{r}_{\mu \nu} + \mathbf{u}_{\mu l \nu} \tag{5.3.3}$$

where $\mathbf{R}_\mu$ represents the origin of the μth mosaic block and the remaining quantities are defined as in Eq. 5.2.2. The lattice vectors $\mathbf{R}_{\mu l}$ are assumed to be given by expressions of the type (A.1) in Appendix A, where the primitive translation vectors of the various blocks are slightly misoriented relative to one another. This implies that we ignore the distortion of the lattice in boundary region between the blocks (see Fig. B.5), an assumption that is justified by the fact that the number of lattice sites in the boundary region is a negligible fraction of the whole.

The scattering amplitude (Eq. 5.3.1) now becomes

$$b_{\mathbf{q}} = \sum_{\mu} b_{\mathbf{q}\mu} \exp(i\mathbf{q} \cdot \mathbf{R}_\mu) \tag{5.3.4}$$

where $b_{\mathbf{q}\mu}$ is the scattering amplitude of the μth mosaic block,

$$b_{\mathbf{q}\mu} = \sum_{l}{}' \sum_{\nu=1}^{n} b_{\mu l \nu} \exp[i\mathbf{q} \cdot (\mathbf{R}_{\mu l} + \mathbf{r}_{\mu \nu} + \mathbf{u}_{\mu l \nu})] \tag{5.3.5}$$

The differential scattering cross section (Eq. 5.0.1) is then given by

$$\frac{d\sigma}{d\Omega} = \sum_{\mu\mu'} \langle b_{\mathbf{q}\mu} \rangle^* \langle b_{\mathbf{q}\mu'} \rangle \exp(i\mathbf{q} \cdot \mathbf{R}_{\mu\mu'}) \tag{5.3.6}$$

where $\mathbf{R}_{\mu\mu'} = \mathbf{R}_{\mu'} - \mathbf{R}_\mu$. If $\mu \neq \mu'$, the quantity $\mathbf{q} \cdot \mathbf{R}_{\mu\mu'}$ is a large number, typically 10^4 to 10^8. Since the mosaic blocks are displaced and misoriented relative to one another in an essentially random way, the complex exponential in Eq. 5.3.6 is a random function of μ and μ'. Thus, it is reasonable to assume that the sum of all the cross terms in Eq. 5.3.6 will vanish even though the Bragg peaks of the individual mosaic blocks may overlap. In other words, the mosaic blocks scatter incoherently, so that

$$\frac{d\sigma}{d\Omega}=\sum_{\mu}\frac{d\sigma_{\mu}}{d\Omega} \tag{5.3.7}$$

where the differential scattering cross section of the μth block is given by

$$\frac{d\sigma_{\mu}}{d\Omega}=|\langle b_{\mathbf{q}\mu}\rangle|^2 \tag{5.3.8}$$

Since each mosaic block is assumed to be a perfect crystal, the cross section (Eq. 5.3.8) can be calculated as in Sec. 5.2 and one finds that

$$\frac{d\sigma_{\mu}}{d\Omega}=N_{\mu}\frac{(2\pi)^3}{V_0}\sum_{h}|F_h|^2\,\Delta_{\mu}(\mathbf{q}-\mathbf{K}_{\mu h}) \tag{5.3.9}$$

The reciprocal lattice vectors $\mathbf{K}_{\mu h}$ depend on μ because the mosaic blocks are slightly misoriented relative to one another. The peak function $\Delta_{\mu}(\mathbf{q})$ also depends on μ because the various mosaic blocks have not only different orientations but also different sizes and shapes. However, the unit-cell structure factor F_h is rigorously independent of μ, since it is an intrinsic property of the crystal structure itself.

The differential scattering cross section (Eq. 5.3.7) can then be expressed as

$$\frac{d\sigma}{d\Omega}=N_0\frac{(2\pi)^3}{V_0}\sum_{h}|F_h|^2\,\Delta(\mathbf{q}-\mathbf{K}_h), \tag{5.3.10}$$

where

$$\Delta(\mathbf{q}-\mathbf{K}_h)=\sum_{\mu}\frac{N_{\mu}}{N_0}\Delta_{\mu}(\mathbf{q}-\mathbf{K}_{\mu h}), \tag{5.3.11}$$

in which, by definition,

$$\mathbf{K}_h=\sum_{\mu}\frac{N_{\mu}}{N_0}\mathbf{K}_{\mu h} \tag{5.3.12}$$

Thus, $\mathbf{K}_h$ is the size-weighted average reciprocal lattice vector and $\Delta(\mathbf{q}-\mathbf{K}_h)$ is the size-weighted average peak function. Alternatively,

$$\Delta(\mathbf{q}_1)=\sum_{\mu}\frac{N_{\mu}}{N_0}\Delta_{\mu}(\mathbf{q}_1-\delta\mathbf{K}_{\mu h}), \tag{5.3.13}$$

where $\mathbf{q}_1=\mathbf{q}-\mathbf{K}_h$ as before (see Eq. 5.2.67) and $\delta\mathbf{K}_{\mu h}=\mathbf{K}_{\mu h}-\mathbf{K}_h$ is the deviation of $\mathbf{K}_{\mu h}$ from the average value. Since $\Delta_{\mu}(\mathbf{q})$ obeys the relations (5.1.14), it follows that

$$\int\Delta(\mathbf{q}_1)\,d\mathbf{q}_1=1, \qquad \int\mathbf{q}_1\,\Delta(\mathbf{q}_1)\,d\mathbf{q}_1=0 \tag{5.3.14}$$

The expression (5.3.10) for the differential scattering cross section of an ideally imperfect mosaic crystal is formally identical to the corresponding expression (Eq. 5.2.18) for a perfect crystal. The only difference is in the detailed shape of the peak function $\Delta(\mathbf{q})$. Consequently, all the results for a perfect crystal in Sec. 5.2 that do not depend explicitly on the shape of the peak function also apply to an ideally imperfect mosaic crystal. For example, the reflectivity is again given by Eq. 5.2.74 in which ρ_h is the same as before. However, the normalized rocking curve $W(\theta_1)$ will now have a different shape because it depends on $\Delta(\mathbf{q}_1)$ according to Eq. 5.2.72.

Normalized rocking curve

For an ideally imperfect mosaic crystal, where $\Delta(\mathbf{q}_1)$ is given by Eq. 5.3.13, the normalized rocking curve (Eq. 5.2.72) becomes

$$W(\theta_1) = \sum_\mu \frac{N_\mu}{N_0} \int \Delta_\mu(\mathbf{q}_2)\, \delta[\theta_1 - \Theta(\mathbf{q}_2 + \delta\mathbf{K}_{\mu h})]\, d\mathbf{q}_2 \tag{5.3.15}$$

where $\mathbf{q}_2 = \mathbf{q}_1 - \delta\mathbf{K}_{\mu h}$. It follows from Eq. 5.2.69 that

$$\Theta(\mathbf{q}_2 + \delta\mathbf{K}_{\mu h}) = \Theta(\mathbf{q}_2) + \theta_\mu \tag{5.3.16}$$

where θ_μ is the angle between $\mathbf{K}_h$ and the projection of $\mathbf{K}_{\mu h}$ on the plane of scattering. In other words, θ_μ is the angle of misorientation of the Bragg planes of the μth mosaic block relative to the average Bragg angle θ_h. Hence,

$$W(\theta_1) = \sum_\mu \frac{N_\mu}{N_0} W_\mu(\theta_1 - \theta_\mu) \tag{5.3.17}$$

where $W_\mu(\theta_1 - \theta_\mu)$ is the normalized rocking curve for the μth mosaic block,

$$W_\mu(\theta_2) = \int \Delta_\mu(\mathbf{q}_2)\, \delta(\theta_2 - \Theta(\mathbf{q}_2))\, d\mathbf{q}_2 \tag{5.3.18}$$

Thus, the normalized rocking curve for an ideally imperfect mosaic crystal is simply the size-weighted average of the normalized rocking curves of the individual mosaic blocks.

The overall width of the normalized rocking curve $W(\theta_1)$ is the combined result of the distribution of the misorientations θ_μ and the intrinsic widths of the $W_\mu(\theta_2)$ distributions themselves. To discuss these two effects, we note first that $W(\theta_1)$ can be expressed alternatively as

$$W(\theta_1) = \int W_1(\theta_1 - \theta) W_2(\theta)\, d\theta + M(\theta_1) \tag{5.3.19}$$

where

$$\begin{aligned} W_1(\theta_1) &= \sum_\mu \frac{N_\mu}{N_0} \delta(\theta_1 - \theta_\mu) \\ W_2(\theta_1) &= \sum_\mu \frac{N_\mu}{N_0} W_\mu(\theta_1) \\ M(\theta_1) &= \sum_\mu \frac{N_\mu}{N_0} [W_\mu(\theta_1 - \theta_\mu) - W_2(\theta_1 - \theta_\mu)] \end{aligned} \tag{5.3.20}$$

We note that

$$\int W_1(\theta_1)\, d\theta_1 = 1$$

$$\int W_2(\theta_1)\, d\theta_1 = 1 \qquad (5.3.21)$$

$$\int M(\theta_1)\, d\theta_1 = 0$$

Thus, $M(\theta_1)$ simply modulates the shape of $W(\theta_1)$ but does not affect the integrated intensity.

Suppose, first of all, that the mosaic blocks are very large, so that the average width of $W_\mu(\theta_1)$ is much less than the average value of θ_μ. Then, in effect, $W_\mu(\theta_1) = \delta(\theta_1)$, so that $W_2(\theta_1) = \delta(\theta_1)$ and $M(\theta_1) = 0$. Hence,

$$W(\theta_1) = W_1(\theta_1) \qquad (5.3.22)$$

which is the size-weighted Bragg-plane misorientation distribution function. When this approximation is valid, the crystal is said to be of type 1 (Zachariasen, 1967).

Suppose, next, that the mosaic blocks are very small, so that the average width of $W_\mu(\theta_1)$ is much greater than the average value of θ_μ. Then, in effect, $\theta_\mu = 0$, so that $W_1(\theta_1) = \delta(\theta_1)$ and $M(\theta_1) = 0$. Hence,

$$W(\theta_1) = W_2(\theta_1) \qquad (5.3.23)$$

which is the size-weighted average of the rocking curves of the individual mosaic blocks. When this approximation is valid, the crystal is said to be of type 2 (Zachariasen, 1967).

Suppose, finally, that the crystal is of intermediate type, but that the mosaic blocks all have roughly the same size and shape, and differ only in their orientations. Then $W_2(\theta_1) = W_\mu(\theta_1)$, independent of μ, so that $M(\theta_1) = 0$ and

$$W(\theta_1) = \int W_1(\theta_1 - \theta) W_2(\theta)\, d\theta \qquad (5.3.24)$$

Thus, it is only when there is a substantial variation in the sizes and shapes of the mosaic blocks that the modulation term $M(\theta_1)$ in Eq. 5.3.19 is significant.

Gaussian approximation

Combining Eqs. 5.3.18 and 5.3.20 we get

$$W_2(\theta_1) = \int \Delta_2(\mathbf{q})\, \delta(\theta_1 - \Theta(\mathbf{q}))\, d\mathbf{q} \qquad (5.3.25)$$

where

$$\Delta_2(\mathbf{q}) = \sum_\mu \frac{N_\mu}{N_0} \Delta_\mu(\mathbf{q}) \tag{5.3.26}$$

In Sec. 5.1 we showed that $\Delta_\mu(\mathbf{q})$ can be represented by an anisotropic Gaussian to an accuracy of about 5 percent. If the crystal as a whole has an approximately isotropic texture and all the mosaic blocks are roughly the same size, then $\Delta_2(\mathbf{q})$ can evidently be represented by an isotropic Gaussian,

$$\Delta_2(\mathbf{q}) = \left(\frac{l}{2\pi}\right)^3 \exp\left[-\frac{1}{\pi}\left(\frac{ql}{2}\right)^2\right] \tag{5.3.27}$$

where l characterizes the linear dimensions of the mosaic blocks. Then $W_2(\theta_1)$ is also Gaussian,

$$W_2(\theta_1) = \frac{1}{\sqrt{2\pi}\,\eta_2} \exp\left[-\frac{1}{2}\left(\frac{\theta_1}{\eta_2}\right)^2\right] \tag{5.3.28}$$

with a standard deviation that is given by

$$\eta_2 = \frac{\lambda}{\sqrt{2\pi}\, l \sin 2\theta_h}, \tag{5.3.29}$$

where λ is the incident-neutron wavelength as before.

Following Darwin (1922) we shall, for simplicity, also assume that the misorientation distribution function is Gaussian,

$$W_1(\theta_1) = \frac{1}{\sqrt{2\pi}\,\eta_1} \exp\left[-\frac{1}{2}\left(\frac{\theta_1}{\eta_1}\right)^2\right] \tag{5.3.30}$$

Here η_1 is the size-weighted rms angle of misorientation of the mosaic blocks,

$$\eta_1^2 = \int \theta_1^2 W_1(\theta_1)\, d\theta_1 = \sum_\mu \frac{N_\mu}{N_0} \theta_\mu^2 \tag{5.3.31}$$

The quantity η_1 can therefore be called the mosaic spread.

It now follows from (Eq. 5.3.24) that the normalized rocking curve is also Gaussian,

$$W(\theta_1) = \frac{1}{\sqrt{2\pi}\,\eta} \exp\left[-\frac{1}{2}\left(\frac{\theta_1}{\eta}\right)^2\right] \tag{5.3.32}$$

where

$$\eta^2 = \eta_1^2 + \eta_2^2 \tag{5.3.33}$$

We shall call η the effective mosaic spread in order to distinguish it from the true mosaic spread η_1. It should be kept in mind that when people speak of the "mosaic spread" they often mean neither η_1 nor η but, rather, the full width at half maximum of the rocking curve, which is given by

$$\Delta\theta = \sqrt{8 \ln 2}\, \eta = 2.3548\eta \tag{5.3.34}$$

It follows from (Eq. 5.3.29) with the typical values $\lambda = 1$ Å, $l = 1\ \mu$m, and $\theta_h = 30$ degrees, that $\eta_2 = 4.6 \times 10^{-5}$ radians $= 0.16$ arc minutes. On the other hand, η_1 is usually 1 to 10 arc minutes. Thus, in most cases $\eta_2 \ll \eta_1$, so that $\eta \simeq \eta_1$ and the crystal is of type 1.

Validity of the kinematical theory

At the end of Sec. 5.2 we showed that the condition for the validity of the kinematical theory of diffraction is that the reflectivity $R \ll 1$. Since R is largest when $\theta = \theta_h$, it follows from Eq. 5.2.74 that the preceding condition requires that $\rho_h W(0) \ll 1$. Since $\rho_h = Q_h t$, where t is the average crystal thickness in the direction of the incident beam, and $W(0) = 1/(\sqrt{2\pi}\, \eta)$ for a mosaic crystal, we therefore require that $t \ll L$, where

$$L = \frac{\sqrt{2\pi}\, \eta}{Q_h} \tag{5.3.35}$$

The quantity Q_h is given by Eq. 5.2.71, from which it follows that for order-of-magnitude purposes,

$$L \simeq \frac{a^3 \eta}{b^2} \tag{5.3.36}$$

where a is the lattice constant and b is the bound coherent scattering length. With the typical values $a \simeq 5$ Å, $b \simeq 5$ fm, and $\eta \simeq 10^{-3}$ radians $\simeq 3$ arc minutes we get $L \simeq 5$ mm. More detailed calculations show that L generally varies from about 1 mm for very strong reflections to about 10 cm for weak reflections. Thus, it is evident that the kinematical theory of neutron diffraction is valid for a mosaic crystal only if its linear dimensions are $\lesssim 1$ mm. For mosaic crystals that are much larger than this, such as monochromator crystals, the reflectivity is dominated by multiple Bragg reflection and must be calculated on the basis of neutron transport theory, as described in Chap. 8.

5.4. Powders

Polycrystals and powders

In crystallographic work the material of interest is often available only as a polycrystal and not as a single crystal. Diffraction measurements on polycrystals are difficult to interpret because the grains are usually neither well-aligned nor randomly oriented. In this case the sample will often be ground into a powder in order to produce a random distribution of grain orientations. This greatly simplifies the theory as we shall see below. The linear dimensions of the grains are normally in the range 0.1 to 10 μm (Azaroff and Buerger, 1958) and, hence, are roughly the same as those of the blocks in a mosaic crystal. If the overall size of the powder sample is $\lesssim 1$ cm, then multiple Bragg reflection will be negligible and the neutron diffraction can be calculated on the basis of the kinematical theory.

Differential scattering cross section

The formal expression (Eq. 5.3.6) for the differential scattering cross section of a mosaic crystal also applies to a powder if the subscript μ is now taken to label the grains in the powder. A similar argument to that before leads to the conclusion that the different grains scatter incoherently and Eq. 5.3.10 then becomes

$$\frac{d\sigma}{d\Omega} = N_0 \frac{(2\pi)^3}{V_0} \sum_h |F_h|^2 G_h(\mathbf{q}) \tag{5.4.1}$$

where

$$G_h(\mathbf{q}) = \sum_\mu \frac{N_\mu}{N_0} \Delta_\mu(\mathbf{q} - \mathbf{K}_{\mu h}) \tag{5.4.2}$$

and has the normalization property

$$\int G_h(\mathbf{q})\, d\mathbf{q} = 1 \tag{5.4.3}$$

For simplicity, we shall ignore the finite width of $\Delta_\mu(\mathbf{q} - \mathbf{K}_{\mu h})$ and approximate this quantity by a delta function. Then

$$G_h(\mathbf{q}) = \sum_\mu \frac{N_\mu}{N_0} \delta(\mathbf{q} - \mathbf{K}_{\mu h}) \tag{5.4.4}$$

The reciprocal lattice vectors $\mathbf{K}_{\mu h}$ of the various grains all have the same magnitude K_h and differ only in their directions. Thus, $G_h(\mathbf{q}) = 0$ unless $q = K_h$ and, hence, is of the form

$$G_h(\mathbf{q}) = C_h\, \delta(q - K_h) \tag{5.4.5}$$

If the grains are randomly oriented then $G_h(\mathbf{q})$ must be independent of the direction of $\mathbf{q}$, and C_h is therefore a constant. It then follows from Eq. 5.4.3 that $C_h = (4\pi K_h^2)^{-1}$ and, hence, that

$$G_h(\mathbf{q}) = \frac{1}{4\pi K_h^2} \delta(q - K_h) \tag{5.4.6}$$

The expression (Eq. 5.4.1) for the differential scattering cross section then becomes

$$\frac{d\sigma}{d\Omega} = N_0 \frac{2\pi^2}{V_0} \sum_h \frac{|F_h|^2}{K_h^2} \delta(q - K_h) \tag{5.4.7}$$

Debye–Scherrer cones

Since the Bragg scattering is elastic, we have $q = 2k \sin(\phi/2)$, where ϕ is the angle of scattering. Thus $q \leq 2k$, and if $2k < K_h$ then $q < K_h$ and $\delta(q - K_h) = 0$. The reciprocal lattice site h can therefore contribute to the differential scattering cross section (Eq. 5.4.7) only if $2k > K_h$. In this case $K_h = 2k \sin \theta_h$, where θ_h is the Bragg angle, and

$$\delta(q - K_h) = \frac{1}{2k} \delta[\sin(\phi/2) - \sin \theta_h] = \frac{1}{k \cos \theta_h} \delta(\phi - 2\theta_h) \tag{5.4.8}$$

Hence, Eq. 5.4.7 becomes

$$\frac{d\sigma}{d\Omega} = \frac{N_0 \pi^2}{V_0 k^3} \sum_h{}' \frac{|F_h|^2}{\sin \theta_h \sin 2\theta_h} \delta(\phi - 2\theta_h) \tag{5.4.9}$$

where the prime means that the summation is restricted to reciprocal lattice sites h for which $2k > K_h$. In other words, the summation extends over all reciprocal lattice sites that lie within a sphere of radius $2k$.

According to Eq. 5.4.9, the differential scattering cross section vanishes unless $\phi = 2\theta_h$ for some h. The Bragg scattering by a powder therefore consists of a set of cones $\phi = 2\theta_h$, each of which is coaxial with the incident beam. These cones are called *Debye–Scherrer cones* in honor of Debye and Scherrer (1917) who, together with Hull (1917), pioneered the powder method in X-ray crystallography. The number of such cones may be $0, 1, 2, \ldots$ depending on the value of the incident-neutron wave vector k. The case in which there are three cones is illustrated in Fig. 5.10. Since $2k = K_h$ when $2\theta_h = 180$ degrees, it is evident that as k increases from 0 to ∞ the hth cone first appears in the backward direction and then gradually folds over into the forward direction.

For a given value of k the series (5.4.9) contains only a finite number of terms. Nevertheless, the number of such terms is very large when k is

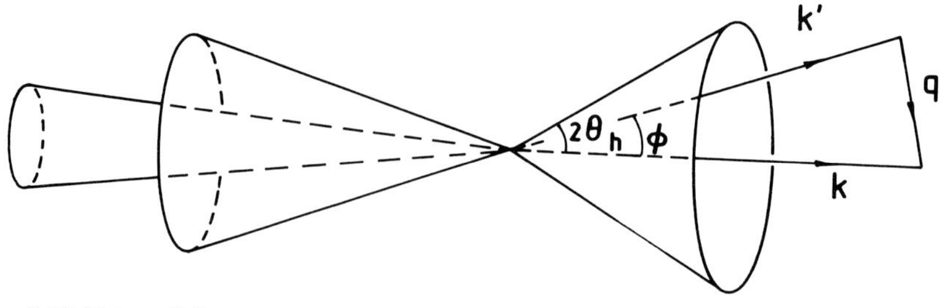

Fig. 5.10 Debye–Scherrer cones.

large. In practice, the amount of labor involved in computing the cross section can be reduced considerably by noting that if h and h' are symmetrically equivalent reciprocal lattice sites then $\theta_h = \theta_{h'}$ and $F_h = F_{h'}$. The multiplicity factor M_h is defined as the number of sites that are symmetrically equivalent to h, including h itself. Values of M_h for cubic crystals are listed in Table 5.2. Thus, Eq. 5.4.9 can be expressed equivalently as

$$\frac{d\sigma}{d\Omega} = \frac{N_0 \pi^2}{V_0 k^3} \sum_h'' \frac{M_h |F_h|^2}{\sin \theta_h \sin 2\theta_h} \delta(\phi - 2\theta_h) \tag{5.4.10}$$

where the double prime means that the summation is restricted to symmetrically inequivalent reciprocal lattice sites h for which $2k > K_h$.

In a powder there are always gaps between neighboring grains, so that the total volume of the grains, $\sum_\mu V_\mu$, is less than the total volume of the powder, V. This effect is expressed in terms of the *packing fraction*,

$$p = \frac{1}{V} \sum_\mu V_\mu \tag{5.4.11}$$

In practice p is typically 50 percent. Since $V_\mu = N_\mu V_0$, it follows from Eq. 5.3.2 that

$$p = \frac{N_0 V_0}{V} \tag{5.4.12}$$

and, hence, that Eq. 5.4.10 can be expressed equivalently as

$$\frac{d\sigma}{d\Omega} = pV \sum_h'' P_h \, \delta(\phi - 2\theta_h) \tag{5.4.13}$$

where

$$P_h = \frac{\lambda^3 M_h |F_h|^2}{8\pi V_0^2 \sin \theta_h \sin 2\theta_h} \tag{5.4.14}$$

The factor $1/(\sin \theta_h \sin 2\theta_h)$ in this expression is the *Lorentz factor* for a powder. The quantity P_h is an intrinsic property of the material, while p and V are sample dependent. In other words, different powder samples of the same material will in general have different values of p and V but P_h will always be the same.

The integrated intensity of the Bragg peak in a powder at $\phi = 2\theta_h$ is proportional to P_h. If the same material were available as a single crystal, the integrated intensity of the rocking curve for the same Bragg reflection would be proportional to Q_h according to Eq. 5.2.75. Comparing Eqs. 5.2.71 and 5.4.14 we see that

$$P_h = \frac{M_h Q_h}{8\pi \sin \theta_h} \tag{5.4.15}$$

In both cases the quantity that is ultimately obtained from the measured integrated intensity of the Bragg peak is $|F_h|^2$.

Grain-size broadening

The Bragg peaks in Eq. 5.4.13 are represented by delta functions because we approximated $\Delta_\mu(\mathbf{q})$ in Eq. 5.4.2 by a delta function. In fact, $\Delta_\mu(\mathbf{q})$ has a finite width $\Delta q \simeq 2\pi/l$, where l characterizes the linear dimensions of the grains (see Sec. 5.1). The corresponding width $\Delta\phi$ of the Bragg peaks can be obtained by differentiating the relation $q = 2k \sin(\phi/2)$. Thus,

$$\frac{\Delta q}{\Delta \phi} = k \cos(\phi/2) \tag{5.4.16}$$

so that

$$\Delta\phi \simeq \frac{\lambda}{l \cos \theta_h} \tag{5.4.17}$$

Thus, the measured widths of the Bragg peaks can be used to determine the grain size (Azaroff and Buerger, 1958). This is most easily done for reflections in the backward direction, $2\theta_h \rightarrow 180$ degrees, where $\cos \theta_h$ is small and $\Delta\phi$ is correspondingly large.

Experimental powder profiles

Figure 5.11 shows the results of neutron diffraction measurements by Armstrong et al. (1985) on a powder sample of K_2PtBr_6 at 296 K with an incident-neutron wavelength $\lambda = 1.407$ Å. This is a typical example of a so-called *powder profile.* Apart from an arbitrary normalization factor,

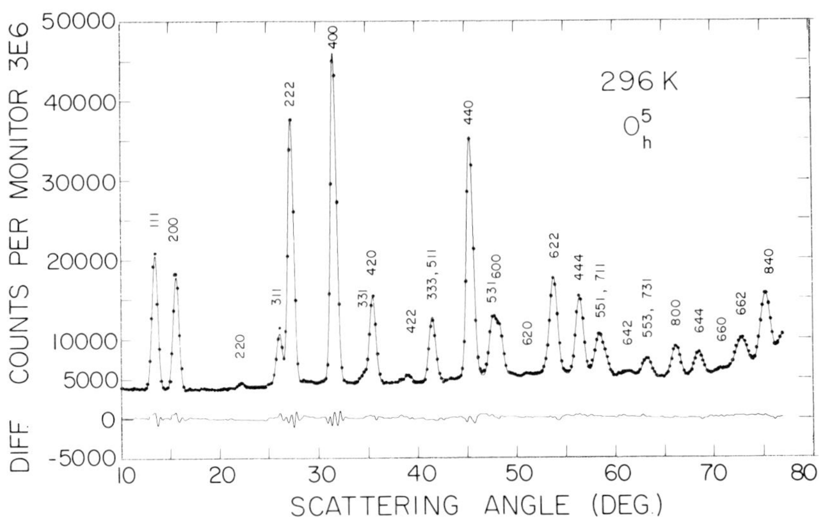

Fig. 5.11 Powder profile of K_2PtBr_6 at 296 K (from Armstrong et al., 1985).

the data point at each angle of scattering gives essentially the differential scattering cross section $d\sigma/d\Omega$, which is seen to consist of a series of Bragg peaks superimposed on a continuous background of diffuse scattering. The widths of the Bragg peaks ($\simeq$1 degree) are due entirely to the finite instrumental resolution. The intrinsic widths given by Eq. 5.4.17 with $l \simeq 1$ μm are two orders of magnitude smaller than the instrumental widths.

The positions of the Bragg peaks in Fig. 5.11 show quite unambiguously that the lattice is face-centered cubic with a lattice constant $a = 10.2854(6)$ Å. The Miller indices hkl for the various peaks are indicated in the figure. The integrated intensities of the Bragg peaks are proportional to $|F_{hkl}|^2$ and these quantities are characteristic of the nature and arrangement of the atoms in the unit cell (see Eq. 5.2.44).

The smooth curve in Fig. 5.11 is the result of a least-squares fit of Eq. 5.4.13 to the data, taking into account the effects of instrumental resolution and the diffuse background, and assuming that the crystal has the antifluorite structure. This is similar to the structure of fluorite (CaF_2) in Table B.1 with K replacing F and $PtBr_6$ replacing Ca. The difference between the observed and fitted profiles is shown by the horizontal line in Fig. 5.11. Apart from random fluctuations due to counting statistics, this difference is essentially zero. The structural parameters obtained from the least-squares fit are the lattice constant a, the Pt–Br bond length, the rms displacements of the K and Pt atoms, and the rms angle of libration of the $PtBr_6$ octahedra.

Total scattering cross section

The total cross section per unit volume for Bragg scattering by a powder is given according to Eq. 5.4.13 by

$$\begin{aligned}\Sigma(\mathbf{k}) &= \frac{1}{V}\int d\sigma \\ &= p\sum_h{}'' P_h \int \delta(\phi - 2\theta_h)\, d\Omega \\ &= 2\pi p \sum_h{}'' P_h \sin 2\theta_h\end{aligned} \tag{5.4.18}$$

This can be expressed with the help of Eq. 5.4.14 in the form

$$\Sigma(\mathbf{k}) = \frac{\lambda^3 p}{4V_0^2}\sum_h{}'' \frac{M_h\,|F_h|^2}{\sin\theta_h} \tag{5.4.19}$$

or, alternatively,

$$\Sigma(\mathbf{k}) = \frac{4\pi^3 p}{V_0^2 k^2}\sum_h{}' \frac{|F_h|^2}{K_h} \tag{5.4.20}$$

a result that was essentially first obtained for neutrons by Halpern et al. (1941).

The summation in Eq. 5.4.20 is restricted to reciprocal lattice sites h for which $K_h < 2k$. Thus,

$$\Sigma(\mathbf{k}) = 0 \qquad \text{if } k < k_c \tag{5.4.21}$$

where k_c is the *Bragg cut-off* and is defined such that $2k_c$ is the magnitude of the shortest reciprocal lattice vector $\mathbf{K}_h$ for which $F_h \neq 0$.

In the limit $k \to \infty$ the number of reciprocal lattice sites h inside a sphere of radius $2k$ also becomes infinite, in which case the summation over h in Eq. 5.4.20 can be replaced by an integral,

$$\frac{(2\pi)^3}{V_0} \sum_h{}' \to \int_0^{2k} d\mathbf{K}_h \tag{5.4.22}$$

If, for simplicity, we confine our attention to a rigid lattice with one atom per primitive unit cell, so that $F_h = b$, we then get

$$\Sigma(\infty) = \rho\sigma_c \tag{5.4.23}$$

where $\rho = p/V_0$ is the number of atoms per unit volume and $\sigma_c = 4\pi\,|b|^2$ is the bound coherent scattering cross section per atom.

Figure 5.12 shows $\Sigma(\mathbf{k})$ as a function of k in reduced units for an fcc crystal with lattice constant a. The discontinuities in $\Sigma(\mathbf{k})$ occur at the thresholds for the appearance of new Debye–Scherrer cones in the backward direction. Between these thresholds $\Sigma(\mathbf{k})$ is proportional to

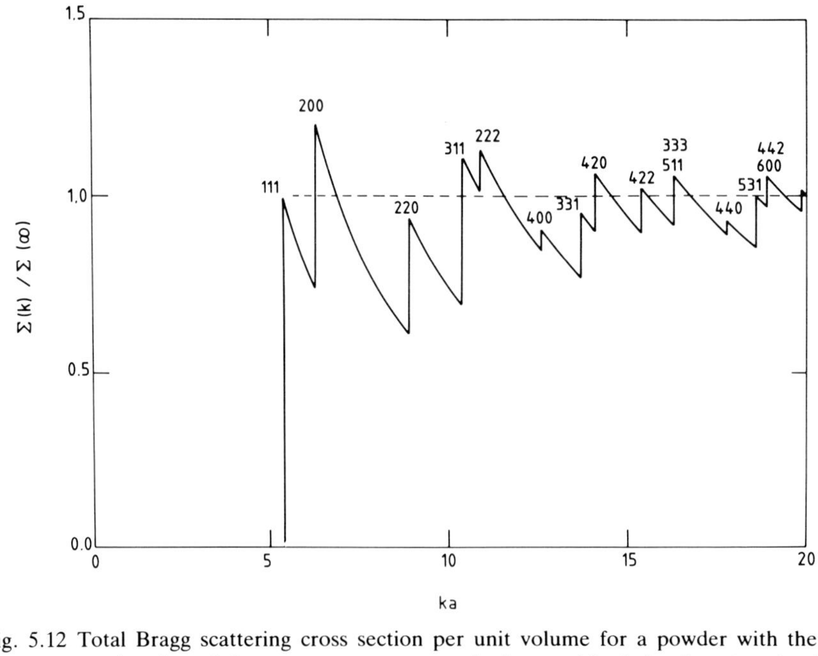

Fig. 5.12 Total Bragg scattering cross section per unit volume for a powder with the fcc structure as a function of the neutron wave vector, both quantities being in reduced units.

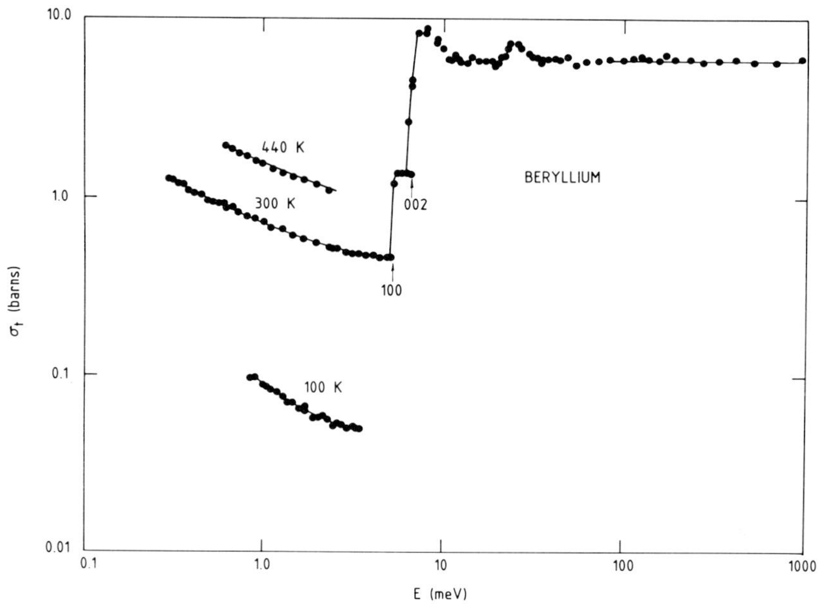

Fig. 5.13 Total collision cross section per atom for powdered beryllium at various temperatures as a function of the neutron energy (from Hughes and Harvey, 1955).

k^{-2}. The shortest reciprocal lattice vector corresponds to the (111) planes, so that $k_c = K_{111}/2 = \sqrt{3}\,\pi/a$.

The measured total collision cross section per atom for a powdered sample of beryllium is shown as a function of the neutron energy in Fig. 5.13. This element has the hexagonal close-packed (hcp) structure with lattice constants $a = 2.29$ Å and $c = 3.58$ Å, so that $\gamma = c/a = 1.57$, which is slightly less than the value $\gamma = \sqrt{8/3} = 1.63$ for the ideal hcp structure. The shortest reciprocal lattice vector is K_{001}. However, $F_{001} = 0$ for an hcp crystal, so that the Bragg cut-off actually corresponds to the (100) reflection. At the Bragg cut-off $E_c = 5.22$ meV, $k_c = 1.59$ Å^{-1}, and $\lambda_c = 2\pi/k_c = 3.96$ Å. The sawtooth structure in the cross section above the Bragg cut-off is largely unresolved in these measurements. Below the Bragg cut-off, where there is no Bragg scattering, the main contribution to σ_t comes from diffuse scattering, which, as is seen in Fig. 5.13, depends strongly on the temperature.

5.5. Magnetic crystals

Magnetic dipole interaction

In the discussion of neutron diffraction in crystals presented thus far we have only included the effect of the strong interaction between the neutron and the nuclei. In magnetic crystals the interaction between the neutron and atomic magnetic dipole moments gives a contribution to the scattering amplitude that is of the same order of magnitude as the nuclear contribution (Bloch, 1936). All other electromagnetic interactions give contributions that are two orders of magnitude smaller than the magnetic dipole interaction (Sears, 1986a) and we shall continue to neglect them.

Let us first consider the magnetic interaction between a neutron at the point $\mathbf{r}$ and a single atom (or ion) centered at the origin. The interaction energy is then given by

$$V(\mathbf{r}) = -\boldsymbol{\mu} \cdot \mathbf{B}(\mathbf{r}) \tag{5.5.1}$$

where $\boldsymbol{\mu}$ is the magnetic dipole moment of the neutron and $\mathbf{B}(\mathbf{r})$ is the magnetic field produced by the atom at the position of the neutron. The magnetic interaction, unlike the nuclear interaction, is a weak long-range interaction and can be treated in the Born approximation. Thus, according to Eq. 2.1.13, $T = V$ and the scattering amplitude operator (Eq. 2.6.21) becomes

$$F = -\frac{4\pi^2 m}{\hbar^2} V \tag{5.5.2}$$

Hence, from Eq. 2.7.10, the corresponding bound scattering length is given by

$$b = -\langle \mathbf{k}' | F | \mathbf{k} \rangle = \frac{4\pi^2 m}{\hbar^2} \int \phi_{\mathbf{k}'}(\mathbf{r})^* V(\mathbf{r}) \phi_{\mathbf{k}}(\mathbf{r}) \, d\mathbf{r} \tag{5.5.3}$$

Since $V(\mathbf{r})$ is independent of the neutron momentum, it then follows from Eq. 2.2.10 that

$$b = \frac{m}{2\pi\hbar^2} V(\mathbf{q}) \tag{5.5.4}$$

where $V(\mathbf{q})$ is the Fourier transform of the interaction energy,

$$V(\mathbf{q}) = \int V(\mathbf{r}) \exp(i\mathbf{q} \cdot \mathbf{r}) \, d\mathbf{r} \tag{5.5.5}$$

and $\mathbf{q} = \mathbf{k} - \mathbf{k}'$ as before. Hence,

$$b = -\frac{m}{2\pi\hbar^2} \boldsymbol{\mu} \cdot \mathbf{B}(\mathbf{q}) \tag{5.5.6}$$

where $\mathbf{B}(\mathbf{q})$ is the corresponding Fourier transform of $\mathbf{B}(\mathbf{r})$.

If the atom remains permanently in its ground electronic state $|0\rangle$ it has an effectively static magnetization $\mathbf{M}$ that will in general contain contributions from both the spins of the electrons and their orbital motion. If, for simplicity, we confine our attention to S-state atoms the latter contribution vanishes and

$$\mathbf{M}(\mathbf{r}) = \left\langle 0 \middle| \sum_j \boldsymbol{\mu}_j \, \delta(\mathbf{r} - \mathbf{r}_j) \middle| 0 \right\rangle \tag{5.5.7}$$

where $\boldsymbol{\mu}_j$ is the magnetic dipole moment and $\mathbf{r}_j$ the position of the jth electron.

The magnetization $\mathbf{M}$ is equivalent to a current density

$$\mathbf{J} = c \operatorname{curl} \mathbf{M} \tag{5.5.8}$$

which, in turn, produces a static magnetic field $\mathbf{B}$ that is governed by the equations

$$\operatorname{curl} \mathbf{B} = \frac{4\pi}{c} \mathbf{J}, \qquad \operatorname{div} \mathbf{B} = 0 \tag{5.5.9}$$

or, equivalently,

$$\mathbf{B} = \operatorname{curl} \mathbf{A}, \qquad \Delta \mathbf{A} = -\frac{4\pi}{c} \mathbf{J} \tag{5.5.10}$$

where $\mathbf{A}$ is the vector potential. The corresponding Fourier-transformed relations are

$$\begin{aligned} \mathbf{J}(\mathbf{q}) &= -ic\mathbf{q} \wedge \mathbf{M}(\mathbf{q}) \\ \mathbf{B}(\mathbf{q}) &= -i\mathbf{q} \wedge \mathbf{A}(\mathbf{q}) \\ \mathbf{A}(\mathbf{q}) &= \frac{4\pi}{cq^2} \mathbf{J}(\mathbf{q}) \end{aligned} \tag{5.1.11}$$

so that

$$\mathbf{B}(\mathbf{q}) = -4\pi\hat{\mathbf{q}} \wedge [\hat{\mathbf{q}} \wedge \mathbf{M}(\mathbf{q})] \tag{5.5.12}$$

where $\hat{\mathbf{q}}$ denotes a unit vector in the direction of $\mathbf{q}$. Hence,

$$\mathbf{B}(\mathbf{q}) = 4\pi \mathbf{M}_\perp(\mathbf{q}) \tag{5.5.13}$$

where

$$\mathbf{M}_\perp(\mathbf{q}) = (\mathbf{1} - \hat{\mathbf{q}}\hat{\mathbf{q}}) \cdot \mathbf{M}(\mathbf{q}) \tag{5.5.14}$$

in which $\mathbf{1}$ denotes the unit tensor. Thus, $\mathbf{M}_\perp(\mathbf{q})$ is the projection of $\mathbf{M}(\mathbf{q})$ on the plane perpendicular to $\mathbf{q}$.

It follows from Eq. 5.5.7 that

$$\mathbf{M}(\mathbf{q}) = \left\langle 0 \left| \sum_j \boldsymbol{\mu}_j \exp(i\mathbf{q} \cdot \mathbf{r}_j) \right| 0 \right\rangle \tag{5.5.15}$$

Since $\boldsymbol{\mu}_j = -2\beta_B \mathbf{s}_j$, where $\mathbf{s}_j$ is the spin of the jth electron and $\beta_B = e\hbar/2m_e c$ is the Bohr magneton, we get

$$\mathbf{M}(\mathbf{q}) = -2\beta_B \mathbf{S} f(\mathbf{q}) \tag{5.5.16}$$

Here $\mathbf{S}$ is the total electron spin,

$$\mathbf{S} = \sum_j \mathbf{s}_j \tag{5.5.17}$$

and $f(\mathbf{q})$ is called the *magnetic form factor,* which is defined such that

$$\mathbf{S}f(\mathbf{q}) = \left\langle 0 \left| \sum_j \mathbf{s}_j \exp(i\mathbf{q}\cdot\mathbf{r}_j) \right| 0 \right\rangle \tag{5.5.18}$$

and therefore has the normalization property $f(0) = 1$. Hence,

$$\mathbf{B}(\mathbf{q}) = -8\pi\beta_B \mathbf{S}_\perp f(\mathbf{q}) \tag{5.5.19}$$

The neutron magnetic dipole moment is given by $\boldsymbol{\mu} = \gamma\beta_N\boldsymbol{\sigma}$, where $\gamma = -1.913$ is the gyromagnetic ratio, $\beta_N = e\hbar/2mc$ is the nuclear magneton, and $\boldsymbol{\sigma}$ the Pauli spin operator so that the neutron spin is $\boldsymbol{\sigma}/2$. Hence, the magnetic scattering length (Eq. 5.5.6) becomes

$$b = \gamma r_e \boldsymbol{\sigma}\cdot\mathbf{S}_\perp f(\mathbf{q}) \tag{5.5.20}$$

where $r_e = e^2/m_e c^2$ is the classical electron radius.

The total scattering length, including both the strong interaction with the nucleus and the magnetic dipole interaction with the electrons, is then given by

$$b = b_N + \gamma r_e \boldsymbol{\sigma}\cdot\mathbf{S}_\perp f(\mathbf{q}) \tag{5.5.21}$$

where b_N is the nuclear scattering length. Since $r_e = 2.818$ fm it follows that $\gamma r_e = -5.391$ fm, which is of the same order of magnitude as a typical value of b_N. Thus, as asserted earlier, the magnetic dipole interaction gives a contribution to b that is of the same order of magnitude as the nuclear contribution.

Let us now assume that the neutron spin is polarized in a direction perpendicular to the plane of scattering. In particular, we choose a system of Cartesian coordinate axes such that the plane of scattering is the xy plane. Then $\mathbf{q}$ is in this plane while the neutron spin is either in the $+z$ direction ($\sigma_z = +1$) or in the $-z$ direction ($\sigma_z = -1$).

We also assume that the atom under consideration is in a crystal that is in thermodynamic equilibrium at some specified temperature and pressure. Then the bound coherent scattering length for given values of the neutron spin is (see Eq. 2.8.6)

$$\langle b_\pm \rangle = \langle b_N \rangle \pm \gamma r_e \langle S_{\perp z} \rangle f(\mathbf{q}) \tag{5.5.22}$$

where the brackets $\langle \cdots \rangle$ denote a thermodynamic average over the degrees of freedom of the crystal as before.

The average magnetic dipole moment of the atom is given by

$$\boldsymbol{\mu}_A = \int \langle \mathbf{M}(\mathbf{r}) \rangle \, d\mathbf{r} = \langle \mathbf{M}(\mathbf{q}=0) \rangle = -2\beta_B \langle \mathbf{S} \rangle \tag{5.5.23}$$

It will be noted that, whereas $\mathbf{S}$ is a quantum-mechanical operator, $\langle \mathbf{S} \rangle$ is a classical vector with a well-defined magnitude and direction. In particular, $\langle \mathbf{S} \rangle$ is antiparallel to $\boldsymbol{\mu}_A$. If $\boldsymbol{\mu}_A$ makes an angle α with the z-axis then

$$\langle S_{\perp z} \rangle = \langle S_z \rangle = -S \cos\alpha \tag{5.5.24}$$

where S is, by definition, the magnitude of $\langle \mathbf{S} \rangle$. Then Eq. 5.5.22 becomes

$$b_{\pm} = b_N \pm b_M \cos \alpha f(\mathbf{q}) \tag{5.5.25}$$

where we now omit the brackets on $b_{\pm}$ and b_N for simplicity, and

$$b_M = -\gamma r_e S \tag{5.5.26}$$

This is the same magnetic scattering length as in Sec. 3.5. Note that since γ is negative, b_M is positive.

Magnetic Bragg scattering

All the relations for Bragg scattering that were obtained in the earlier sections of this chapter also apply to magnetic crystals when one takes into account the contribution of the magnetic dipole interaction to the total scattering length. Thus, for example, the expression (Eq. 5.2.18) for the differential scattering cross section becomes

$$\frac{d\sigma_{\pm}}{d\Omega} = N_0 \frac{(2\pi)^3}{V_0} \sum_h |F_{h\pm}|^2 \, \Delta(\mathbf{q} - \mathbf{K}_h) \tag{5.5.27}$$

where the sign refers to incident neutrons that are polarized parallel (+) or antiparallel (−) to the z direction which, by assumption, is perpendicular to the plane of scattering.

The unit-cell structure factor (Eq. 5.2.17) becomes

$$F_{h\pm} = \sum_{\nu=1}^{n} b_{\nu\pm} \exp(-W_{\nu h}) \exp(i\mathbf{K}_h \cdot \mathbf{r}_\nu) \tag{5.5.28}$$

where

$$b_{\nu\pm} = b_{N\nu} \pm b_{M\nu} \cos \alpha_\nu f_{\nu h} \tag{5.5.29}$$

Here $b_{N\nu}$ is the nuclear bound coherent scattering length of the νth atom in the unit cell, and $b_{M\nu} = -\gamma r_e S_\nu$ is the corresponding magnetic scattering length. Also, α_ν is the angle that the average magnetic dipole moment of the νth atom makes with the z-axis, and $f_{\nu h} \equiv f_\nu(\mathbf{K}_h)$ is the magnetic form factor for this atom.

If the νth atom is nonmagnetic then $b_{\nu\pm} = b_{N\nu}$, since the magnetic dipole interaction vanishes in this case. If the νth atom is magnetic but the crystal is in its high-temperature paramagnetic phase, then S_ν, the magnitude of $\langle \mathbf{S}_\nu \rangle$, vanishes in the absence of an external magnetic field, in which case $b_{M\nu} = 0$ and $b_{\nu\pm} = b_{N\nu}$ once more. Thus, magnetic Bragg scattering occurs only in ordered magnetic phases, i.e., below the Curie temperature T_C in ferromagnets or below the Néel temperature T_N in antiferromagnets.

The results of polarized-neutron diffraction experiments are often expressed in terms of the *flipping ratio,*

$$r = \frac{d\sigma_+}{d\sigma_-} \tag{5.5.30}$$

If $\mathbf{q} \simeq \mathbf{K}_h$ then $d\sigma_\pm$ is proportional to $|F_{h\pm}|^2$ and the flipping ratio becomes

$$r_h = \left|\frac{F_{h+}}{F_{h-}}\right|^2 \tag{5.5.31}$$

Let us now suppose that the incident-neutron beam is unpolarized. Such a beam can be regarded as an incoherent mixture of equal numbers of up and down spins. In this case the differential scattering cross section is given by

$$\frac{d\sigma}{d\Omega} = \frac{1}{2}\left[\frac{d\sigma_+}{d\Omega} + \frac{d\sigma_-}{d\Omega}\right] \tag{5.5.32}$$

This is then identical to Eq. 5.2.18 if we put

$$|F_h|^2 = \tfrac{1}{2}\{|F_{h+}|^2 + |F_{h-}|^2\} \tag{5.5.33}$$

Even though the incident-neutron beam is unpolarized, the Bragg-reflected beam may in general be partially polarized with a polarization that is given by (see Eq. 3.5.23)

$$p'_h = \frac{R_{h+} - R_{h-}}{R_{h+} + R_{h-}} \tag{5.5.34}$$

Since the reflectivity $R_{h\pm}$ is proportional to $|F_{h\pm}|^2$ according to Eq. 5.2.74, we see that

$$p'_h = \frac{|F_{h+}|^2 - |F_{h-}|^2}{|F_{h+}|^2 + |F_{h-}|^2} \tag{5.5.35}$$

or,

$$p'_h = \frac{r_h - 1}{r_h + 1} \tag{5.5.36}$$

where r_h is the flipping ratio (Eq. 5.5.31). Conversely,

$$r_h = \frac{1 + p'_h}{1 - p'_h} \tag{5.5.37}$$

so that the flipping ratio can also be obtained from an experiment with unpolarized incident neutrons by measuring the polarization p'_h of the Bragg-reflected neutrons.

Ferromagnets

Consider a ferromagnet with one atom per primitive unit cell. An example of such a ferromagnet is iron which has the bcc structure (Fig. B.3a). Then

$$F_{h\pm} = b_\pm \exp(-W_h) \tag{5.5.38}$$

where

$$b_{\pm} = b_N \pm b_M \cos\alpha f_h \tag{5.5.39}$$

Alternatively,

$$b_{\pm} = b_N(1 \pm \chi_h \cos\alpha) \tag{5.5.40}$$

where

$$\chi_h = \frac{b_M}{b_N} f_h = \left(\frac{-\gamma r_e}{b_N}\right) S f_h \tag{5.5.41}$$

The quantity χ_h is the fundamental quantity of interest for Bragg reflection in ferromagnets. By measuring χ_h as a function of h, and using the normalization condition $f_0 = 1$, one can determine S and f_h separately. The value of S then gives the average magnetic dipole moment per atom, $\mu_A = 2\beta_B S$, while the magnetic form factor f_h determines the spin-density distribution in the atom. Let us now consider how χ_h is determined in various kinds of experiments.

Unpolarized neutrons

If the incident-neutron beam is unpolarized, then the reflectivity R_h is proportional to $|F_h|^2$ where, according to Eq. 5.5.33,

$$\begin{aligned} |F_h|^2 &= \tfrac{1}{2}(b_+^2 + b_-^2)\exp(-2W_h) \\ &= b_N^2(1 + \chi_h^2 \cos^2\alpha)\exp(-2W_h) \end{aligned} \tag{5.5.42}$$

At temperatures above the Curie point ($T > T_C$), where S and, hence, χ_h vanish, we have

$$|F_h|^2 = b_N^2 \exp(-2W_h) \tag{5.5.43}$$

Ignoring the effect of the temperature dependence of the Debye–Waller factor, we then get

$$\frac{R_h(T < T_C)}{R_h(T > T_C)} = 1 + \chi_h^2 \cos^2\alpha \tag{5.5.44}$$

In the preceding discussion we have assumed that all the atomic magnetic dipole moments make the same (average) angle α with the z-axis when $T < T_C$. In the absence of an external magnetic field a ferromagnetic crystal has a natural domain structure. Within each domain the atomic magnetic dipole moments are parallel to one another but their direction varies from one domain to the next in such a way that the crystal as a whole usually has little or no net magnetization. If the distribution of directions were completely random, the average value of the factor $\cos^2\alpha$ in Eq. 5.5.44 would be 1/3. However, as a result of crystal-field effects, this is usually not a good approximation, especially in noncubic crystals.

If a sufficiently strong external magnetic field **H** is applied, the domain structure breaks down and the atomic magnetic dipole moments will all be aligned parallel to **H**. In particular, if **H** is parallel to the z-axis then $\alpha = 0$ and $\cos\alpha = 1$, while if **H** is perpendicular to the z-axis then $\alpha = 90$ degrees and $\cos\alpha = 0$. Thus,

$$\frac{R_h(\mathbf{H} \parallel \mathbf{z})}{R_h(\mathbf{H} \perp \mathbf{z})} = 1 + \chi_h^2 \tag{5.5.45}$$

In this way χ_h^2 can be obtained directly from the reflectivity ratio, and the previous uncertainty in the $\cos^2\alpha$ factor in Eq. 5.5.44 is eliminated, as is also the need to correct for the temperature variation of the Debye–Waller factor that was neglected in Eq. 5.5.44.

It is evident from Eq. 5.5.45 that χ_h can be determined accurately from diffraction measurements with unpolarized neutrons only if χ_h is of order unity, i.e., only if b_M and b_N are of roughly the same order of magnitude. If b_M is an order of magnitude smaller than b_N, then $\chi_h \lesssim 0.1$ and $\chi_h^2 \lesssim 0.01$, in which case an accurate determination of χ_h is very difficult.

Polarized neutrons

The flipping ratio (Eq. 5.5.31) can be obtained directly from experiments with polarized incident neutrons. According to Eq. 5.5.38,

$$r_h = \left(\frac{b_+}{b_-}\right)^2 = \left(\frac{1 + \chi_h \cos\alpha}{1 - \chi_h \cos\alpha}\right)^2 \tag{5.5.46}$$

If the atomic magnetic dipole moments are aligned by a magnetic field in the z direction, so that $\alpha = 0$, then

$$r_h = \left(\frac{1 + \chi_h}{1 - \chi_h}\right)^2 \tag{5.5.47}$$

It is clear from this expression that the use of polarization techniques is especially advantageous when χ_h is small. For example, if $\chi_h = 0.1$ then $r_h = 1.49$ while $1 + \chi_h^2 = 1.01$. In this case the magnetic dipole interaction gives a 50 percent contribution to the flipping ratio but only a 1 percent contribution to the reflectivity ratio (Eq. 5.5.45) that is obtained from measurements with unpolarized neutrons.

The values of χ_h^2 and r_h for the 3d ferromagnets iron, cobalt, and nickel are given in Table 5.4 assuming $f_h = 0.75$. It is evident that χ_h can be measured for cobalt using either polarized or unpolarized neutrons, but that an accurate measurement of this quantity for iron and especially for nickel requires the use of polarized neutrons.

A ferromagnetic monochromator crystal can also be used to produce the collimated monoenergetic beam of polarized neutrons that is required for polarized-neutron diffraction measurements. The polarization of the

Table 5.4 Properties of Bragg reflection in the 3d ferromagnets. It is assumed that $f_h = 0.75$ and that the external magnetic field is in the z direction

Quantity	Fe	Co	Ni
b_N (fm)	9.54	2.50	10.3
b_M (fm)	5.98	4.64	1.62
χ_h	0.470	1.39	0.118
χ_h^2	0.221	1.94	0.0139
r_h	7.70	37.2	1.61
p'_h	0.770	0.948	0.233

Bragg-reflected beam is given, according to Eq. 5.5.35 and 5.5.38, by

$$p'_h = \frac{b_+^2 - b_-^2}{b_+^2 + b_-^2} = \frac{2\chi_h \cos \alpha}{1 + \chi_h^2 \cos^2 \alpha} \tag{5.5.48}$$

Here α is the angle that the external magnetic field $\mathbf{H}$ makes with the z axis. If $\chi_h > 1$ one can make $p'_h = +1$ by tuning the direction of $\mathbf{H}$ such that $\cos \alpha = \chi_h^{-1}$. For some ferromagnetic monochromator crystals discussed in Sec. 5.6 the nuclear and magnetic scattering lengths are matched in the sense that $b_N = b_M f_h$ for one particular value of h. In this case $\chi_h = 1$, so that $p'_h = +1$ when $\alpha = 0$, i.e., a 100 percent polarized beam is obtained when the external magnetic field is in the z direction.

Antiferromagnets

Although the formal expression (Eq. 5.5.28) for $F_{h\pm}$ applies to both ferromagnets and antiferromagnets, the neutron diffraction by these two kinds of magnetic material is, in fact, qualitatively very different. This is best illustrated by means of a specific example. Let us consider an antiferromagnet with the bcc structure such as chromium (see Fig. B.3b). In the paramagnetic phase ($T > T_N$), where the atomic magnetic dipole moments are randomly oriented, the lattice is bcc with one atom per primitive unit cell. In the antiferromagnetic phase ($T < T_N$) the lattice is simple cubic with two atoms per primitive unit cell, the magnetic dipole moments of which are aligned antiparallel to one another. The structure of the antiferromagnetic phase can, therefore, be thought of as consisting of two interpenetrating simple cubic sublattices of oppositely directed magnetic dipole moments.

The structural parameters that characterize the antiferromagnetic phase are listed in Table 5.5. Hence, from Eq. 5.5.28,

$$F_{h\pm} = \{b_N[1 + \exp(i\mathbf{K}_h \cdot \mathbf{r})] \pm b_M \cos \alpha f_h[1 - \exp(i\mathbf{K}_h \cdot \mathbf{r})]\} \exp(-W_h) \tag{5.5.49}$$

Table 5.5 Structural parameters for a bcc antiferromagnet

Quantity	$\nu = 1$	$\nu = 2$
$\mathbf{r}_\nu$	0	$\mathbf{r}$
(x_ν, y_ν, z_ν)	(0, 0, 0)	$(\frac{1}{2}, \frac{1}{2}, \frac{1}{2})$
α_ν	α	$\pi - \alpha$
$\cos \alpha_\nu$	$\cos \alpha$	$-\cos \alpha$
$b_{\nu\pm}$	$b_N \pm b_M \cos \alpha f_h$	$b_N \mp b_M \cos \alpha f_h$

Since

$$\begin{aligned} \exp(i\mathbf{K}_h \cdot \mathbf{r}) &= \exp[2\pi i(hx_2 + ky_2 + lz_2)] \\ &= \exp[i\pi(h + k + l)] \\ &= (-1)^{h+k+l} \end{aligned} \tag{5.5.50}$$

we get

$$F_{h\pm} = \begin{cases} 2b_N \exp(-W_h), & h + k + l = \text{even} \\ \pm 2b_M \cos \alpha f_h \exp(-W_h), & h + k + l = \text{odd} \end{cases} \tag{5.5.51}$$

Thus, the Bragg scattering is purely nuclear when $h + k + l$ is even and purely magnetic when $h + k + l$ is odd. In the paramagnetic phase, where $b_M = 0$, only the nuclear reflections occur. Here $h + k + l$ is even, which corresponds to the reciprocal lattice sites of the bcc lattice (see Eq. A.39). The reciprocal lattice sites where $h + k + l$ is odd, at which the magnetic reflections occur in the antiferromagnetic phase, represent the so-called *superlattice sites*.

The reflectivity is proportional to $|F_{h\pm}|^2$ and, hence, is independent of the incident-neutron polarization. This is as one would expect, since the antiferromagnetic structure looks the same to an up spin as it does to a down spin. Thus, the flipping ratio $r_h = 1$. Also, if the incident-neutron beam is unpolarized, so is the Bragg-reflected beam, $p'_h = 0$. The Bragg reflection by an antiferromagnet is therefore very different from that by a ferromagnet.

5.6. Applications

The most well-known and widely applied neutron optical effect is without doubt Fraunhofer diffraction by crystal lattices. In this final section we shall briefly review the historical development of this subject with emphasis on those applications that can be understood, at least qualitatively, within the kinematical theory of diffraction already given.

At the time the neutron was discovered (Chadwick, 1932), the Bragg reflection of X-rays by crystals was already well established as an important tool for the determination of crystal structures. The Bragg reflection of electrons by crystals had also been demonstrated by Davisson and Germer (1927) in an important experiment that helped establish the validity of the new quantum theory. Following the subsequent discovery of neutron thermalization by Fermi et al. (1934), and the realization that thermal neutrons have wavelengths of the same order of magnitude as the interatomic distances in crystals, it was natural for Elsasser (1936) to suggest that thermal neutrons might also be Bragg-reflected by crystals. This was soon demonstrated by Halban and Preiswerk (1936) in powdered iron and, independently, by Mitchell and Powers (1936) in single crystals of MgO. The effect of the Bragg-reflection of neutrons in crystals was also seen indirectly through the observation by Whitaker and Beyer (1939) that single crystals are much more transparent to thermal-neutron beams than are polycrystals and powders.

The Rn–Be neutron sources used in the work mentioned were far too weak to do any more than simply demonstrate the existence of the diffraction of thermal neutrons by crystals. It was only after the development of thermal fission reactors by Fermi and others during the Second World War that it became possible to produce the intense, highly collimated beams of thermal neutrons that were needed for serious practical applications of neutron diffraction.

Crystal spectrometers

The first practical application of the Bragg reflection of thermal neutrons was in the development of crystal spectrometers similar to those that had originally been developed for X-rays by Bragg and Bragg (1913). Since the thermal-neutron flux from a fission reactor is much less than the corresponding flux from an X-ray source, the sizes of the beams and samples must be much larger for neutrons than for X-rays (i.e., of the order of centimeters rather than millimeters). To obtain comparable angular resolution, the physical size of a neutron spectrometer is therefore much larger than that of an X-ray spectrometer. Otherwise, the two types of instrument are very similar.

The *single-axis crystal spectrometer* (Zinn, 1947) is illustrated schematically in Fig. 5.14a. Here a collimated beam of neutrons from a thermal fission reactor (R) is incident on a monochromator crystal (M). Those neutrons with wavelength

$$\lambda = 2d_M \sin \theta_M \tag{5.6.1}$$

where d_M is the spacing of the relevant Bragg planes, will be Bragg-reflected through the angle $2\theta_M$ so that a collimated monoenergetic beam of neutrons is incident on the sample (S), and the subsequent transmitted beam then enters the detector (D). The ratio of the detector count rates in the presence and absence of the sample gives the transmissivity (Eq. 3.2.10) and, hence, the total collision cross section σ_t.

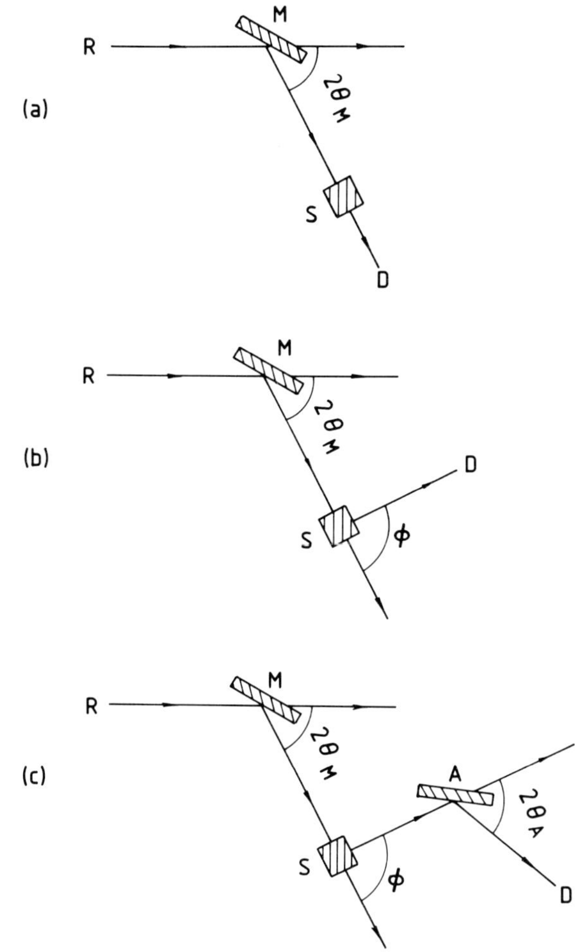

Fig. 5.14 Crystal spectrometers: (a) single-axis, (b) double-axis, (c) triple-axis.

With the development of the single-axis crystal spectrometer it became possible to measure σ_t as a function of the neutron energy throughout the thermal region at high resolution. One of the first applications (Zinn, 1947) was to verify explicitly that the neutron absorption in boron follows very accurately the $1/v$ law (Eq. 2.4.43), and to measure the (n, γ) resonance in ^{113}Cd at 178 meV and show that it is well described by the formula of Breit and Wigner (1936). These were important results, not only from the point of view of nuclear physics, but also because the validity of the $1/v$ law was assumed in calculating the efficiency of the BF_3 proportional counters that were used as neutron detectors in those days. In addition, cadmium shutters, which absorb thermal neutrons but are transparent to fast neutrons (Amaldi et al., 1934), are used to measure fast-neutron background levels. Subsequent work has helped determine the thermal-neutron cross sections and resonance parameters of all the elements and their isotopes (Mughabghab et al., 1981; Mughabghab, 1984), quantities that are important not only for thermal-neutron scattering experiments in condensed matter, but also for nuclear physics and reactor engineering.

In the *double-axis crystal spectrometer* (Fermi and Marshall, 1947), which is illustrated in Fig. 5.14b, the detector is able to revolve about an axis through the sample. This enables one to measure the angular distribution of the scattered neutrons, and the detector count rate is essentially proportional to the differential scattering cross section $d\sigma/d\Omega$. With this instrument it became possible to measure not only the Bragg scattering from crystalline materials, as discussed later in this section, but also the angular distribution of the diffuse scattering in gases (Alcock and Hurst, 1949) and liquids (Chamberlain, 1950), which provides information on the static atomic pair correlations in the system (see Eq. 2.9.34).

As the neutron flux available from thermal fission reactors increased, it eventually became practicable to construct and use the *triple-axis crystal spectrometer* (Brockhouse, 1955, 1961) that is illustrated in Fig. 5.14c. Here an analyser crystal (A) is located between the sample and the detector so that only those neutrons with wavelength

$$\lambda' = 2d_A \sin \theta_A, \qquad (5.6.2)$$

where d_A is the spacing of the relevant Bragg planes, are Bragg reflected through the angle $2\theta_A$ into the detector. This enables one to measure both the angular and energy distribution of the scattered neutrons. The detector count rate is essentially proportional to the double-differential scattering cross section $d^2\sigma/d\Omega\, d\varepsilon'$, which provides information on both the spatial and temporal correlations in the system (see Eq. 2.9.2).

With the development of the triple-axis crystal spectrometer it became possible to study not only the static structure of matter on the atomic level but also the dynamics. Thus, for example, one could now measure phonon dispersion relations in crystals (Brockhouse and Stewart, 1955) so as to obtain information on interatomic forces, and magnon dispersion relations in magnetic crystals (Brockhouse, 1957) so as to obtain information on magnetic interactions. The discussion of neutron inelastic scattering is beyond the scope of this book and we refer the interested reader to, for example, the books by Squires (1978) or Lovesey (1984).

In measurements of neutron scattering by crystals with a double-axis crystal spectrometer, the observed scattering consists of a series of Bragg peaks superimposed on a continuous background of diffuse scattering (see Fig. 5.11). If the measurements are performed instead on a triple-axis crystal spectrometer with the analyser set at the elastic position ($\lambda' = \lambda$), then the Bragg scattering is the same as before (because it is elastic) but the diffuse scattering is largely suppressed (because it is almost entirely inelastic in most materials).

In Sec. 5.5 we saw that it is often advantageous to use polarized neutrons to study neutron diffraction in magnetic materials and that polarized-neutron beams can be produced by Bragg reflection from a ferromagnetic monochromator crystal. A double-axis *polarized-beam spectrometer* was first developed by Shull et al. (1951). Following the development of high-flux reactors it later became practicable to use triple-axis polarized-beam spectrometers too (Moon *et al.*, 1969).

Monochromators

The single most important requirement of a neutron monochromator is that it be available as a large single crystal with a large homogeneous mosaic spread. Beyond that, it is desirable that it have small incoherent scattering and absorption and that any radioactive absorption products be short-lived so as to avoid handling problems when the monochromator is removed from the spectrometer.

Suitable monochromator crystals are typically 10 to 15-cm long and 5-cm wide. The optimum thickness depends on a number of variables (as discussed later in Chap. 8) but is usually in the range 0.2 to 2.0 cm. The mosaic spread, as given by the full width at half maximum of the rocking curve, is typically 0.2 to 0.5 degree for most materials, but for pyrolytic graphite it is somewhat larger (0.4 to 0.8 degree, say). A number of other properties of the most common monochromator crystals are listed in Table 5.6.

In most crystal spectrometers the working range of $2\theta_M$ is from about 20 to 90 degrees. The corresponding wavelength range that is then available for a number of monochromator reflections is shown in Fig. 5.15. It is evident that crystal monochromators can produce neutron

Table 5.6 Properties of monochromator crystals

Structure	Usual reflections	Material	Lattice constants[b] (Å)	b_c (fm)	σ_c (b)	σ_i (b)	σ_a^c (b)
Fcc	111 200 220 311	Cu	$a = 3.6147$	7.718(4)	7.486(8)	0.52(4)	3.78(2)
		Al	$a = 4.0495$	3.449(5)	1.495(4)	0.0092(7)	0.231(3)
		Pb	$a = 4.9502$	9.4017(20)	11.108(5)	0.0030(7)	0.171(2)
Bcc	110 200 211 220	Fe	$a = 2.8664$	9.54(6)	11.44(14)	0.39(3)	2.56(3)
Diamond	111 220 311	Si	$a = 5.4309$	4.149(1)	2.163(1)	0.015(2)	0.171(3)
		Ge	$a = 5.6575$	8.1929(17)	8.435(4)	0.17(6)	2.3(2)
Hcp	100 002	Be	$a = 2.2856$ $c = 3.5832$	7.79(1)	7.63(2)	0.005(1)	0.0076(8)
Hexagonal layer	002	PG[a]	$a = 2.461$ $c = 6.708$	6.6460(12)	5.551(2)	0.000(4)	0.00350(7)

[a] Pyrolytic graphite.
[b] At 300 K.
[c] At 2200 m s^{-1}.

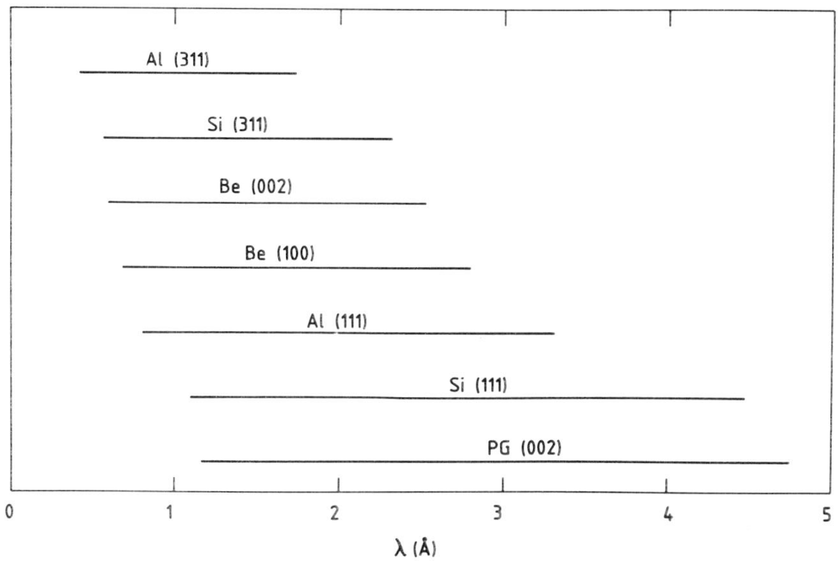

Fig. 5.15 Wavelength range of various monochromator reflections for $20 \leq 2\theta_M \leq 90$ degrees.

wavelengths from about 0.5 to 5.0 Å. The corresponding neutron energies vary from about 330 to 3.3 meV.

For many years the preferred monochromator for polarized-beam spectrometers was the ferromagnetic alloy $Co_{0.92}Fe_{0.08}$, for which the nuclear and magnetic scattering amplitudes are approximately matched for both the (111) and (200) reflections (Nathans et al., 1959). One disadvantage of this material is the large absorption cross section of Cobalt (37 b) and the fact that the absorption product (^{60}Co) is radioactive with a long half-life (5.26 years). As a result, only thin crystals can be used and this limits the intensity of the polarized beam. It is now more common to use Heusler alloy (Cu_2MnAl), for which the matched reflection is (111) (Delapalme *et al.*, 1971). Since the absorption in this material produces no long-lived radioactive isotopes, much thicker crystals can be used with a gain of a factor of 4 or more in the intensity of the polarized beam.

It was pointed out at the end of Sec. 5.3 that the reflectivity of large mosaic crystals such as neutron monochromators is dominated by multiple Bragg reflection and must be calculated on the basis of neutron transport theory as described in Chap. 8. We shall therefore defer the further discussion of monochromator crystals until that time. Additional information on this subject can also be found in the review article by Freund and Forsyth (1979).

Filters

It was first noted by Anderson et al. (1946) that the effective temperature of a thermal-neutron beam, as measured by its average kinetic energy, is reduced from 300 K to 18 K when it is filtered through 23 cm of polycrystalline graphite. To see why this happens we note first that the transmissivity of a slab of material of thickness d for a beam at normal incidence is given according to Eq. 3.2.10 by

$$T = \exp(-\rho\sigma_t d) \tag{5.6.3}$$

where ρ is the number density and σ_t the total collision cross section per atom. In a polycrystal or powder having small incoherent scattering and absorption there is a discontinuity of one to two orders of magnitude in the value of σ_t at the Bragg cut-off λ_c (see Fig. 5.13). Thus, it is possible to choose d such that

$$\rho\sigma_t d \begin{cases} \gg 1, & \lambda < \lambda_c \\ \ll 1, & \lambda > \lambda_c \end{cases} \tag{5.6.4}$$

and, hence, that

$$T \simeq \begin{cases} 0, & \lambda < \lambda_c \\ 1, & \lambda > \lambda_c \end{cases} \tag{5.6.5}$$

The material will then transmit only neutrons with $\lambda > \lambda_c$ while those with $\lambda < \lambda_c$ will be Bragg-reflected in some other direction.

An important application of such *polycrystalline filters* is in the elimination of order contamination in crystal spectrometers. In general, when a monochromator crystal is set to Bragg-reflect neutrons of wavelength λ, it will also produce higher-order reflections at wavelengths $\lambda/2$, $\lambda/3, \ldots$ (see Eq. 5.2.24). If a filter is available with $\lambda/2 < \lambda_c < \lambda$, then the first-order neutrons will be transmitted while all the higher-order neutrons will be removed from the beam.

The cut-off wavelength λ_c has the value 6.70 Å for graphite (Anderson et al., 1946) and 3.96 Å for beryllium (Fermi et al., 1947). All other suitable filter materials (Fe, BeO, CaO, etc.) have values of λ_c between the above two values. Thus, polycrystalline filters can be used to eliminate order contamination only if $\lambda \gtrsim 4.0$ Å. More information on polycrystalline filters can be found in the review article by Egelstaff and Pease (1954).

In neutron crystallography one always uses small wavelengths $\lambda \simeq 1$ Å in order to be able to cover an adequately large region of the reciprocal lattice and, also, to minimize the effect of secondary extinction, as will be discussed in Chap. 8. In this case the higher-order wavelengths are in the region $\lesssim 0.5$ Å, where suitable polycrystalline filters are unavailable. In this region the order contamination can, if necessary, be eliminated by the use of *resonance filters* (Freund and Forsyth, 1979), i.e., materials such as ^{239}Pu that have a sharp (n, γ) resonance at the required wavelength.

It was first pointed out by Brockhouse (1959) that large perfect single crystals can be used to remove fast neutrons and γ-rays from a neutron beam while still having a high transmission for thermal neutrons. (In this respect, *single-crystal filters* have the opposite effect of *cadmium filters,* which absorb thermal neutrons but transmit fast neutrons and γ-rays.) Single-crystal filters are especially useful in reducing shielding requirements for spectrometers that are located at the reactor face rather than at the end of a long guide tube.

To understand the behavior of a single-crystal filter, we note first that, when a collimated polyenergetic beam of neutrons is incident on an almost perfect single crystal, only a small fraction of the neutrons will have wavelengths for which Bragg scattering can occur (see Fig. 5.2a). If the material has small absorption and incoherent scattering, and is cooled to liquid-nitrogen temperature to minimize inelastic scattering, then the total collision cross section σ_t will be very small for thermal neutrons. For fast neutrons ($E \gtrsim 1$ eV, say), on the other hand, inelastic scattering will not be negligible and σ_t will have essentially its free-atom value, which is very much larger than its value at thermal-neutron energies. Thus, it will be possible, as before, to choose a crystal thickness d for which the transmissivity is essentially zero for fast neutrons and 100 percent for thermal neutrons.

Materials commonly used as single-crystal filters are silicon, quartz (SiO_2), and sapphire (Al_2O_3). The total filter thickness is typically 30 cm. More information on single-crystal filters can be found in the article by Nieman et al. (1980).

Scattering length measurement

The first application of the double-axis crystal spectrometer was in the measurement of neutron scattering lengths (Fermi and Marshall, 1947). It is evident from Eq. 5.2.44 that, if the crystal structure is already known, the values of $|F_{hkl}|$ obtained from the measured integrated intensities of a series of Bragg reflections (hkl) can be used to determine the bound coherent scattering lengths of all the atoms in the unit cell. To do this, the sign of the scattering length of at least one atom in the unit cell must already be known. This can be obtained from, for example, mirror reflection measurements, as discussed earlier in Sec. 3.4.

The initial experiments of Fermi and Marshall (1947) were done with single crystals. It was later shown by Wollan and Shull (1948) that the scattering lengths could be determined more accurately with powder samples, for which the effect of extinction on the measured values of $|F_{hkl}|$ is negligible. Within a few years, values of the bound coherent scattering lengths of some 50 elements had been determined by the neutron powder diffraction method (Shull and Wollan, 1951).

The values obtained by Shull and Wollan had an average accuracy of about 5 percent. Even under the most favorable conditions it is difficult to obtain better than 1 percent accuracy with the powder diffraction method. As a result, this method cannot compete with modern neutron-optical methods (see Table 3.3) and is no longer used today for measuring bound coherent scattering lengths.

The neutron powder diffraction method is, however, still used for measuring bound incoherent scattering lengths. To see how this is done, we note first that, according to Eq. 5.2.6, the quantity "b" in the expression (Eq. 5.2.44) for F_{hkl} is actually $\langle b \rangle$. It is then evident from Eq. 3.5.30 that measurements of $\langle b \rangle$ with polarized neutrons and polarized nuclei can be used to determine the bound incoherent scattering length. This was first demonstrated by Shull and Ferrier (1963). The polarized-neutron diffraction method has since been used by Ito and Shull (1969), Abragam et al. (1972), and many others to measure the incoherent scattering lengths of isotopes with nonvanishing spin.

Crystal structure determination

Another early application of the double-axis crystal spectrometer was for crystal structure determinations. To put this subject in perspective, we note first that for an infinite perfect crystal Eqs. 5.2.3 and 5.2.8 can be combined to give

$$\langle b_{\mathbf{q}} \rangle = \frac{(2\pi)^3}{V_0} \sum_h F_h \, \delta(\mathbf{q} - \mathbf{K}_h) \tag{5.6.6}$$

The average bound scattering length density can then be obtained from the inverse of Eq. 5.1.3,

$$\begin{aligned} \langle b(\mathbf{r}) \rangle &= \frac{1}{(2\pi)^3} \int \langle b_{\mathbf{q}} \rangle \exp(-i\mathbf{q} \cdot \mathbf{r}) \, d\mathbf{q} \\ &= \frac{1}{V_0} \sum_h F_h \exp(-i\mathbf{K}_h \cdot \mathbf{r}) \end{aligned} \tag{5.6.7}$$

Alternatively, this quantity can be expressed with the help of (Eq. 5.1.4) in the form

$$\langle b(\mathbf{r}) \rangle = \sum_{l\nu} b_\nu P_\nu(\mathbf{r} - \mathbf{R}_l - \mathbf{r}_\nu) \tag{5.6.8}$$

where b_ν is the bound coherent scattering length of the νth atom in the unit cell (see Eq. 5.2.6) and

$$P_\nu(\mathbf{u}) = \langle \delta(\mathbf{u} - \mathbf{u}_{l\nu}) \rangle \tag{5.6.9}$$

Thus, $P_\nu(\mathbf{u}) \, d\mathbf{u}$ is the normalized probability that the νth atom in a unit cell is in $d\mathbf{u}$. Since the rms displacements of the atoms are small in comparison with the lattice constant, the function $\langle b(\mathbf{r}) \rangle$ is sharply peaked at the equilibrium positions of the atoms and the integrated intensity of the peak at $\mathbf{R}_l + \mathbf{r}_\nu$ is equal to b_ν.

If one could somehow measure the unit-cell structure factor F_h and the reciprocal lattice vector $\mathbf{K}_h$ for all h, then one could, in principle, calculate $\langle b(\mathbf{r})\rangle$ from Eq. 5.6.7. The expression (5.6.8) could then be used to determine the equilibrium position $\mathbf{r}_\nu$ and bound coherent scattering length b_ν of each atom in the unit cell. The chemical composition of each atom could then be identified from a table of scattering lengths of the elements. In this case it is clear that the determination of crystal structures would be a purely straightforward process.

It is evident from the expression (5.2.18) for the differential scattering cross section that the measured positions of the Bragg peaks determine $\mathbf{K}_h$, but the integrated intensities of these peaks give only $|F_h|$. In general,

$$F_h = |F_h| \exp(i\Phi_h) \tag{5.6.10}$$

and the phase Φ_h cannot be measured in a neutron diffraction experiment. As a result of this so-called *phase problem* the neutron diffraction data alone are not sufficient to determine the structure of the crystal and additional information is needed.

In practice, two pieces of additional information that are essential are a knowledge of the chemical composition and bulk density of the crystal. Since the size of the unit cell can be obtained from the diffraction data, this additional information then determines the number of formula units in the unit cell, and all that remains to be determined are the positions of the atoms in the unit cell.

If it is found that there are only a few atoms in the primitive unit cell, and the measured values of $|F_h|$ show that the crystal has high symmetry, then the number of possible structures is very small and the correct structure can easily be determined by the "trial-and-error" method of discovering which structure agrees with the measured values of $|F_h|$. On the other hand, if the crystal has low symmetry and a large number of atoms in the unit cell, then more systematic methods are required (see, for example, Buerger [1960], Bragg [1962], or Vainshtein [1981]).

Following the development of crystal spectrometers for neutrons, it was soon demonstrated that neutron diffraction had two important advantages over X-ray diffraction, namely, in the location of hydrogen (Shull et al., 1948) and in the discrimination between neighboring atoms in the periodic table (Shull and Siegel, 1949).

The initial work on the use of neutron diffraction for crystal structure determinations employed the powder method. Later it was shown by Peterson and Levy (1951) that the single-crystal method could also be used provided the size of the crystal was small enough ($\simeq$2 mm) to avoid extinction effects. This was important because it allowed neutron diffraction to be used to determine structures with a large number of atoms per unit cell. When the unit cell in the direct lattice is very large, the corresponding unit cell in the reciprocal lattice is very small, with the result that the high-index Bragg peaks in the powder profile become crowded together and difficult to resolve. This problem is avoided in the single-crystal method.

Another important tool in crystallography is *Laue photography,* which was first demonstrated with neutrons by Wollan et al. (1948). For more information on the use of neutron diffraction in crystal structure determinations, we refer the reader to Bacon (1963, 1975), Willis (1970), and Dachs (1978).

Magnetic structure determination

We saw earlier, in Sec. 5.5, that neutron diffraction measurements in magnetic crystals can be used to determine atomic magnetic dipole moments and magnetic form factors as well as the spin arrangements in ordered magnetic phases. Until quite recently (Gibbs et al., 1985) this type of information could not be obtained by X-ray diffraction.

The use of neutron diffraction measurements for magnetic structure determinations began with the work of Shull and Smart (1949), whose determination of the structure of MnO represented the first direct observation of antiferromagnetism. This was followed by systematic investigations of the magnetic structure of the transition-metal elements and their various alloys and compounds. More recent work has focused on one- and two-dimensional magnetic systems and on systems having both magnetically-ordered and superconducting phases. For more information on this subject, see Nathans and Pickart (1963), Izyumov and Ozerov (1970), Willis (1970), Bacon (1975), Dachs (1978), and Lovesey (1984).

Materials science applications

In Sec. 5.1 we showed that a body with linear dimension L produces a Fraunhofer diffraction peak in the forward direction with an angular width $\phi \simeq \lambda/L$, where λ is the neutron wavelength. In heterogeneous materials the variation in the scattering length density at different positions in the sample gives rise to analogous small-angle scattering.

One of the earliest applications of small-angle scattering was for the measurement of coherent scattering lengths (Weiss, 1951) and, when used in conjunction with the Christiansen filter technique, this is still an important method today (see Sec. 3.4).

In more recent years, instruments have been built specifically for small-angle neutron scattering which, with long-wavelength neutrons (up to 20 Å), enable one to study inhomogeneities with linear dimensions up to 10^4 Å or more. With small-angle scattering one can therefore study precipitates in alloys, defects such as voids in metals, magnetic inhomogeneities, polymer configurations, biological macromolecules, etc. More information on neutron small-angle scattering can be found in the review articles by Schmatz et al. (1974) and Kostorz (1979).

Another materials-science application that might be mentioned is *neutron diffraction topography* (Doi et al., 1971; Schlenker and Shull, 1973), which is similar to ordinary radiography except that one records a Bragg-reflected beam rather than the transmitted beam. This technique is especially useful for studying twin boundaries in crystals and domain boundaries in ferromagnets and antiferromagnets.

Finally, a method of nondestructive testing that is of growing commercial importance is the use of neutron diffraction for *residual-stress measurements* (MacEwen et al., 1983; Allen et al., 1985). Here the residual stresses in polycrystalline materials are determined from neutron diffraction measurements of the variations in the lattice constant throughout the sample. Information on the texture of the sample can also be obtained from measurements of the grain-orientation distribution function.

6

DYNAMICAL THEORY OF NEUTRON DIFFRACTION

The term *dynamical theory of diffraction* was originally introduced by Ewald (1917a) to distinguish his approach to the theory of X-ray diffraction in crystals from the "kinematical" theory of Laue (Friedrich *et al.*, 1912), which ignored multiple scattering, and from the primitive "geometrical" theory of Bragg (1913). Although the importance of multiple scattering was recognized in earlier work by Darwin (1914a,b), who gave an approximate theory of these effects, Ewald (1917a,b) was the first to treat the theory of diffraction in a completely self-consistent way. In a later theory, Laue (1931) showed that the local-field effects that complicate the Ewald method can be bypassed if one simply assumes at the outset that the coherent wave in a crystal satisfies Maxwell's equations with a complex periodic dielectric constant. This approach proved extremely popular among those who were not interested in the theory of dispersion per se but only in the application of dynamical diffraction theory to specific problems. More recently, accounts of Ewald's theory have been given by Ewald (1965, 1969) and Wagenfeld (1968). Reviews of the Laue theory and its subsequent extensions have been given by, for example, Zachariasen (1945), James (1963), Batterman and Cole (1964), Cowley (1975), and Pinsker (1978).

The dynamical theory of neutron diffraction in crystals has been developed in analogy with the corresponding X-ray theories. Thus, those authors (Lax, 1951, 1952; Dederichs, 1972) who are primarily interested in the theory of dispersion have used the Ewald method, while those (Goldberger and Seitz, 1947; Sears, 1978; Rauch and Petrascheck, 1978; Klein and Werner, 1983; Werner and Klein, 1986) interested in the theory of reflection, refraction, and diffraction have adopted the Laue method.

In Chap. 2 we showed that the coherent wave $\psi(\mathbf{r})$, which describes all neutron optical phenomena, satisfies the one-body Schrödinger equation (2.8.32),

$$\left\{-\frac{\hbar^2}{2m}\Delta + v(\mathbf{r})\right\}\psi(\mathbf{r}) = E\psi(\mathbf{r}) \tag{6.0.1}$$

Here $v(\mathbf{r})$ is the complex optical potential that represents the effective interaction of the neutron with the medium and is given in the elementary theory of dispersion by Eq. 2.8.37,

$$v(\mathbf{r}) = \frac{2\pi\hbar^2}{m}\langle b(\mathbf{r})\rangle, \tag{6.0.2}$$

where $b(\mathbf{r})$ denotes the bound scattering length density (Eq. 2.8.38).

In the kinematical theory of neutron diffraction one neglects multiple scattering by solving Eq. 6.0.1 in the Born approximation. In the dynamical theory of neutron diffraction, on the other hand, the analogue of *Laue's method* consists in finding an essentially exact solution of Eq. 6.0.1 at all points $\mathbf{r}$ inside and outside the crystal subject to the requirement that $\psi(\mathbf{r})$ and $\nabla\psi(\mathbf{r})$ be continuous at the boundary of the crystal and that $\psi(\mathbf{r})$ have the appropriate asymptotic behavior at infinity.

The accuracy of the results obtained by the Laue method is limited by the accuracy of the elementary expression (Eq. 6.0.2) for the optical potential, which neglects local-field effects. This limitation is overcome in the *Ewald method,* which is not based on Eqs. 6.0.1 and 6.0.2 but consists instead in finding a systematic self-consistent solution of the Ewald equations (4.1.11) and (4.1.13) when the atomic positions $\mathbf{r}_i$ form a regular crystalline lattice. In the Ewald method the local-field effects are taken rigorously into account. If these effects are neglected, however, the Ewald method is equivalent to the Laue method.

In Chap. 4 we gave an account of Ewald's method for a homogeneous medium and showed that the local-field correction to $v(\mathbf{r})$ is, in fact, very small for neutrons, typically $\lesssim 10^{-4}$. Thus, the Laue method, which we shall use in the present chapter, is adequate for most practical purposes.

In our use of the term *dynamical diffraction* we shall include all diffraction phenomena that lie outside the kinematical theory, i.e., all situations in which the Born approximation gives the wrong answer when applied to the solution of Eq. 6.0.1. In this case, the dynamical theory of neutron diffraction includes not only Bragg reflection in large perfect crystals but also, for example, the classical optical phenomena of Fraunhofer and Fresnel diffraction by apertures in opaque screens. Weiss' (1951) treatment of small-angle neutron scattering by a sphere is also a dynamical diffraction theory in this sense.

The distinction between the kinematical and dynamical theories of diffraction is useful only for X-rays and neutrons for which the index of refraction n is such that $|n-1| \lesssim 10^{-5}$. In ordinary optics, where $n-1 = O(1)$, there are no practical situations in which the kinematical theory is valid, so that all diffraction phenomena are of the dynamical type.

6.1. Coherent wave in a perfect crystal

Optical potential

According to Eq. 5.6.7, the average bound scattering length density for an infinite perfect crystal is given by

$$\langle b(\mathbf{r})\rangle = \frac{1}{V_0}\sum_h F_h \exp(-i\mathbf{K}_h \cdot \mathbf{r}) \tag{6.1.1}$$

Alternatively, this can be expressed with the help of Eq. 5.2.19 in the form

$$\langle b(\mathbf{r})\rangle = \rho b \sum_h \hat{F}_h \exp(-i\mathbf{K}_h \cdot \mathbf{r}) \tag{6.1.2}$$

where ρ is the number density, b is the average bound coherent scattering length per atom, and $\hat{F}_h = F_h/F_0$ is the normalized unit-cell structure factor.

The optical potential (Eq. 6.0.2) then becomes

$$v(\mathbf{r}) = v_0 \sum_h \hat{F}_h \exp(-i\mathbf{K}_h \cdot \mathbf{r}) \tag{6.1.3}$$

where

$$v_0 = \frac{2\pi\hbar^2}{m}\rho b \tag{6.1.4}$$

as before (see Eqs. 2.8.43 or 3.1.6). It is evident that $v(\mathbf{r})$ is a periodic function, or lattice function (see Appendix A),

$$v(\mathbf{r} + \mathbf{R}_l) = v(\mathbf{r}), \qquad \text{all } l \tag{6.1.5}$$

It follows from Eq. A.15 that v_0 is the average value of $v(\mathbf{r})$ within a unit cell,

$$v_0 = \frac{1}{V_0}\int_{V_0} v(\mathbf{r})\, d\mathbf{r} \tag{6.1.6}$$

For a finite but otherwise perfect crystal of volume V we have

$$v(\mathbf{r}) = \begin{cases} v_0 \sum_h \hat{F}_h \exp(-i\mathbf{K}_h \cdot \mathbf{r}), & \text{inside } V \\ 0, & \text{outside } V \end{cases} \tag{6.1.7}$$

Since $v(\mathbf{r})$ is discontinuous at the surface of V, we must find separate solutions of (Eq. 6.0.1) inside and outside V and then match these solutions by requiring that $\psi(\mathbf{r})$ and $\nabla\psi(\mathbf{r})$ be continuous at the boundary. One might, perhaps, question whether it is permissible to treat the boundary as a well-defined geometrical surface, since the wavelength of a thermal neutron is typically of the order of 1Å and, hence, of the same order of magnitude as the interatomic distance in a crystal. However, the important parameter here is not the wavelength but the extinction length Δ_h. In Sec. 5.2 we saw that Δ_h is typically 50 μm, which is very much greater than the interatomic distance. Thus, as long as the scale of any surface roughness is much less than Δ_h, the surface is optically smooth and Eq. 6.1.7 is valid.

Exterior solution

At points $\mathbf{r}$ outside V the general solution of Eq. 6.0.1 can be expressed as a superposition of plane waves,

$$\psi(\mathbf{r}) = \sum a \exp(i\mathbf{k} \cdot \mathbf{r}) \tag{6.1.8}$$

in which the wave vectors $\mathbf{k}$ all have the same magnitude, which is determined by the incident-neutron energy,

$$E = \frac{(\hbar k)^2}{2m} \tag{6.1.9}$$

The values of a and the directions of $\mathbf{k}$ are determined by the boundary conditions, as will be discussed in Sec. 6.2.

Interior solution

At points $\mathbf{r}$ inside V, the optical potential is a periodic function, so that the general solution of Eq. 6.0.1 can be expressed as a superposition of *Bloch waves,*

$$\psi(\mathbf{r}) = \sum A(\mathbf{r}) \exp(i\mathbf{K} \cdot \mathbf{r}) \tag{6.1.10}$$

in which $A(\mathbf{r})$ is also periodic,

$$A(\mathbf{r} + \mathbf{R}_l) = A(\mathbf{r}), \qquad \text{all } l \tag{6.1.11}$$

and, hence, can be expanded in the form

$$A(\mathbf{r}) = \sum_h A_h \exp(-i\mathbf{K}_h \cdot \mathbf{r}) \tag{6.1.12}$$

The coefficients A_h satisfy the equations

$$D_h A_h = \xi \sum_{h'} \hat{F}_{h-h'} A_{h'} \tag{6.1.13}$$

in which $h - h'$ refers to the reciprocal lattice vector $\mathbf{K}_{h-h'} \equiv \mathbf{K}_h - \mathbf{K}_{h'}$ and

$$D_h = 1 - \frac{(\mathbf{K} - \mathbf{K}_h)^2}{k^2} \tag{6.1.14}$$

Finally,

$$\xi = \frac{v_0}{E} = \frac{4\pi}{k^2}\rho b = \frac{\lambda^2}{\pi}\rho b \tag{6.1.15}$$

as before (see Eq. 3.1.7). Thus, to each value of $\mathbf{K}$ in Eq. 6.1.10 there corresponds a set of coefficients A_h satisfying Eq. 6.1.13. The values of these quantities are determined by the boundary conditions, as will be discussed in Sec. 6.2. In the meantime we note that a necessary condition for the system of homogeneous linear equations (6.1.13) to possess a solution is that the determinant of the coefficients vanish,

$$\text{Det}\|\xi\hat{F}_{h-h'} - D_h\delta_{hh'}\| = 0 \tag{6.1.16}$$

The Schrödinger equation for the coherent wave inside the crystal is formally the same as for an electron in the band theory of solids. However, the boundary conditions are quite different. In the band theory of solids the allowed values of $\mathbf{K}$ are determined by requiring that $\psi(\mathbf{r})$

satisfy periodic boundary conditions. For each such value of $\mathbf{K}$ one then solves Eq. 6.1.16 to determine the possible values of k and, hence, the band energies (Eq. 6.1.9). In the dynamical theory of neutron diffraction, on the other hand, k is determined by the incident-neutron energy (Eq. 6.1.9), and Eq. 6.1.16 is then solved to determine the so-called *dispersion surface*, $\mathbf{K} = \mathbf{K}(k)$. The allowed values of $\mathbf{K}$ on the dispersion surface are called *tie points* and are determined by the boundary conditions for the particular problem under consideration.

The solution of the present problem is facilitated by the fact that the dimensionless parameter ξ, which measures the magnitude of the optical potential relative to the incident-neutron energy, is very small at thermal-neutron energies. Typically, $\xi \simeq 10^{-5}$ when $\lambda = 4.0$ Å and $\xi \simeq 10^{-6}$ when $\lambda = 1.3$ Å. Thus, ξ is a useful parameter in terms of which to make a perturbation expansion.

We shall show that

$$\mathbf{K} = \mathbf{k}[1 + O(\xi)] \tag{6.1.17}$$

Hence, from Eq. 6.1.14,

$$D_h = \frac{2\mathbf{k} \cdot \mathbf{K}_h - K_h^2}{k^2} + O(\xi) \tag{6.1.18}$$

and from Eq. 6.1.13,

$$D_h A_h = O(\xi) \tag{6.1.19}$$

Since $\mathbf{K}_0 = 0$ we therefore have

$$D_0 = O(\xi), \qquad A_0 = O(1) \tag{6.1.20}$$

If the reciprocal lattice site h lies on the surface of the Ewald sphere (see Chap. 5) then

$$2\mathbf{k} \cdot \mathbf{K}_h = K_h^2 \tag{6.1.21}$$

which is equivalent to Bragg's law (Eq. 5.2.24), and

$$D_h = O(\xi), \qquad A_h = O(1) \tag{6.1.22}$$

On the other hand, if Bragg's law is not satisfied (at least to order ξ) then

$$D_h = O(1), \qquad A_h = O(\xi) \tag{6.1.23}$$

Thus, it is clear that the nature of the solution depends on the number of reciprocal lattice vectors $\mathbf{K}_h$ for which Bragg's law is simultaneously satisfied for the given incident-neutron wave vector $\mathbf{k}$.

One-wave case

Suppose $\mathbf{k}$ does not satisfy Bragg's law for any $\mathbf{K}_h$. Then,

$$D_h = \begin{cases} O(\xi), & h = 0 \\ O(1), & h \neq 0 \end{cases} \tag{6.1.24}$$

and

$$A_h = \begin{cases} O(1), & h = 0 \\ O(\xi), & h \neq 0 \end{cases} \tag{6.1.25}$$

Hence, from Eq. 6.1.13,

$$D_h A_h = \xi \hat{F}_h A_0 + O(\xi^2) \tag{6.1.26}$$

Keeping only the leading term, we see that $D_0 = \xi$. On the other hand, Eq. 6.1.14 gives $D_0 = 1 - n^2$, where $n = K/k$ is the index of refraction. Thus, $n^2 = 1 - \xi$ as before (see Eq. 3.1.4). Also, with $A_0 \equiv A$, it follows from Eq. 6.1.10 that

$$\psi(\mathbf{r}) = \sum A\left\{\exp(i\mathbf{K}\cdot\mathbf{r}) + \xi \sum_{h\neq 0} \frac{\hat{F}_h}{D_h} \exp[i(\mathbf{K}-\mathbf{K}_h)\cdot\mathbf{r}] + O(\xi^2)\right\} \tag{6.1.27}$$

Thus, $\psi(\mathbf{r})$ is a superposition of the internal transmitted wave, with wave vector $\mathbf{K}$, plus a Bragg-reflected wave, with wave vector $\mathbf{K} - \mathbf{K}_h$, from each reciprocal lattice site $h \neq 0$. Whereas all possible Bragg reflections are excited, the amplitude of each Bragg-reflected wave relative to that of the transmitted wave is of order ξ, so that the relative intensity of the Bragg reflection is of order $\xi^2 \lesssim 10^{-10}$, which is negligibly small. Thus, for all practical purposes we can put

$$\psi(\mathbf{r}) = \sum A \exp(i\mathbf{K}\cdot\mathbf{r}) \tag{6.1.28}$$

which is identical to Eq. 3.1.1. In other words, when Bragg's law is not satisfied the coherent wave inside the crystal is the same as in a homogeneous body of the same composition and density. Consequently, all the results that were derived for a homogeneous body in Chap. 3 also apply to a crystal as long as the incident-neutron wave vector does not satisfy Bragg's law.

Two-wave case

Let us now suppose that $\mathbf{k}$ satisfies the condition for Bragg reflection via one particular reciprocal lattice vector $\mathbf{K}_h$ so that

$$D_{h'} = \begin{cases} O(\xi), & h' = 0, h \\ O(1), & \text{otherwise} \end{cases} \tag{6.1.29}$$

and, hence, that

$$A_{h'} = \begin{cases} O(1), & h' = 0, h \\ O(\xi), & \text{otherwise} \end{cases} \tag{6.1.30}$$

In this case,

$$D_{h'} A_{h'} = \xi(\hat{F}_{h'} A_0 + \hat{F}_{h'-h} A_h) + O(\xi^2) \tag{6.1.31}$$

Keeping only the leading term, we see that

$$\begin{aligned}(D_0 - \xi)A_0 - \xi\hat{F}_{-h}A_h &= 0 \\ -\xi\hat{F}_h A_0 + (D_h - \xi)A_h &= 0\end{aligned} \tag{6.1.32}$$

These equations describe the multiple Bragg reflection of the neutron inside the crystal and have a solution only if

$$(D_0 - \xi)(D_h - \xi) - \xi^2\hat{F}_h\hat{F}_{-h} = 0 \tag{6.1.33}$$

in which case

$$X \equiv \frac{A_h}{A_0} = \frac{D_0 - \xi}{\xi\hat{F}_{-h}} = \frac{\xi\hat{F}_h}{D_h - \xi} \tag{6.1.34}$$

With $A_0 \equiv A$, the coherent wave then becomes

$$\begin{aligned}\psi(\mathbf{r}) = \sum A\Big\{&\exp(i\mathbf{K}\cdot\mathbf{r}) + X\exp[i(\mathbf{K} - \mathbf{K}_h)\cdot\mathbf{r}] \\ &+ \xi \sum_{h'\neq 0,h} \frac{\hat{F}_{h'} + X\hat{F}_{h'-h}}{D_{h'}}\exp[i(\mathbf{K} - \mathbf{K}_{h'})\cdot\mathbf{r}] + O(\xi^2)\Big\}\end{aligned} \tag{6.1.35}$$

Again, $\psi(\mathbf{r})$ is a superposition of the internal transmitted wave plus a Bragg-reflected wave from each reciprocal lattice site $h' \neq 0$. However, only the Bragg-reflected wave from h has an amplitude comparable with that of the transmitted wave. All the others have relative amplitudes of order ξ and, hence, are negligible as before. Thus, for all practical purposes we can put

$$\psi(\mathbf{r}) = \sum A\{\exp(i\mathbf{K}\cdot\mathbf{r}) + X\exp(i\mathbf{K}'\cdot\mathbf{r})\} \tag{6.1.36}$$

where

$$\mathbf{K}' = \mathbf{K} - \mathbf{K}_h \tag{6.1.37}$$

which is analogous to the Laue equation (5.2.15).

The theory can easily be extended to the case in which $\mathbf{k}$ satisfies the condition for Bragg reflection for two or more values of h simultaneously. This leads to interesting interference phenomena such as the *Renninger effect* (Renninger, 1937; James, 1963). However, in what follows we shall confine our attention to the two-wave case.

Dispersion surface

We now consider the solution of Eq. 6.1.33 that determines the dispersion surface in the two-wave case. Let us begin by introducing the indices of refraction for the internal transmitted and Bragg-reflected waves,

$$n = \frac{K}{k}, \qquad n' = \frac{K'}{k} \tag{6.1.38}$$

With

$$D_0 = \xi p, \qquad D_h = \xi p' \tag{6.1.39}$$

it then follows from Eq. 6.1.14 that

$$n^2 = 1 - \xi p, \qquad n'^2 = 1 - \xi p' \tag{6.1.40}$$

and from Eq. 6.1.33 that

$$(p-1)(p'-1) = \hat{F}_h \hat{F}_{-h} \tag{6.1.41}$$

Let us suppose that the incident wave enters the crystal through a plane boundary, which we shall take to be the $z = 0$ plane. It will be shown later than the boundary conditions require that the quantity $\boldsymbol{\kappa} \equiv \mathbf{K} - \mathbf{k}$ be perpendicular to the boundary. Thus, we can put $\boldsymbol{\kappa} = \kappa_z \hat{\mathbf{z}}$ where $\hat{\mathbf{z}}$ denotes a unit vector in the z direction. Hence, it follows that to first order in ξ,

$$\kappa_z = -\varepsilon p, \qquad \varepsilon = \frac{\xi k^2}{2k_z} \tag{6.1.42}$$

Also, Eq. 6.1.37 implies that

$$p' = \alpha p + \beta \tag{6.1.43}$$

where

$$\begin{aligned} \alpha &= 1 - \frac{K_{hz}}{k_z} \\ \beta &= \frac{2\mathbf{k} \cdot \mathbf{K}_h - K_h^2}{\xi k^2} \end{aligned} \tag{6.1.44}$$

Thus, Eq. 6.1.41 is a quadratic equation in p with two roots,

$$p_\pm = 1 + \gamma \pm (\gamma^2 + \delta)^{1/2} \tag{6.1.45}$$

in which

$$\gamma = \frac{1 - \alpha - \beta}{2\alpha}, \qquad \delta = \frac{\hat{F}_h \hat{F}_{-h}}{\alpha} \tag{6.1.46}$$

Since p and, hence, p' are double-valued, so are the refractive indices n and n'. Thus, in the simplest case, the coherent wave (Eq. 6.1.36) is a superposition of four plane waves,

$$\begin{aligned} \psi(\mathbf{r}) = {} & A_+\{\exp(i\mathbf{K}_+ \cdot \mathbf{r}) + X_+ \exp(i\mathbf{K}'_+ \cdot \mathbf{r})\} \\ & + A_-\{\exp(i\mathbf{K}_- \cdot \mathbf{r}) + X_- \exp(i\mathbf{K}' \cdot \mathbf{r})\} \end{aligned} \tag{6.1.47}$$

in which

$$\begin{aligned} \mathbf{K}_\pm &= \mathbf{k} + \boldsymbol{\kappa}_\pm \\ \mathbf{K}'_\pm &= \mathbf{K}_\pm - \mathbf{K}_h \\ X_\pm &= \frac{p_\pm - 1}{\hat{F}_{-h}} \end{aligned} \tag{6.1.48}$$

and

$$\boldsymbol{\kappa}_\pm = \kappa_{z\pm} \hat{\mathbf{z}}, \qquad \kappa_{z\pm} = -\varepsilon p_\pm \tag{6.1.49}$$

Thus, given $\mathbf{k}$ and $\mathbf{K}_h$, the only undetermined quantities in the wave function (Eq. 6.1.47) are $A_{\pm}$. These will be determined from the boundary conditions in Sec. 6.2.

The essential difference between the kinematical and dynamical theories of neutron diffraction in crystals is now apparent. In the kinematical theory, refraction is neglected so that there is no distinction between the internal and external waves. Consequently, there is only one transmitted wave and one Bragg-reflected wave. In the dynamical theory, on the other hand, refraction is taken into account and the refractive indices for both the transmitted and Bragg-reflected waves are found to be double-valued. As a result, the external incident wave produces two internal transmitted waves with wave vectors $\mathbf{K}_{\pm}$ and two internal Bragg-reflected waves with wave vectors $\mathbf{K}'_{\pm}$. In Sec. 6.2 it will be shown that these waves subsequently emerge from the crystal as a single external transmitted wave and a single external Bragg-reflected wave.

Since the refractive indices are generally both complex and double-valued, the dynamical theory contains a number of effects that are completely absent in the kinematical theory. These effects include extinction, Pendellösung interference, and anomalous absorption, and will be discussed in Sec. 6.3.

6.2. Bragg reflection by a plane slab

We saw in Sec. 5.2 that in the kinematical theory of neutron diffraction in a perfect crystal the positions of the Bragg peaks are determined by the crystal structure but the intrinsic shapes of these peaks are determined by the shape of the crystal and are independent of the crystal structure. This is also true in the dynamical theory of diffraction. Whereas the kinematical theory can be carried through for a crystal of arbitrary shape, this cannot be done in the dynamical theory because the solution of the boundary-value problem depends explicitly on the detailed shape of the crystal.

The crystals used in dynamical diffraction experiments are always in the form of plane slabs or wedges. This is partly because such shapes are the easiest to fabricate but mainly because these are the only shapes for which the boundary-value problem is easily solved. In what follows we shall assume that the crystal takes the form of a plane slab of thickness d and infinite lateral extent, and that the incident wave is likewise of infinite width. The results obtained on the basis of these assumptions will be valid in practice provided the actual widths of the crystal and the incident wave are much greater than the thickness of the crystal.

Boundary-value problem

Let us choose a system of Cartesian coordinate axes such that the faces of the slab are the planes $z = 0$ and $z = d$. If the incident wave enters the

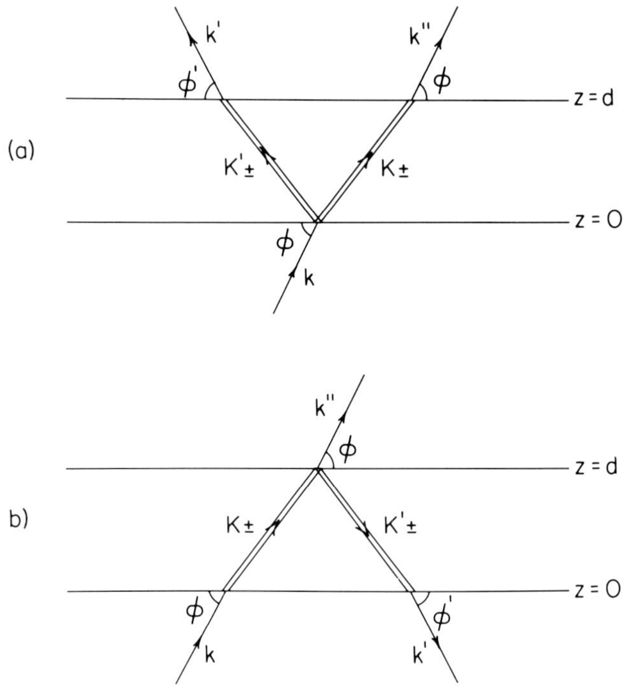

Fig. 6.1 Wave vector relations for the dynamical theory of Bragg reflection in a plane slab for (a) the Laue case and (b) the Bragg case.

crystal through the face $z = 0$, then the Bragg-reflected wave may emerge from the crystal through either the face $z = d$ or the face $z = 0$ as, illustrated in Fig. 6.1. The former is the case of *transmission geometry* which, following Ewald (1917b), is usually referred to as the *Laue case*. The latter is the case of *reflection geometry* and is called the *Bragg case*.

In general, a wave that is incident on a boundary between two media with different refractive indices will be partially reflected and refracted at the boundary. However, we saw in Sec. 3.1 that reflection is appreciable only for glancing incidence where $\phi \lesssim 0.2$ degree. In what follows we shall exclude the case of glancing incidence, so that reflection can be neglected. This case has been discussed in detail by Stassis and Oberteuffer (1974) and Zeilinger and Beatty (1983).

If reflection is neglected, the external wave function (Eq. 6.1.8) is given in the Laue case by

$$\psi(\mathbf{r}) = \begin{cases} a \exp(i\mathbf{k} \cdot \mathbf{r}), & z < 0 \\ a' \exp(i\mathbf{k}' \cdot \mathbf{r}) + a'' \exp(i\mathbf{k}'' \cdot \mathbf{r}), & z > d \end{cases} \tag{6.2.1}$$

and in the Bragg case by

$$\psi(\mathbf{r}) = \begin{cases} a \exp(i\mathbf{k} \cdot \mathbf{r}) + a' \exp(i\mathbf{k}' \cdot \mathbf{r}), & z < 0 \\ a'' \exp(i\mathbf{k}'' \cdot \mathbf{r}), & z > d \end{cases} \tag{6.2.2}$$

For $0<z<d$ the internal wave function is given by Eq. 6.1.47 for both the Laue and Bragg cases.

In general, the internal and external wave functions would be matched by requiring that $\psi(\mathbf{r})$ and $\nabla\psi(\mathbf{r})$ be continuous at the boundary. However, since we have neglected reflection we must, for consistency, also ignore the requirement that $\nabla\psi(\mathbf{r})$ be continuous at the boundary. The requirement that $\psi(\mathbf{r})$ be continuous at $z=0$ and $z=d$ then gives, for both the Laue and Bragg cases,

$$\mathbf{k}'=\mathbf{k}-\mathbf{K}_h-\mathbf{q}_1, \qquad \mathbf{k}''=\mathbf{k} \tag{6.2.3}$$

where, in the notation of Sec. 6.1,

$$\mathbf{q}_1=-\left(\frac{\varepsilon\beta}{\alpha}\right)\hat{\mathbf{z}} \tag{6.2.4}$$

The amplitudes are given for the Laue case by

$$\begin{aligned}
A_+&=\left(\frac{-X_-}{X_+-X_-}\right)a\\
A_-&=\left(\frac{X_+}{X_+-X_-}\right)a\\
a'&=-\left(\frac{X_+X_-(Y_+-Y_-)}{X_+-X_-}\right)a\exp(iq_{1z}d)\\
a''&=\left(\frac{X_+Y_--X_-Y_+}{X_+-X_-}\right)a
\end{aligned} \tag{6.2.5}$$

and for the Bragg case by

$$\begin{aligned}
A_+&=\left(\frac{-X_-Y_-}{X_+Y_+-X_-Y_-}\right)a\\
A_-&=\left(\frac{X_+Y_+}{X_+Y_+-X_-Y_-}\right)a\\
a'&=\left(\frac{X_+X_-(Y_+-Y_-)}{X_+Y_+-X_-Y_-}\right)a\\
a''&=\left(\frac{Y_+Y_-(X_+-X_-)}{X_+Y_+-X_-Y_-}\right)a
\end{aligned} \tag{6.2.6}$$

In both cases,

$$Y_\pm=\exp(i\kappa_{z\pm}d) \tag{6.2.7}$$

where $\kappa_{z\pm}$ is given by Eq. 6.1.49 as before. Thus we see that, given a, $\mathbf{k}$, and $\mathbf{K}_h$, all the remaining quantities in both the internal and external wave functions are uniquely determined.

Reflectivity and transmissivity

The reflectivity and transmissivity are given by

$$R = \frac{j' \sin \phi'}{j \sin \phi}$$
$$T = \frac{j'' \sin \phi}{j \sin \phi} \tag{6.2.8}$$

where j, j', and j'' are the neutron current densities in the incident, Bragg reflected, and transmitted waves respectively (see Eq. 3.1.27). Since

$$\frac{\sin \phi'}{\sin \phi} = |\alpha| \tag{6.2.9}$$

to first order in ξ for both the Laue and Bragg cases, it follows that

$$R = |\alpha| \left|\frac{a'}{a}\right|^2$$
$$T = \left|\frac{a''}{a}\right|^2 \tag{6.2.10}$$

Thus, for the Laue case,

$$R = |\alpha| \left|\frac{X_+ X_-(Y_+ - Y_-)}{X_+ - X_-}\right|^2$$
$$T = \left|\frac{X_+ Y_- - X_- Y_+}{X_+ - X_-}\right|^2 \tag{6.2.11}$$

and for the Bragg case,

$$R = |\alpha| \left|\frac{X_+ X_-(Y_+ - Y_-)}{X_+ Y_+ - X_- Y_-}\right|^2$$
$$T = \left|\frac{Y_+ Y_-(X_+ - X_-)}{X_+ Y_+ - X_- Y_-}\right|^2 \tag{6.2.12}$$

Case of zero absorption

Let us first consider the case of zero absorption where b is real and the unit-cell structure factor satisfies Friedel's law (Eq. 5.2.46). In this case the quantities α, β, γ, and δ in Sec. 6.1 are all real and it then follows that for both the Laue and Bragg cases

$$R + T = 1 \tag{6.2.13}$$

This is a statement of neutron conservation since, in the case of zero absorption, the incident neutron is either Bragg reflected or transmitted.

Thus, we need only give the expression for R which, for the Laue case, is

$$R(x, y) = \frac{\sin^2[y(x^2+1)^{1/2}]}{x^2+1} \tag{6.2.14}$$

and, for the Bragg case, is

$$R(x, y) = \begin{cases} \dfrac{\sinh^2[y(1-x^2)^{1/2}]}{1-x^2+\sinh^2[y(1-x^2)^{1/2}]}, & x^2<1 \\ \dfrac{\sin^2[y(x^2-1)^{1/2}]}{x^2-1+\sin^2[y(x^2-1)^{1/2}]}, & x^2>1 \end{cases} \tag{6.2.15}$$

Here,

$$x^2 = \frac{\gamma^2}{|\delta|} \qquad y^2 = (\varepsilon d)^2\,|\delta| \tag{6.2.16}$$

and the signs of x and y are arbitrary.

For the rotating-crystal method (Sec. 5.2) it follows that

$$x = \frac{\theta - \theta_h}{\chi_h} + \psi_h, \qquad y = \frac{\pi d}{\Delta_h} \tag{6.2.17}$$

where θ_h is the Bragg angle, χ_h is given by Eq. 5.2.86, Δ_h by Eq. 5.2.87, and

$$\psi_h = \frac{1}{2\,|\hat{F}_h|}\left[\pm\left(\frac{\sin\phi'}{\sin\phi}\right)^{1/2} \pm \left(\frac{\sin\phi}{\sin\phi'}\right)^{1/2}\right], \tag{6.2.18}$$

in which the sign assignment is:

$+\,-$ for the Laue case with $b>0$
$-\,+$ for the Laue case with $b<0$
$-\,-$ for the Bragg case with $b>0$
$+\,+$ for the Bragg case with $b<0$

Also, it follows from Eq. 6.2.4 that

$$q_{1z} = \pm\left(\frac{k\sin 2\theta_h}{\sin\phi'}\right)(\theta - \theta_h) \tag{6.2.19}$$

in which the plus sign refers to the Bragg case and the minus sign to the Laue case. This relation is identical to Eq. 5.2.81. Consequently, the expressions (5.2.82) and (6.2.3) for the scattered-neutron wave vector $\mathbf{k}'$ are also identical. In other words, $\mathbf{k}'$ is given correctly by the kinematical theory. However, the reflectivity is not given correctly by the kinematical theory. In the dynamical theory, R is not only a different function of x and y from that in the kinematical theory (see Eq. 5.2.84) but the dynamical expression (6.2.17) for x contains an additional term ψ_h that is absent in the corresponding kinematical expression (5.2.85).

Dynamical diffraction effects are usually studied experimentally for symmetric scattering geometries, where the Bragg planes of interest are either parallel or perpendicular to the face of the slab (Fig. 5.8). The values of χ_h and Δ_h for the symmetric Laue and Bragg cases are listed in Table 5.3. For symmetric reflections Eq. 6.2.18 reduces to

$$\psi_h = \begin{cases} 0, & \text{Laue case} \\ \pm|\hat{F}_h|^{-1}, & \text{Bragg case} \end{cases} \tag{6.2.20}$$

where the plus sign applies when $b < 0$ and the minus sign when $b > 0$. The reflectivity $R(x, y)$ is an even function of x for both the Laue and Bragg cases. However, it is only in the symmetric Laue case, where $\psi_h = 0$, that the reflectivity is an even function of $\theta - \theta_h$. In all other cases $\psi_h \neq 0$ and the rocking curve is not symmetric about the Bragg angle.

Finally, it will be noted that in both the Laue and Bragg cases $R(x, y) \leq 1$ for all x and y. Thus, the dynamical theory of neutron diffraction, unlike the kinematical theory, satisfies the requirement of neutron conservation.

Thin crystals

Let us first consider a thin crystal, i.e., one for which $d \ll \Delta_h$. Then $y \ll 1$, so that $|x| \gg 1$ for virtually all values of x for which $R(x, y)$ is appreciably different from zero. In this case the term ψ_h in the expression (6.2.17) for x is negligible, and Eqs. 6.2.14 and 6.2.15 each reduce to

$$R(x, y) = \left(\frac{\sin xy}{x}\right)^2 \tag{6.2.21}$$

which is simply the kinematical result (Eq. 5.2.84). Thus, as argued earlier in Sec. 5.2, the kinematical theory of neutron diffraction is asymptotically correct in the thin-crystal limit.

Thick crystals

For a thick crystal, on the other hand, where $d \gg \Delta_h$ and $y \gg 1$, the reflectivity is appreciably different from zero only if $|x| \lesssim 1$, i.e., only if $|\theta - \theta_h| \lesssim \chi_h$. Since $\chi_h \simeq \xi$ and $\xi \simeq 10^{-6}$ radians = 0.2 arc seconds for 1.5 Å neutrons, it is evident that the rocking curve for a thick crystal is extremely narrow and, within the peak, $R(x, y)$ is a rapidly oscillating function of x.

This is illustrated in Fig. 6.2, which shows $R(x, y)$ as a function of x for $y = 10\pi$ as calculated for the Laue case from Eq. 6.2.14 and for the Bragg case from Eq. 6.2.15. This value of y corresponds to a crystal thickness $d = 10\Delta_h$. Since the extinction length Δ_h is typically 50 μm = 0.05 mm, this means that $d \simeq 0.5$ mm. While such a crystal is thick in the optical

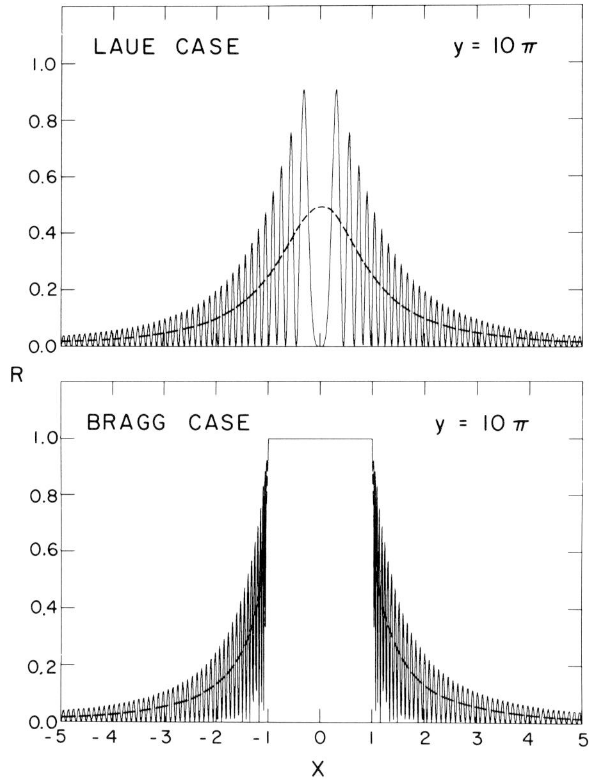

Fig. 6.2 The reflectivity $R(x, y)$ for Bragg reflection in a plane slab in the case of zero absorption according to the dynamical theory of neutron diffraction. These curves were calculated for the Laue case from Eq. 6.2.14 and for the Bragg case from Eq. 6.2.15, and the results are shown as a function of x for $y = 10\pi$. The dashed curves show the corresponding mean reflectivity $\bar{R}(x)$ calculated from Eqs. 6.2.22 and 6.2.23.

sense it is still, of course, very thin from a practical point of view. As d and, hence, y increase, the fringe separation in Fig. 6.2 becomes proportionately smaller.

In practice, the fine structure in the reflectivity curves is not observed owing to the effect of the finite instrumental resolution. Consequently, the quantity that is measured is essentially the mean reflectivity $\bar{R}$, which is obtained from R by averaging over the oscillations. In the thick crystal limit, $y \to \infty$, this quantity is given in the Laue case by

$$\bar{R}(x) = \frac{1}{2(x^2+1)} \tag{6.2.22}$$

and in the Bragg case by

$$\bar{R}(x) = \begin{cases} 1, & x^2 < 1 \\ 1-(1-x^{-2})^{1/2}, & x^2 > 1 \end{cases} \tag{6.2.23}$$

These results were first obtained by Ewald (1917b) in the context of X-ray diffraction and are represented by the dashed curves in Fig. 6.2. The Bragg-case expression (Eq. 6.2.23) is qualitatively similar to an approximate result that had been obtained earlier by Darwin (1914b). Consequently, the region $x^2 < 1$ is called the *Darwin gap* and the value $\bar{R}(x) = 1$ within this gap is called the *Darwin plateau.*

Figure 6.3 shows the rocking curves for both the Bragg-reflected and the transmitted beams that were measured by Sippel et al. (1964) for a symmetric (220) Laue-case reflection in a 0.91-mm-thick crystal of germanium with 1.18 Å neutrons. It will be noted that the reflectivity has the Lorentzian shape that is characteristic of Laue-case reflections when absorption is negligible.

Figure 6.4 shows the rocking curve that was measured by Kikuta et al. (1975) for a symmetric (111) Bragg-case reflection in a 13-mm thick crystal of silicon. The crystal was such that $\mu d \simeq 10^{-3}$, so that absorption is negligible, and $y \simeq 10^3$, so that the fine structure in the reflectivity is not observable. The angular divergence of the incident beam was 0.065 arc seconds, which is an order of magnitude smaller than the width of the rocking curve, so that Fig. 6.4 represents, essentially, the intrinsic reflectivity (Eq. 6.2.23). In this figure I_i is the incident intensity and I_h the reflected intensity, so that the peak reflectivity is 100 percent as it should be according to Eq. 6.2.23.

In the work represented in Fig. 6.4, the absolute angle of incidence θ was not measured, so that it was not possible to detect the peak shift

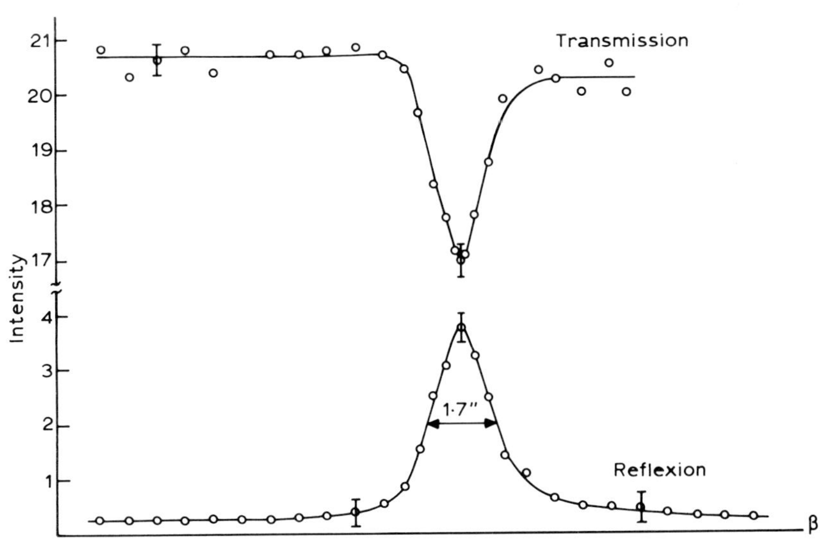

Fig. 6.3 Rocking curves for both the Bragg-reflected and transmitted beams measured by Sippel et al. (1964) for the symmetric Laue case in germanium.

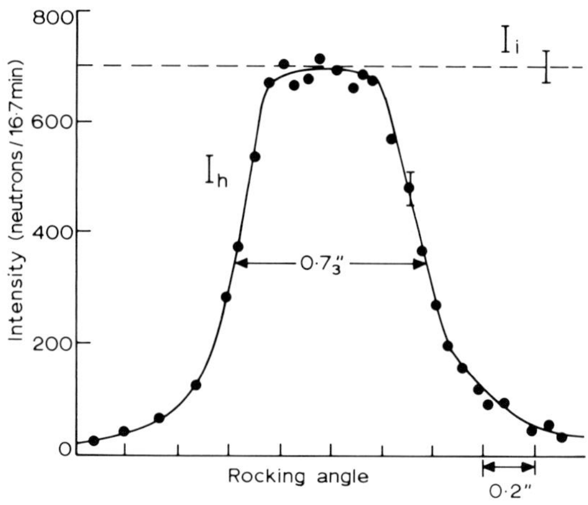

Fig. 6.4 Rocking curve measured by Kikuta et al. (1975) for the symmetric Bragg case in silicon.

arising from the ψ_h term in Eq. 6.2.17. This shift has, however, recently been observed by Treimer et al. (1985) in an asymmetric (111) Bragg-case reflection in silicon. From the measured value of the peak shift these authors were able to obtain a value 4.15(6) fm for the bound coherent scattering length of silicon.

6.3. Dynamical diffraction effects in Bragg reflection

We now consider some of the dynamical diffraction effects that are characteristic of Bragg reflection in a perfect crystal. These effects include extinction, Pendellösung interference, and anomalous absorption, all of which are absent in the kinematical theory of diffraction.

Extinction

We begin by considering the integrated reflectivity

$$\rho(y) = \int_{-\infty}^{\infty} R(x, y)\, dx \tag{6.3.1}$$

In the case of zero absorption it follows from Eqs. 6.2.14 and 6.2.15 that

$$\rho(y) = \begin{cases} \dfrac{\pi}{2} \displaystyle\int_0^{2y} J_0(t)\, dt, & \text{(Laue case)} \\ \pi \tanh y, & \text{(Bragg case)} \end{cases} \tag{6.3.2}$$

where $J_0(t)$ denotes the zeroth-order Bessel function. These results are

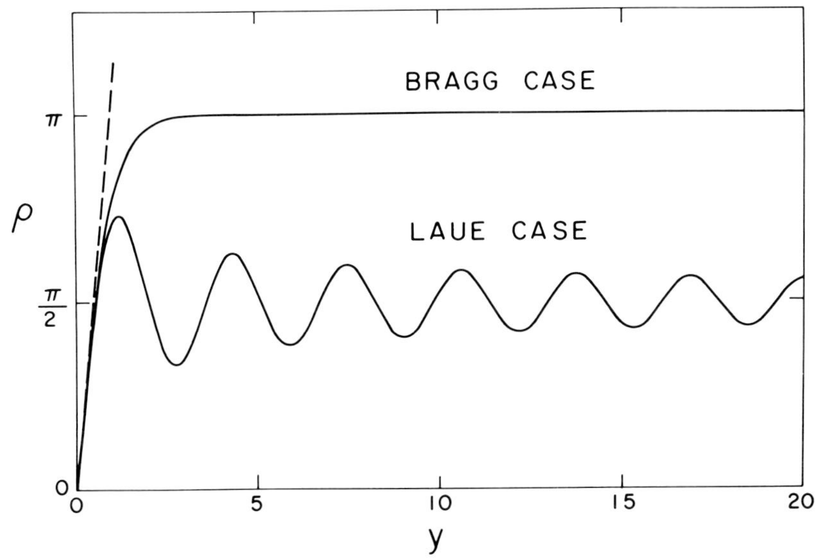

Fig. 6.5 Integrated reflectivity ρ as a function of y in the case of zero absorption for both the Bragg and Laue cases as calculated from Eq. 6.3.2. The dashed line shows the kinematical result Eq. 6.3.3.

illustrated in Fig. 6.5. In the thin-crystal limit ($y \to 0$) the preceding expressions for both the Laue and Bragg cases reduce to the kinematical result (Eq. 5.2.89),

$$\rho_k(y) = \pi y \tag{6.3.3}$$

This is represented by the dashed line in Fig. 6.5. In the thick-crystal limit, on the other hand,

$$\rho(\infty) = \begin{cases} \pi/2, & \text{(Laue case)} \\ \pi, & \text{(Bragg case)} \end{cases} \tag{6.3.4}$$

Figure 6.6 shows the integrated reflectivity measured by Sippel et al. (1965) for symmetric (220) Laue-case reflections from silicon crystals of various thicknesses with 1.18 Å neutrons. The data are in excellent agreement with the theoretical result (Eq. 6.3.2), which is represented by the smooth curve in this figure.

It is evident from Fig. 6.5 that $\rho(y) < \rho_k(y)$. This reduction in the integrated reflectivity is called *extinction* and is due to the effect of the multiple Bragg reflection that is present in the dynamical theory (see Eq. 6.1.32) but is neglected in the kinematical theory. Writing

$$\rho(y) = \rho_k(y)\, E(y) \tag{6.3.5}$$

we can then call $E(y)$ the *extinction correction.* If the expressions (6.3.2) are expanded in powers of y, it then follows with the help of Eq. 6.3.3

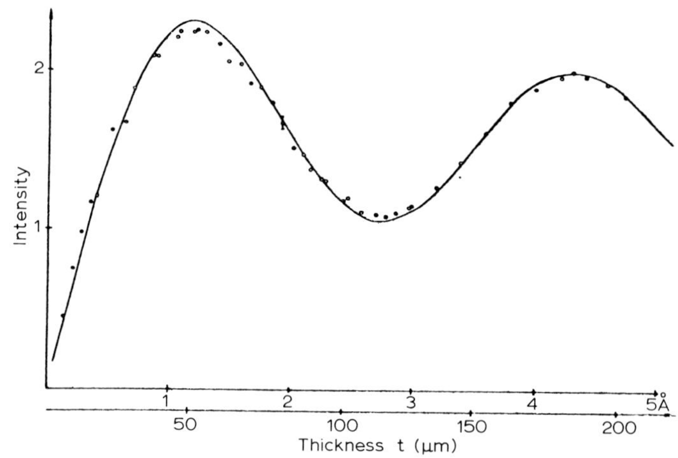

Fig. 6.6 Integrated reflectivity as a function of crystal thickness measured by Sippel et al. (1965) for the symmetric Laue case in silicon. The smooth curve shows the theoretical result (Eq. 6.3.2).

that for a thin crystal ($y \ll 1$),

$$E(y) = \begin{cases} 1 - \frac{1}{3}y^2 + \frac{1}{20}y^4 - \cdots, & \text{(Laue case)} \\ 1 - \frac{1}{3}y^2 + \frac{2}{15}y^4 - \cdots, & \text{(Bragg case)} \end{cases} \tag{6.3.6}$$

The lowest-order correction term is seen to be the same in both cases. These results apply only in the case of zero absorption. More generally, $\rho(y)$ will also be reduced by the effect of absorption and diffuse scattering. However, except in very strongly absorbing crystals, extinction is by far the dominant effect.

In mosaic crystals it is necessary to distinguish between the *primary* extinction arising from multiple Bragg reflection within a single mosaic block and the *secondary* extinction arising from multiple Bragg reflection between different mosaic blocks. Since each mosaic block is essentially a perfect crystal, primary extinction is a dynamical diffraction effect. In contrast, secondary extinction is a transport effect (see Chap. 8) because the different mosaic blocks scatter incoherently.

For order-of-magnitude purposes the *primary extinction correction* can be taken to be

$$E_p = 1 - \frac{1}{3}\left(\frac{\pi l}{\Delta_h}\right)^2 \tag{6.3.7}$$

where l characterizes the linear dimensions of the mosaic blocks. In most cases $l \lesssim 1\ \mu\text{m}$ and we saw earlier in Sec. 5.2 that the extinction length $\Delta_h \gtrsim 10\ \mu m$, so that $0.97 \lesssim E_p < 1.00$. Thus, it is only when $l \gg 1\ \mu\text{m}$ that primary extinction is more than a few percent for even the strongest reflections.

Pendellösung interference

This effect occurs in its simplest form in the symmetric Laue case with $\theta = \theta_h$. Then $\psi_h = 0$ and $x = 0$ so that, according to Eqs. 6.2.13 and 6.2.14,

$$R(0, y) = \sin^2 y, \qquad T(0, y) = \cos^2 y \tag{6.3.8}$$

This oscillatory behavior of R and T as a function of y describes the coherent transfer of neutron flux back and forth between the transmitted and Bragg-reflected beams with a period $y = \pi$ or

$$d = \Delta_h = \frac{\pi V_0 \cos \theta_h}{\lambda \, |F_h|} \tag{6.3.9}$$

This is analogous to the transfer of energy between two coupled pendulums. Consequently, Ewald (1917a) called this result the *pendulum solution* (i.e., *Pendellösung*). The extinction length Δ_h is therefore often called the Pendellösung period. The Pendellösung effect arises from the interference between the $\pm$ components in the internal wave function (Eq. 6.1.47), which produces a spatial modulation of the amplitudes of the transmitted and Bragg-reflected waves inside the crystal. The Pendellösung interference effect is also present in the integrated reflectivity for the Laue case (see Figs. 6.5 and 6.6).

Figure 6.7 shows the Pendellösung interference fringes measured by Shull (1968) for symmetric (111) Laue-case reflections from three silicon crystals with different thicknesses d. The data points give the reflected intensity at $\theta = \theta_h$ and the fringes were produced by varying the neutron wavelength λ and, hence, Δ_h. The minima are labeled by the fringe order number $y/\pi = d/\Delta_h$. (We might add that in these experiments the incident-neutron beam was limited by a narrow slit on the entrance face of the crystal, as a result of which the appropriate expression for the reflectivity differs from Eq. 6.2.14 when $x \neq 0$.)

Pendellösung interference measurements of the type described permit an absolute determination of the Pendellösung period Δ_h and, hence, of the unit-cell structure factor $|F_h|$. The latter quantity can then be used to obtain the value of the bound coherent scattering length b. The use of this method requires that the material of interest be available as an essentially perfect crystal because the presence of dislocations in any appreciable concentration would wash out the interference fringes. As a result, values of b have so far been determined by this method only for silicon (Shull, 1968; Shull and Oberteuffer, 1972) and germanium (Shull and Shaw, 1973). These values of b have an accuracy of about 0.03 percent, which is comparable with that of the best neutron-optical methods (see Table 3.3).

Pendellösung interference can also be observed at a fixed neutron wavelength by varying the effective thickness of the crystal. This can be done by rotating the crystal slab about the normal to the reflecting planes

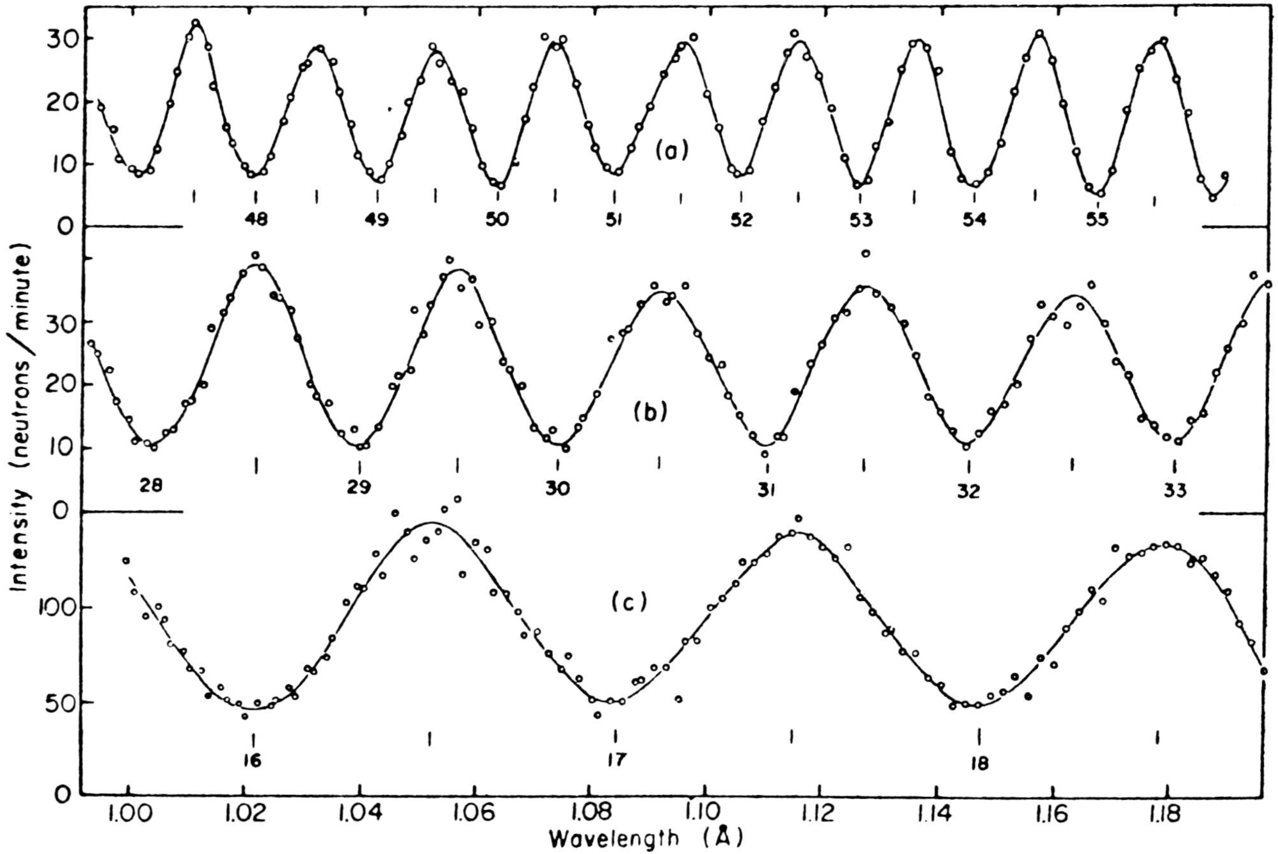

Fig. 6.7 Pendellösung interference fringes measured by Shull (1968) for symmetric Laue-case reflections from three silicon crystals with thicknesses (a) $d = 1.0000$ cm, (b) 0.5939 cm, and (c) 0.3315 cm.

(Somenkov et al., 1978). Alternatively, one can use a wedge-shaped crystal and record the Pendellösung fringes by neutron-diffraction topography (Kikuta et al., 1971; Malgrange et al., 1976).

Pendellösung interference has also been studied by Baruchel et al. (1982) in the ferromagnetic crystal yttrium iron garnet ($Y_3Fe_5O_{12}$) by polarized neutron diffraction. Pendellösung oscillations were observed in both the integrated reflectivity of the Bragg peaks and the flipping ratio as a function of the neutron wavelength.

Finally, we might mention the interesting theoretical investigation by Mendiratta (1976) on the inelastic scattering of thermal neutrons by a perfect crystal under Bragg-reflecting conditions. He showed on the basis of a distorted-wave Born approximation that each of the four components in the internal coherent wave (Eq. 6.1.47) is attenuated by inelastic scattering. The inelastically scattered waves then interfere to produce Pendellösung effects is the intensitites of the coherent one-phonon inelastic peaks.

Anomalous absorption

When the effect of absorption is taken into account, the neutron conservation condition (Eq. 6.2.13) becomes

$$R + T + A = 1 \tag{6.3.10}$$

where the *absorptivity* A is the fraction of incident neutrons that are absorbed. If the incident-neutron wave vector does not satisfy the condition for Bragg reflection, then the reflectivity $R = 0$, and the transmissivity T is given by Lambert's law (Eq. 3.2.10), so that

$$A = 1 - \exp(-\mu D) \tag{6.3.11}$$

where μ is the absorption coefficient and $D = d \operatorname{cosec} \phi$ is the thickness of the slab in the direction of the incident beam. When the crystal is rocked through the Bragg angle, however, deviations from Eq. 6.3.11 occur that are referred to as *anomalous absorption*.

Anomalous absorption occurs in its simplest form in the symmetric Laue case with $\theta = \theta_h$. For simplicity we shall also assume a rigid lattice with one atom per primitive unit cell. Then $F_h = b$ and (see Sec. 6.1)

$$\alpha = 1, \qquad \beta = 0, \qquad \gamma = 0, \qquad \delta = 1 \tag{6.3.12}$$

so that

$$\begin{aligned} p_+ &= 2, \qquad X_+ = 1, \qquad Y_+ = \exp(-2iy) \\ p_- &= 0, \qquad X_- = -1, \qquad Y_- = 1 \end{aligned} \tag{6.3.13}$$

where (see Eq. 6.2.16)

$$y = \varepsilon d = \frac{\lambda \rho b d}{\cos \theta_h} \tag{6.3.14}$$

Hence, it follows from Eq. 6.2.11 that

$$\begin{aligned} R &= \tfrac{1}{4}\,|1 - \exp(-2iy)|^2 \\ T &= \tfrac{1}{4}\,|1 + \exp(-2iy)|^2 \end{aligned} \tag{6.3.15}$$

Since b is in general complex, so is y:

$$b = b' - ib'', \qquad y = y' - iy'' \tag{6.3.16}$$

The real part of y can be expressed in the form of Eq. 6.2.17,

$$y' = \frac{\pi d}{\Delta_h} \tag{6.3.17}$$

where the extinction length is given by

$$\Delta_h = \frac{\pi \cos \theta_h}{\lambda \rho b'} \tag{6.3.18}$$

Since $\rho = 1/V_0$, this agrees with our previous result (Eq. 6.3.9) for the symmetric Laue case. The imaginary part of y can be expressed as

$$y'' = \tfrac{1}{2}\mu D \tag{6.3.19}$$

where μ is the absorption coefficient (Eq. 3.2.12) and $D = d \sec \theta_h = d \operatorname{cosec} \phi$ as before.

The relations (6.3.15) now become

$$\begin{aligned} R &= \tfrac{1}{4}[1 + \exp(-4y'') - 2\exp(-2y'')\cos(2y')] \\ T &= \tfrac{1}{4}[1 + \exp(-4y'') + 2\exp(-2y'')\cos(2y')] \end{aligned} \tag{6.3.20}$$

The terms containing $\cos(2y')$ represent the Pendellösung interference effect for finite absorption. In the case of zero absorption, where $\mu = 0$ and $y'' = 0$, the expressions (6.3.20) reduce to Eq. 6.3.8 as they should. Combining Eqs. 6.3.10 and 6.3.20, we get

$$R + T = 1 - A = \tfrac{1}{2}[1 + \exp(-4y'')] \tag{6.3.21}$$

The Pendellösung interference effect vanishes in the sum $R + T$ and, hence, in A. This is as one would expect, since the Pendellösung interference simply describes the coherent transfer of neutron flux back and forth between the transmitted and Bragg-reflected beams.

The absorptivity is therefore given by

$$A = \tfrac{1}{2}[1 - \exp(-2\mu D)] \tag{6.3.22}$$

Figure 6.8 shows A as a function of μD as calculated for normal absorption from Eq. 6.3.11 and for anomalous absorption from Eq. 6.3.22. Normal absorption occurs when the angle of incidence θ is sufficiently different from the Bragg angle θ_h that Bragg reflection is negligible, while anomalous absorption occurs under Bragg-reflecting conditions where $\theta \simeq \theta_h$. Equation 6.3.22 refers, in particular, to a rigid

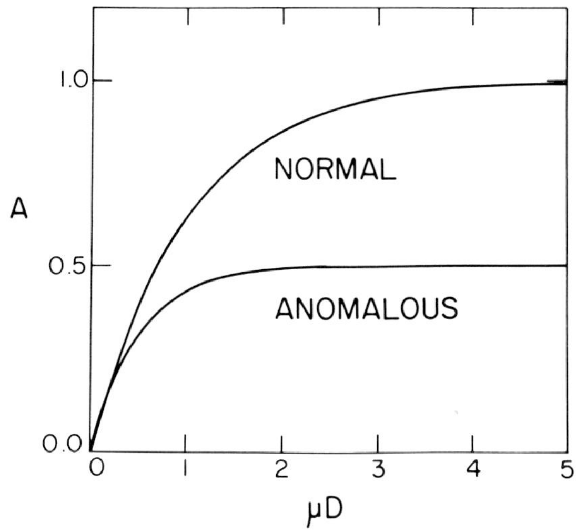

Fig. 6.8 Absorptivity A as a function of μD for normal and anomalous absorption.

lattice with one atom per primitive unit cell for the symmetric Laue case with $\theta = \theta_h$ exactly.

If $\mu D \ll 1$ then Eqs. 6.3.11 and 6.3.22 both reduce to $A = \mu D$, so that the anomalous absorption is appreciable only if $\mu D \gtrsim 1$. If $\mu D \gg 1$, then, for normal absorption, $A = 1$ and all the incident neutrons are absorbed, while, for anomalous absorption, $A = \frac{1}{2}$ and only 50 percent of the incident neutrons are absorbed. This is a remarkable result, since it means that no matter how large the absorption (even $\mu \to \infty$) or how thick the crystal (even $D \to \infty$), at least 50 percent of the incident neutrons are propagated through the crystal and out the other side when $\theta = \theta_h$.

The origin of anomalous absorption can be understood by considering the nature of the coherent wave (Eq. 6.1.47) at the positions $\mathbf{R}_l$ of the absorbing nuclei. In general,

$$\psi(\mathbf{r}) = \psi_+(\mathbf{r}) + \psi_-(\mathbf{r}) \tag{6.3.23}$$

and one then finds with the help of Eqs. 6.3.13 and A.11 that when $\theta = \theta_h$,

$$\psi_+(\mathbf{R}_l) = 2A_+ \exp(i\mathbf{K}_+ \cdot \mathbf{R}_l), \qquad \psi_-(\mathbf{R}_l) = 0 \tag{6.3.24}$$

for all l. Thus, when $\theta = \theta_h$ the component $\psi_-(\mathbf{r})$ has nodes at the positions of the absorbing nuclei and, hence, cannot be attenuated by absorption.

Anomalous absorption is often referred to as the *Borrmann effect*, since it was first observed by Borrmann (1941, 1950) in X-ray diffraction by perfect crystals. The anomalous absorption of neutrons was first observed by Knowles (1956), who measured the change in the (n, γ) yield that occurred when a calcite crystal was rocked through the Bragg position. The anomalous absorption of neutrons has also been observed indirectly by Sippel et al. (1962, 1964) from simultaneous measurements of R and T for a symmetric (220) Laue-case reflection from a 7.2-mm thick crystal of InSb. Their results are shown in Fig. 6.9, where it is seen that at the position of the maximum in R (i.e., at $\theta = \theta_h$) the sum $R + T$ is anomalously large, so that A is anomalously small. For the 1.18 Å neutrons used in these experiments, $\mu = 1.95\ \text{cm}^{-1}$ and $\mu d = 1.4$, which is sufficiently large that anomalous absorption is quite pronounced for this crystal. For the germanium crystal in Fig. 6.3, on the other hand, $\mu d = 0.007$ and anomalous absorption is negligible.

In Sec. 4.3 we showed that the absorption coefficient μ contains contributions from both true neutron absorption by the nuclei and diffuse scattering. Thus, in a perfect crystal under Bragg-reflecting conditions one would expect to observe not only anomalous absorption but also anomalous diffuse scattering. This has, in fact, been done by Sippel and Eichhorn (1968), who observed anomalous incoherent scattering by hydrogen atoms for a symmetric (004) Laue-case reflection in a nearly perfect crystal of KDP (i.e., KH_2PO_4). Effects associated with the anomalous time of flight of neutrons through a perfect crystal under Bragg-reflecting conditions have been observed by Shull et al. (1980).

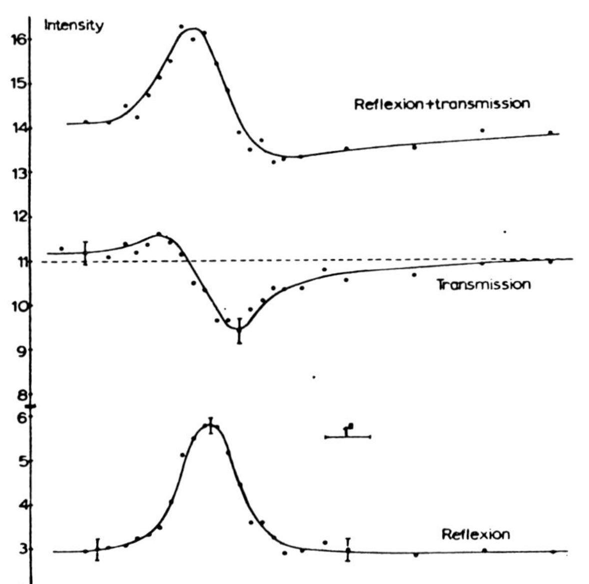

Fig. 6.9 Rocking curves for R and T measured by Sippel et al. (1964) for the symmetric Laue case in InSb. The $R + T$ curve shows the effect of anomalous absorption at the Bragg position.

Other effects

The dynamical theory of diffraction in perfect crystals is a rich source of interesting wave phenomena and it is beyond the scope of this book to attempt to treat them all in detail. We shall simply indicate here two situations in which some of these other phenomena arise. For more details we refer the reader to the general references cited at the beginning of this chapter.

In a perfect crystal under diffracting conditions one consequence of the intense multiple Bragg reflection is that the net neutron current tends to be parallel to the Bragg planes. This effect can be demonstrated most dramatically by masking the entrance face of the crystal with a narrow absorbing slit, as illustrated schematically in Fig. 6.10. To obtain detailed agreement with experiment in situations like this, it is found (Shull and Oberteuffer, 1972; Shull and Shaw, 1973; Shull, 1973) that a spherical-wave theory must be used.

Another class of phenomena arises when one considers the effect of external forces such as gravity or an inhomogeneous magnetic field on the dynamical diffraction of neutrons by a perfect crystal. In the presence of such external forces the total potential energy is no longer a strictly periodic function inside the crystal and the coherent wave is no longer a simple Bloch wave. Werner (1980) has investigated this problem in detail and has shown that it can be treated using the same methods that were developed for discussing X-ray diffraction in strained crystals, i.e., the

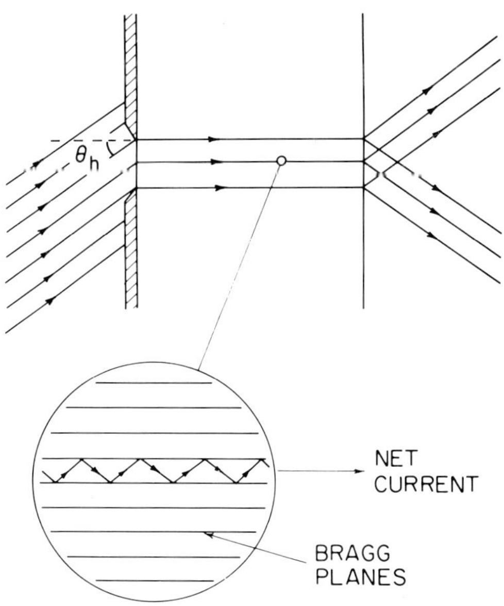

Fig. 6.10 Net neutron current in a perfect crystal under diffracting conditions where $\theta = \theta_h$ exactly.

eikonal method and the Takagi–Taupin method. The effect of external magnetic fields on Bragg reflection in perfect silicon crystals has been studied experimentally by Zeilinger and Shull (1979) and Zeilinger et al. (1986).

6.4. Diffraction by apertures

We now turn to a completely different type of dynamical diffraction phenomenon, namely, the diffraction of thermal neutrons by an opening, or aperture, in a plane opaque screen. In practice, such a screen is made from a material such as cadmium or gadolinium that has a very large absorption cross section for thermal neutrons, so that even though the screen may be relatively thin ($\simeq 1$ mm, say) it is still thick enough that essentially all the incident neutrons would be absorbed in the absence of the aperture.

We have previously discussed the diffraction of neutrons by transparent bodies and have shown that, for both crystalline and amorphous materials, the intrinsic shape of the diffraction peaks is characteristic of the three-dimensional shape of the body. For the diffraction of neutrons by an aperature in an opaque screen, we shall find that the diffraction pattern is characteristic of the two-dimensional shape of the aperture.

In general, the diffraction pattern depends not only on the diffracting element but also on the experimental conditions, i.e., on the position of the source and the detector relative to the diffracting element. If the source and the detector are both effectively at infinity, so that the incident wave at the diffracting element and the diffracted wave at the detector are both essentially plane waves, one obtains the case of *Fraunhofer diffraction*. Otherwise, it is called *Fresnel diffraction*. All the diffraction phenomena that we have considered earlier are of the Fraunhofer type. In the present discussion of neutron diffraction by apertures we shall consider both Fraunhofer and Fresnel diffraction.

In a rigorous theory of neutron diffraction by apertures one would have to solve the Schrödinger equation (6.0.1) for the coherent wave ψ at all points inside and outside the screen subject to the requirement that ψ and $\nabla\psi$ be continuous at the surface of the screen and that ψ have the asymptotic behavior at infinity appropriate to the given experimental conditions. In ordinary optics (Born and Wolf, 1975), such a rigorous theory of diffraction was first given by Sommerfeld in 1896 for the Fresnel diffraction of light by a straight edge.

In neutron optics, as also in most cases of practical importance in ordinary optics, the wavelength is many orders of magnitude smaller than the linear dimensions of the aperture so that one can, with negligible error, use the approximate theory of diffraction that was given by Fresnel in 1818. In developing this theory we shall employ the elegant Green-function method that was first used in this context by Kirchhoff in 1882.

Helmholtz–Kirchhoff theorem

In free space, where the optical potential vanishes, the Schrödinger equation (6.0.1) reduces to Helmholtz' equation,

$$(\Delta + k^2)\psi(\mathbf{r}) = 0 \tag{6.4.1}$$

where $E = (\hbar k)^2/2m$ as before. According to Eqs. 4.1.7 and 4.1.20, the corresponding Green function,

$$G(\mathbf{r}) = \frac{1}{r}\exp(ikr) \tag{6.4.2}$$

satisfies the equation

$$(\Delta + k^2)G(\mathbf{r}) = -4\pi\,\delta(\mathbf{r}) \tag{6.4.3}$$

It then follows from Gauss' theorem (Eq. 2.4.8) that

$$\begin{aligned}\int_S (G\,\nabla\psi - \psi\,\nabla G)\cdot\hat{\mathbf{n}}\,dS &= \int_V \nabla\cdot(G\,\nabla\psi - \psi\,\nabla G)\,d\mathbf{r}\\ &= \int_V (G\,\Delta\psi - \psi\,\Delta G)\,d\mathbf{r}\\ &= \int_V 4\pi\psi\,\delta(\mathbf{r})\,d\mathbf{r}\\ &= 4\pi\psi_P\end{aligned} \tag{6.4.4}$$

where the origin $\mathbf{r} = 0$ is taken to be at some point P inside the surface S that bounds the volume V. Thus, we obtain the *Helmholtz–Kirchhoff theorem,*

$$\psi_P = \frac{1}{4\pi}\int_S (G\,\nabla\psi - \psi\,\nabla G)\cdot\hat{\mathbf{n}}\,dS \tag{6.4.5}$$

This relation expresses ψ at any point P inside V in terms of the values of ψ and $\nabla\psi$ on the boundary S. It is an exact relation for any scalar field ψ that satisfies the Helmholtz equation throughout V.

Kirchhoff's formula

Let us now consider the diffraction of neutrons by an aperture in an infinite plane opaque screen, as illustrated schematically in Fig. 6.11. We want to calculate the coherent wave ψ_P at a point P on the opposite side of the screen from the point source P_0. Since the surface S encloses a volume of free space, ψ_P is given exactly by the Helmholtz–Kirchhoff theorem (Eq. 6.4.5).

We note that S consists of three parts: A, B, and C. Suppose the source is switched on at time t_0 and we consider ψ_P at some later time t after the wave front has passed through P but before it reaches C. Then ψ and $\nabla\psi$ vanish identically on C at time t. These quantities also vanish

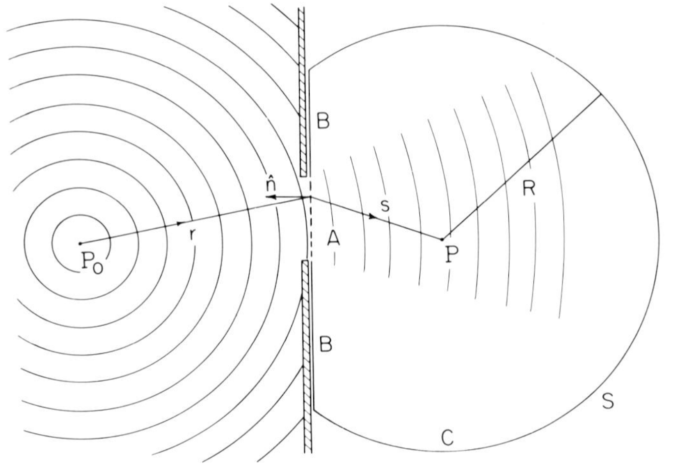

Fig. 6.11 Diffraction by an aperture in a plane opaque screen.

on B for all t, since the screen is assumed to be ideally opaque to thermal neutrons. In this case Eq. 6.4.5 reduces to

$$\psi_P = \frac{1}{4\pi} \int_A (G\,\nabla\psi - \psi\,\nabla G) \cdot \hat{\mathbf{n}}\, dS \tag{6.4.6}$$

where the integration is now restricted to the aperture A. To complete the argument, we first let $R \to \infty$ and then let $t_0 \to -\infty$. In this case ψ_P, as constructed earlier, can be assumed to be a stationary state as is required for the validity of the Helmholtz–Kirchhoff theorem, and Eq. 6.4.6 is then valid for all t.

It will be noted that the surface A is to some extent arbitrary. It might be taken to coincide with the plane of the aperture as in Fig. 6.11. Alternatively, it might be taken to coincide (as much as possible) with the spherical wave front of the incident wave from P_0. The only restriction on A is that it completely cover the aperture.

Since

$$G = \frac{1}{s} \exp(iks) \tag{6.4.7}$$

we now only need to know ψ and $\nabla\psi$ on A in order to determine ψ_P. In practice, the linear dimensions of the aperture are many orders of magnitude larger than the neutron wavelength. Thus, it is reasonable to assume that, except in the neighborhood of the edge of the aperture, ψ will be the same as in the absence of the screen, so that, apart from an arbitrary normalization factor,

$$\psi = \frac{1}{r}\exp(ikr) \qquad \text{on } A \tag{6.4.8}$$

This approximation is called the *Kirchhoff boundary condition.*

If $(\hat{\mathbf{n}}, \mathbf{r})$ denotes the angle between $\hat{\mathbf{n}}$ and $\mathbf{r}$ in Fig. 6.11, then

$$\begin{aligned}\hat{\mathbf{n}} \cdot \nabla\psi &= \frac{d\psi}{dr}\cos(\hat{\mathbf{n}}, \mathbf{r}) \\ &= \left(ik - \frac{1}{r}\right)\psi\cos(\hat{\mathbf{n}}, \mathbf{r}) \\ &\simeq ik\psi\cos(\hat{\mathbf{n}}, \mathbf{r})\end{aligned} \tag{6.4.9}$$

In the last line we have used the fact that $r \gg \lambda$, where $\lambda = 2\pi/k$ is the neutron wavelength. In the derivation of the Helmholtz–Kirchhoff theorem it was assumed that the origin $\mathbf{s} = 0$ is at P. However, in Fig. 6.11 the origin of $\mathbf{s}$ is on the aperture A. In this case,

$$\begin{aligned}\hat{\mathbf{n}} \cdot \nabla G &= \frac{dG}{ds}\cos(\hat{\mathbf{n}}, -\mathbf{s}) \\ &\simeq -ikG\cos(\hat{\mathbf{n}}, \mathbf{s})\end{aligned} \tag{6.4.10}$$

if $s \gg \lambda$. Hence, Eq. 6.4.6 becomes

$$\psi_P = \frac{i}{2\lambda}\int_A \frac{\exp[ik(r+s)]}{rs}[\cos(\hat{\mathbf{n}}, \mathbf{r}) + \cos(\hat{\mathbf{n}}, \mathbf{s})]\, dS \tag{6.4.11}$$

which is *Kirchhoff's formula.* Unlike the Helmholtz–Kirchhoff theorem (Eq. 6.4.5), this is not an exact result but depends on the validity of the Kirchhoff boundary conditon (Eq. 6.4.8). However, the error can be expected to be negligible as long as $\lambda \ll d$, r, and s, where d denotes the linear dimension of the aperture.

Fresnel's formula and Huygens' principle

If the surface A is taken to be the wave front at the aperture then r and, hence, ψ are constant on A. Also, $(\hat{\mathbf{n}}, \mathbf{r}) = \pi$ and $(\hat{\mathbf{n}}, \mathbf{s}) = \pi - (\mathbf{r}, \mathbf{s})$, so that Eq. 6.4.11 reduces to *Fresnel's formula,*

$$\psi_P = \psi\int_A GK\, dS \tag{6.4.12}$$

in which

$$K = -\frac{i}{2\lambda}[1 + \cos(\mathbf{r}, \mathbf{s})] \tag{6.4.13}$$

In 1678 Huygens derived the laws of reflection and refraction by assuming that light consists of waves, that waves have the property that each point on a wave front is the origin of a secondary spherical wavelet, and that the wavefront at any later instant is given by the envelope of all these wavelets. Later, in 1818, Fresnel showed that the diffraction of light could be explained simply by combining *Huygens' principle* with the *principle of superposition.* In fact, Fresnel's formula (Eq. 6.4.12) simply asserts that each point on the wave front A is the origin of a secondary spherical wavelet G (Huygens' principle) and that the resulting wave ψ_P at P is a coherent superposition of all these wavelets (principle of superposition). Fresnel simply postulated the expression (6.4.12) on the basis of physical intuition and was not able to derive the inclination factor K. However, this did not matter, because K is essentially constant in most experimental arrangements. The expression (6.4.13) for K was, in fact, first obtained by Stokes in 1849.

The present derivation of Fresnel's formula from the Helmholtz–Kirchhoff theorem was first given by Kirchhoff in 1882 and shows that Huygens' "secondary wavelet" is simply the Green function for the Helmholtz equation.

6.5. Fraunhofer and Fresnel diffraction

Experimental conditions

In actual diffraction experiments the conditions are such that a simplification of Kirchhoff's formula (Eq. 6.4.11) occurs. Let the surface A be the plane of the aperture and write

$$\mathbf{r} = \mathbf{a} + \mathbf{u}, \qquad \mathbf{s} = \mathbf{b} - \mathbf{u} \tag{6.5.1}$$

as illustrated in Fig. 6.12. In practice, a and b are many orders of magnitude larger than u. Thus, $r \simeq a$ and $s \simeq b$, so that we can put $rs = ab$ in the denominator of Eq. 6.4.11. In the exponent we put

$$k(r+s) = k(a+b) + \chi(\mathbf{u}) \tag{6.5.2}$$

so that $\chi(0) = 0$. The distinction between Fraunhofer and Fresnel diffraction depends on the nature of the phase shift $\chi(\mathbf{u})$, as we shall see subsequently. In most experimental arrangements the points P_0 and P in Fig. 6.12 are sufficiently near the axis of the aperture that we can put

$$(\hat{\mathbf{n}}, \mathbf{r}) = (\hat{\mathbf{n}}, \mathbf{s}) = \pi \tag{6.5.3}$$

Finally, $dS = d\mathbf{u}$ since $\mathbf{u}$ is a two-dimensional vector in the plane of the aperture.

With the preceding simplifications Kirchhoff's formula (6.4.11) reduces to

$$\psi = f\frac{\exp[ik(a+b)]}{ab} \tag{6.5.4}$$

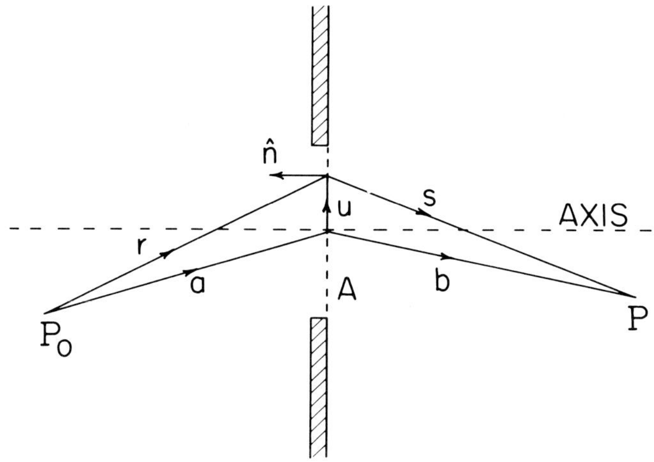

Fig. 6.12 Diffraction geometry.

where we now omit the subscript P, and f is the scattering amplitude,

$$f = -\frac{i}{\lambda}\int_A \exp[i\chi(\mathbf{u})]\,d\mathbf{u} \tag{6.5.5}$$

According to Eq. 6.5.4,

$$\psi = \psi_i f \frac{\exp(ikb)}{b} \tag{6.5.6}$$

where ψ_i is the incident wave at the center of the aperture,

$$\psi_i = \frac{1}{a}\exp(ika) \tag{6.5.7}$$

The corresponding neutron fluxes are given as in Sec. 2.4 by

$$J = |\psi|^2\left(\frac{\hbar k}{m}\right), \qquad J_i = |\psi_i|^2\left(\frac{\hbar k}{m}\right) \tag{6.5.8}$$

so that

$$J = J_i\left|\frac{f}{b}\right|^2 \tag{6.5.9}$$

This is formally the same as Eq. 2.4.18 and a similar argument then shows that the differential scattering cross section is given by

$$\frac{d\sigma}{d\Omega} = |f|^2 \tag{6.5.10}$$

Alternatively,

$$\psi = \psi_0 g \tag{6.5.11}$$

where ψ_0 is the coherent wave at P in the absence of the screen,

$$\psi_0 = \frac{\exp[ik(a+b)]}{a+b} \tag{6.5.12}$$

and

$$g = \left(\frac{a+b}{ab}\right)f \tag{6.5.13}$$

The corresponding flux ratio is then given by

$$\frac{J}{J_0} = \left|\frac{\psi}{\psi_0}\right|^2 = |g|^2 \tag{6.5.14}$$

We shall see that the differential scattering cross section (Eq. 6.5.10) is the natural quantity for describing Fraunhofer diffraction, while the flux ratio (Eq. 6.5.14) is most useful for Fresnel diffraction.

Fraunhofer diffraction

Let us write

$$\mathbf{a} = a\hat{\mathbf{a}}, \qquad \mathbf{b} = b\hat{\mathbf{b}} \tag{6.5.15}$$

where $\hat{\mathbf{a}}$ and $\hat{\mathbf{b}}$ are unit vectors that define the directions of $\mathbf{a}$ and $\mathbf{b}$ respectively. The term *Fraunhofer diffraction* refers to the limit $a \to \infty$ and $b \to \infty$ for fixed directions $\hat{\mathbf{a}}$ and $\hat{\mathbf{b}}$. Thus, for Fraunhofer diffraction, the source (P_0) and the detector (P) are both effectively at infinity. The incident wave at the aperture is then, in effect, a plane wave with wave vector $\mathbf{k} = k\hat{\mathbf{a}}$ and the diffracted wave at the detector is effectively a plane wave with wave vector $\mathbf{k}' = k\hat{\mathbf{b}}$.

According to Eq. 6.5.1,

$$r = a + \hat{\mathbf{a}} \cdot \mathbf{u} + O(1/a), \qquad s = b - \hat{\mathbf{b}} \cdot \mathbf{u} + O(1/b) \tag{6.5.16}$$

Thus, in the Fraunhofer limit,

$$r + s = a + b + (\hat{\mathbf{a}} - \hat{\mathbf{b}}) \cdot \mathbf{u} \tag{6.5.17}$$

so that from Eq. 6.5.2,

$$\chi(\mathbf{u}) = k(\hat{\mathbf{a}} - \hat{\mathbf{b}}) \cdot \mathbf{u} = \mathbf{q} \cdot \mathbf{u} \tag{6.5.18}$$

where $\mathbf{q} = \mathbf{k} - \mathbf{k}'$ as before (see Eq. 2.5.5). The scattering amplitude (Eq. 6.5.5) now becomes

$$f = -\frac{i}{\lambda} \int_A \exp(i\mathbf{q} \cdot \mathbf{u})\, d\mathbf{u} \tag{6.5.19}$$

Alternatively,

$$f = -\frac{i}{\lambda} \int S(\mathbf{u}) \exp(i\mathbf{q} \cdot \mathbf{u})\, d\mathbf{u} \tag{6.5.20}$$

where the integration is now unrestricted and $S(\mathbf{u})$ is the shape function,

$$S(\mathbf{u}) = \begin{cases} 1, & \text{inside } A \\ 0, & \text{outside } A \end{cases} \tag{6.5.21}$$

Thus, the Fraunhofer diffraction of neutrons by an aperture in a plane opaque screen is determined by the Fourier transform of the two-dimensional shape function of the aperture. In contrast, the Fraunhofer diffraction of neutrons by a transparent body is, as we saw in Chap. 5, determined in the kinematical theory by the Fourier transform of the three-dimensional shape function of the body.

For Fraunhofer diffraction, where a and b are infinite, we see from Eq. 6.5.13 that $g \rightarrow 0$ and, hence, that the flux ratio $J/J_0 \rightarrow 0$. Thus, the differential scattering cross section (Eq. 6.5.10) is the natural quantity for describing Fraunhofer diffraction.

Let us choose a system of Cartesian coordinate axes such that the plane of the aperture is the yz plane, as illustrated in Fig. 6.13. Suppose that $\mathbf{k}$ is in the x direction,

$$\mathbf{k} = (k, 0, 0) \tag{6.5.22}$$

Then, to first order in α and β,

$$\mathbf{k}' = (k, k\alpha, k\beta) \tag{6.5.23}$$

and

$$\mathbf{q} = (0, -k\alpha, -k\beta) \tag{6.5.24}$$

Thus, $\mathbf{q}$ is effectively a two-dimensional vector in the plane of the aperture and

$$d\mathbf{q} = k^2\, d\alpha\, d\beta = k^2\, d\Omega \tag{6.5.25}$$

Hence, it follows from (Eq. 6.5.10) that the total scattering cross section

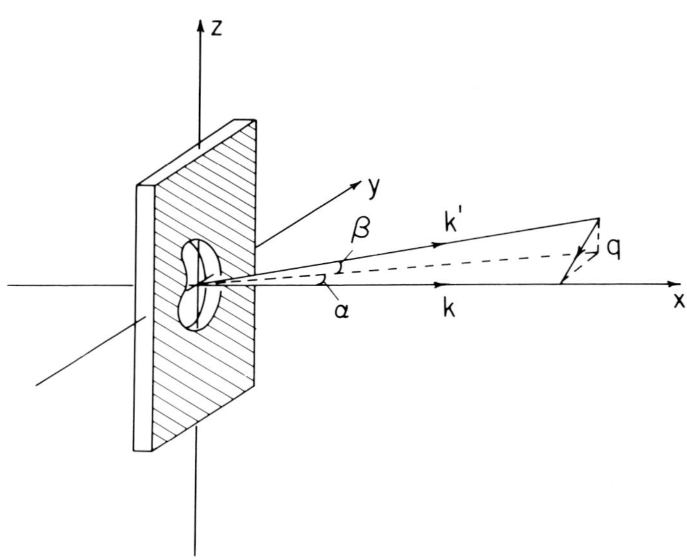

Fig. 6.13 Fraunhofer diffraction by an aperture.

is given by

$$\begin{aligned}\sigma &= \int d\sigma \\ &= \int |f|^2 \, d\Omega \\ &= \frac{1}{k^2} \int |f|^2 \, d\mathbf{q}\end{aligned} \tag{6.5.26}$$

The expression (6.5.20) for the scattering amplitude then gives

$$\sigma = \int S(\mathbf{u})^2 \, d\mathbf{u} \tag{6.5.27}$$

and, since $S(\mathbf{u})^2 = S(\mathbf{u})$, we get

$$\sigma = \int S(\mathbf{u}) \, d\mathbf{u} \equiv A \tag{6.5.28}$$

Thus, the total scattering cross section is equal to the area of the aperture, A. This is, of course, what one would expect classically. The preceding derivation shows that this is also true quantum-mechanically as long as the neutron wavelength is much less than the linear dimensions of the aperture.

Rectangular aperture

Since $\mathbf{u} = (0, y, z)$ the scattering amplitude (Eq. 6.5.19) becomes

$$f = -\frac{i}{\lambda} \iint_A \exp[-ik(\alpha y + \beta z)] \, dy \, dz \tag{6.5.29}$$

For a rectangular aperture with a width w in the y direction and a height h in the z direction, we then get

$$f = \frac{-i\lambda}{\pi^2 \alpha\beta} \sin\left(\frac{\pi\alpha}{\alpha_0}\right) \sin\left(\frac{\pi\beta}{\beta_0}\right) \tag{6.5.30}$$

where

$$\alpha_0 = \frac{\lambda}{w}, \qquad \beta_0 = \frac{\lambda}{h} \tag{6.5.31}$$

The differential scattering cross section (Eq. 6.5.10) can then be expressed as

$$\frac{d\sigma}{d\Omega} = A \, \Delta(\alpha) \, \Delta(\beta) \tag{6.5.32}$$

where $A = wh$ is the area of the aperture,

$$\Delta(\alpha) = \frac{1}{\alpha_0} \left[\frac{\sin(\pi\alpha/\alpha_0)}{\pi\alpha/\alpha_0}\right]^2 \tag{6.5.33}$$

and $\Delta(\beta)$ is given by a similar expression. Since

$$\int \Delta(\alpha)\, d\alpha = 1 \tag{6.5.34}$$

it follows that the total scattering cross section $\sigma = A$, as it should according to Eq. 6.5.28.

The quantities α_0 and β_0 are approximately equal to the full widths at half maximum of the diffraction peak in the y and z directions respectively. In the limit of a *slit*, where $w \ll h$ and $\alpha_0 \gg \beta_0$, the quantity $\Delta(\beta)$ is effectively a delta function and

$$\frac{d\sigma}{d\Omega} = A\, \Delta(\alpha)\, \delta(\beta) \tag{6.5.35}$$

In this case the diffraction is essentially confined to the xy plane.

If one were to discuss the diffraction of neutrons by an aperture in classical language, one would say that the neutrons that are incident on the screen are absorbed while those that are incident on the aperture are diffracted. In other words, the neutrons that are diffracted are the ones that do not collide with the atoms in the screen. This paradoxical conclusion simply indicates the inadequacy of classical language in describing what is, in fact, a purely quantum-mechanical phenomenon.

The diffraction of neutrons by an aperture in an opaque screen is a manifestation of the *Heisenberg principle.* Thus, for example, a neutron that is observed to have passed through a slit has an uncertainty $\Delta y \simeq w$ in its position and a corresponding uncertainty $\Delta p_y \simeq \hbar k \alpha_0$ in its momentum, so that $\Delta y\, \Delta p_y \simeq \hbar$. Thus, measurements of the Fraunhofer diffraction of thermal neutrons by a slit provide direct experimental confirmation of the Heisenberg principle.

Such measurements have been performed by Shull (1969) with 4.43 Å neutrons. The results obtained when slits of various widths were introduced between the two perfect silicon crystals in a double-crystal spectrometer are shown in Fig. 6.14. The upper curve was obtained with a large width $w = 770\ \mu$m and represents the instrumental resolution function of the spectrometer. The lower curves, corresponding to $w = 5.6$ and $4.1\ \mu$m, show that broadening produced by Fraunhofer diffraction. Data were also obtained for $w = 21.0\ \mu$m but are not shown in the figure. The measured intrinsic widths of the distributions are all in excellent agreement with the theoretical expression (Eq. 6.5.31).

The Fraunhofer diffraction by ruled gratings has also been observed using thermal neutrons (Kurz and Rauch, 1969), cold neutrons (Graf et al., 1979), and ultracold neutrons (Scheckenhofer and Steyerl, 1977).

Fresnel diffraction

Fresnel diffraction occurs in its simplest form when the points P_0 and P in Fig. 6.12 are both on the axis of the aperture. Then **a** and **b** are both

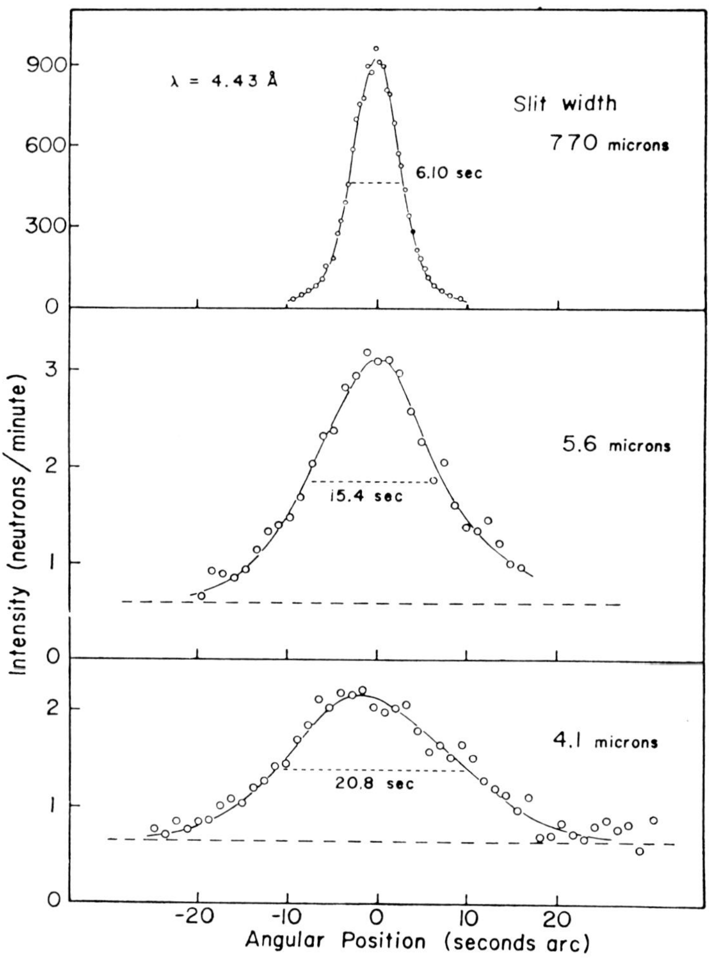

Fig. 6.14 Fraunhofer diffraction of 4.43 Å neutrons measured by Shull (1969) for three different slit widths.

orthogonal to **u** and Eq. 6.5.1 gives

$$r = a + \frac{u^2}{2a}, \qquad s = b + \frac{u^2}{2b} \tag{6.5.36}$$

to lowest order in u. Hence, from Eq. 6.5.2,

$$\chi(\mathbf{u}) = \frac{k}{2}\left(\frac{1}{a} + \frac{1}{b}\right)u^2 \tag{6.5.37}$$

The linear term (Eq. 6.5.18) is absent because $\hat{\mathbf{a}} = \hat{\mathbf{b}}$ and, hence, $\mathbf{q} = 0$ in the present case. Alternatively,

$$\chi(\mathbf{u}) = \pi\left(\frac{u}{l}\right)^2 \tag{6.5.38}$$

where

$$l = \left(\frac{\lambda ab}{a+b}\right)^{1/2} \tag{6.5.39}$$

Hence, it follows from Eq. 6.5.5 and 6.5.13 that

$$g = -\frac{i}{l^2}\int_A \exp\left[i\pi\left(\frac{u}{l}\right)^2\right] d\mathbf{u} \tag{6.5.40}$$

The ratio of the neutron flux at the point P in the presence of the screen to its value in the absence of the screen is given, according to Eq. 6.5.14, by $|g|^2$. We see that this quantity does not depend on λ, a, and b separately, but only on the characteristic length l, which we shall call the *Fresnel length*. In ordinary optics, where $\lambda \simeq 5000$ Å and $a \simeq b \simeq 50$ cm, say, we get $l \simeq 350\ \mu$m. In a typical neutron optics experiment, on the other hand, $\lambda \simeq 20$ Å and $a \simeq b \simeq 5.0$ m, so that $l \simeq 70\ \mu$m. Thus, in practice, the Fresnel length is only a factor of 5 or so smaller for neutrons than it is for light.

Fraunhofer diffraction is characterized by the wave vector transfer $\mathbf{q}$, and Fresnel diffraction by the Fresnel length l. The condition for Fraunhofer diffraction is that

$$qR \gg \pi\left(\frac{R}{l}\right)^2 \tag{6.5.41}$$

where R is the radius of the aperture. In general, $q \simeq k\alpha$ and, for Fraunhofer diffraction, the angle of scattering $\alpha \simeq \lambda/2R$. Thus, $qR \simeq \pi$ and the preceding condition becomes $R \ll l$. Conversely, Fresnel diffraction dominates when $R \gg l$.

Circular aperture

For a circular aperture of radius R, Eq. 6.5.40 gives

$$g = 1 - \exp(i\pi N) \tag{6.5.42}$$

where

$$N = \left(\frac{R}{l}\right)^2 \tag{6.5.43}$$

Hence, the flux ratio (Eq. 6.5.14) becomes

$$\frac{J}{J_0} = 4\sin^2\left(\frac{\pi}{2}N\right) \tag{6.5.44}$$

The significance of this result can be understood with the help of the *Fresnel zone* construction shown in Fig. 6.15a. It is evident that all the zones have the same area πl^2 and, since $R = \sqrt{N}\, l$, the quantity N is the number of zones inside the aperture. If N is an integer then

$$\frac{J}{J_0} = \begin{cases} 4, & N = 1, 3, 5, \ldots \\ 0, & N = 2, 4, 6, \ldots \end{cases} \tag{6.5.45}$$

This is a consequence of the fact, which can be verified from Eq. 6.5.40, that the waves from adjacent zones are 180 degrees out of phase and, hence, cancel. Thus, if N is even, J/J_0 vanishes while, if N is odd, J/J_0 equals the contribution from a single zone.

Zone plates

If we mask off zones $2, 4, 6, \ldots$ with an opaque layer, as illustrated in Fig. 6.15b then, since the waves from zones $1, 3, 5, \ldots$ are all in phase,

$$\frac{J}{J_0} = 4\left(\frac{N}{2}\right)^2 = N^2 \tag{6.5.46}$$

Better still, if we coat zones $2, 4, 6, \ldots$ with a transparent layer that changes the phase of the coherent wave by π, then the waves from all the zones will be in phase and

$$\frac{J}{J_0} = 4N^2 \tag{6.5.47}$$

In either case, if $N \gg 1$ then $J \gg J_0$ and the zone plate acts as a lens.

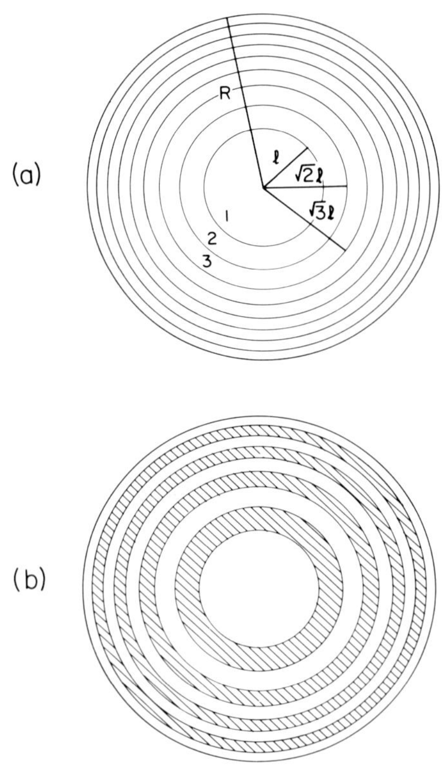

Fig. 6.15 (a) Fresnel zones for a circular aperture of radius R; (b) zone plate.

According to Eq. 6.5.39,

$$\frac{1}{a}+\frac{1}{b}=\frac{\lambda}{l^2}=\frac{1}{f} \tag{6.5.48}$$

where f is now, by definition, the focal length of the zone plate (i.e., $b=f$ when $a=\infty$). This is the same as the relation (Eq. 3.1.42) that describes the focal properties of thin lenses and spherical mirrors in Gaussian optics. The corresponding focal lengths are given by Eqs. 3.1.39 and 3.1.43 so that, in summary,

$$f=\begin{cases} r/2\,, & \text{spherical mirror} \\ \dfrac{\pi r}{\lambda^2 \rho b}\,, & \text{thin lens} \\ \dfrac{R^2}{N\lambda}\,, & \text{zone plate} \end{cases} \tag{6.5.49}$$

The physical phenomena underlying the focal properties of these devices are reflection, refraction, and diffraction, respectively.

The zone plates used in neutron optics are always of the phase-reversal type and are produced by using photolithographic techniques to coat alternate zones with a uniform layer of material such as copper that changes the phase of the coherent wave by π. The required layer thickness d is given, according to Eq. 3.2.6, by

$$d=\frac{\lambda}{2(n-1)}=\frac{\lambda}{\xi} \tag{6.5.50}$$

Thus, for example, if $\lambda = 20$ Å one finds that $\xi = 8.3\times10^{-4}$ for copper and, hence, that $d = 2.4\ \mu$m. For light, on the other hand, $n \simeq 1.5$ and $d \simeq \lambda \simeq 0.5\ \mu$m which is about a factor of 5 smaller than for neutrons.

Figure 6.16 is a schematic illustration of the experiment of Kearney et

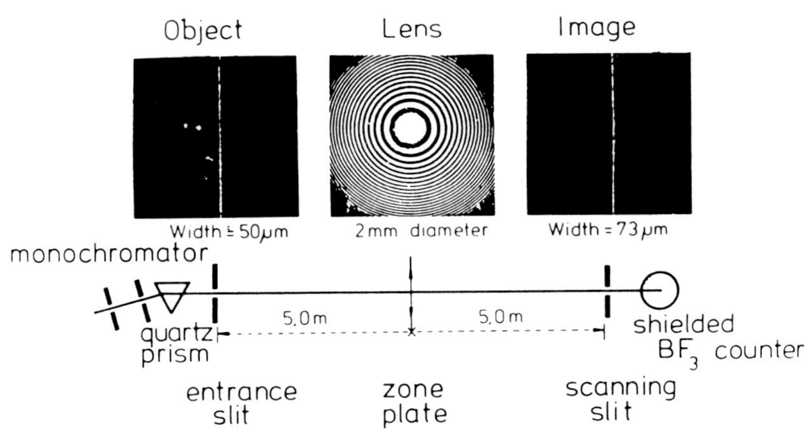

Fig. 6.16 Focusing of 20 Å neutrons by a Fresnel zone plate (Kearney et al. 1980).

al. (1980) that provided the first demonstration of the use of a zone plate for the focusing of a neutron beam. Since a zone plate is nonachromatic (i.e., f depends on λ), monoenergetic neutrons must be used to obtain a sharp image of the entrance slit at the position of the scanning slit. Here, a quartz prism is used to produce a beam of 20 Å neutrons. The zone plate has a radius $R = 1.0$ mm and the total number of zones $N = 200$. Thus, according to Eq. 6.5.47, the neutron flux at the detector is ideally a factor $4N^2 = 1.6 \times 10^5$ larger than it would have been in the absence of the zone plate. The radius of the first zone is equal to the Fresnel length $l = R/\sqrt{N} = 70.7$ μm and the width of the last zone is $l/2\sqrt{N} = 2.5$ μm, which is the limit of resolution of the plating technique. The focal length of the zone plate is $f = R^2/N\lambda = 2.5$ m and the f-number is $f/2R = 1250$ or, as a photographer would say, the aperture ($2R$) is $f/1250$. Of course, this is very poor in comparison with lenses used with visible light but is comparable with those used in electron optics.

In later work the same authors (Klein et al., 1981) used a cylindrical zone plate to focus 20 Å neutrons. A concave Fresnel zone mirror has also been used by Schütz et al. (1980) to focus ultracold neutrons. Finally, we might mention the work of Beynon and Pink (1980), who used a zone plate, not for imaging, but as a shadowing device in a demonstration of *neutron holography*. More information on the use of zone plates in neutron optics can be found in the review article by Klein and Zeilinger (1984).

Straight edge

Let us now consider the Fresnel diffraction of neutrons by a straight edge that is orthogonal to the line P_0P and at a distance d below it, as illustrated in Fig. 6.17. Then Eq. 6.5.40 gives

$$g = -iF(\infty)[F(w) + F(\infty)] \tag{6.5.51}$$

where $w = \sqrt{2}\, d/l$ and $F(w)$ is the *Fresnel integral*,

$$F(w) = C(w) + iS(w) = \int_0^w \exp\left(i\frac{\pi}{2}t^2\right) dt \tag{6.5.52}$$

Since $F(\infty) = (1+i)/2$ the flux ratio (Eq. 6.5.14) becomes

$$\frac{J}{J_0} = \frac{1}{2}\left\{\left[C(w) + \frac{1}{2}\right]^2 + \left[S(w) + \frac{1}{2}\right]^2\right\} \tag{6.5.53}$$

In particular,

$$\frac{J}{J_0} = \begin{cases} 1, & w = \infty \\ 1/4, & w = 0 \\ 0, & w = -\infty \end{cases} \tag{6.5.54}$$

where we have used the fact that $F(0) = 0$ and $F(-w) = -F(w)$.

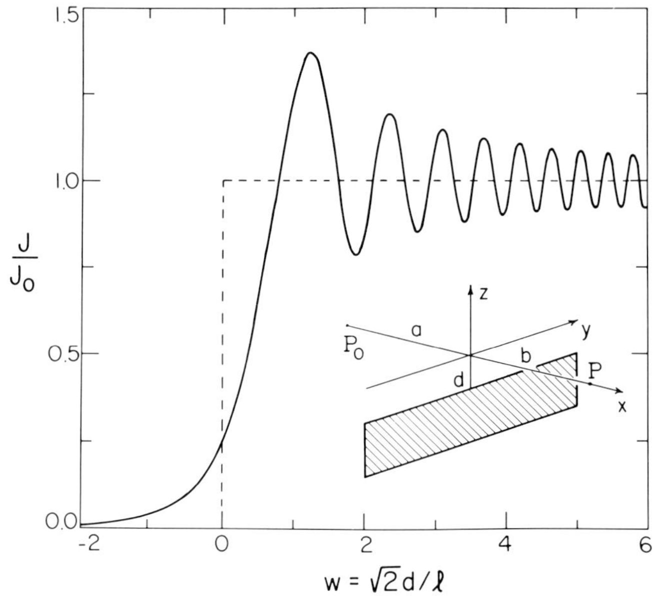

Fig. 6.17 Fresnel diffraction by a straight edge calculated from Eq. 6.5.53. The dashed line shows the geometrical optics result (Eq. 6.5.55).

In 1874 Cornu gave an ingeneous graphical method of calculating Eq. 6.5.53 based on the well-known *Cornu spiral* (Born and Wolf, 1975). However, this method is now only of historical interest and for modern computational purposes the approximate analytical expressions for $C(w)$ and $S(w)$ given by Abramowitz and Stegun (1965) are far more useful (Klein and Martin, 1977). Alternatively, one can use the values of the Fresnel integrals given in tables such as those of Jahnke and Emde (1945).

Figure 6.17 shows J/J_0 as a function of w as calculated from Eq. 6.5.53. The first maximum occurs at $w = 1.22$, where $d = 0.86l$, and has the value $J/J_0 = 1.37$. The dashed line represents the corresponding result for geometrical optics (i.e., classical mechanics),

$$\frac{J}{J_0} = \begin{cases} 1, & w > 0 \\ 0, & w < 0 \end{cases} \tag{6.5.55}$$

Two features are especially worth noting. First, the coherent wave penetrates well into the region of the *geometrical shadow* ($w < 0$) and, second, the oscillations at $w > 0$ are only very slowly damped.

The Fresnel diffraction of neutrons by an opaque straight edge was first measured by Landkammer (1966) and later by Klein et al. (1977) and Gähler et al. (1981). The results of the last-named authors for 20 Å neutrons are shown in Fig. 6.18. The oscillations are somewhat damped

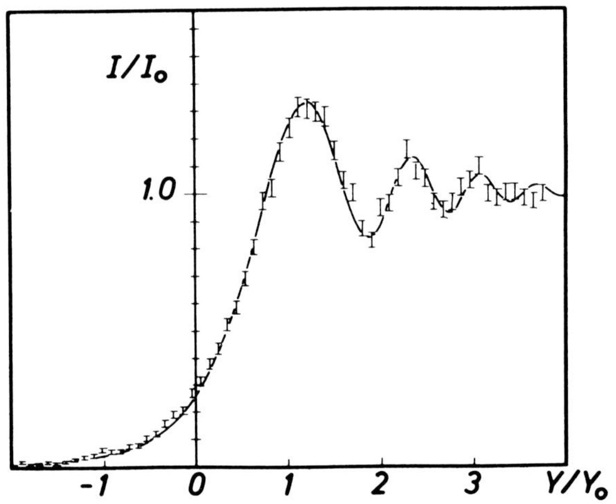

Fig. 6.18 Fresnel diffraction of 20 Å neutrons by a straight edge measured by Gähler et al. (1981).

by the effect of the finite instrumental resolution and counting statistics, but the first three maxima are clearly resolved. These are very impressive results when one considers that the first maximum occurs at $d = 61$ μm while $a = b = 5$ m (see Fig. 6.17). These authors also measured the Fresnel diffraction by a wire. Finally, Klein and Opat (1975, 1976) have demonstrated the properties of spinor rotation by measuring the Fresnel diffraction of neutrons by a ferromagnetic domain boundary.

7

NEUTRON INTERFEROMETRY

Although the absolute phase of a wave function ψ is unobservable, the relative phase of the two components in a superposition $\psi_1 + \psi_2$ is, in fact, observable since measurable quantities (expectation values, transition probabilities, etc.) all depend on the relative phase. Neutron interferometry provides a method of constructing superpositions of plane-wave states with a controllable relative phase. This enables one to perform experiments involving the interference between waves that provide information that is often difficult or impossible to obtain in other ways. In such experiments the incident wave is divided into two coherent parts that are made to travel different paths and are then superimposed prior to observation. The observed interference fringes provide a direct measurement of the relative phase, or phase shift, between the two components, as we shall show.

In neutron interferometry, as in ordinary interferometry with visible light, the phase shift arises from a difference in the optical path lengths of the two components. This, in turn, may be due either to a difference in the actual path lengths or to a difference in the indices of refraction of the media through which the beams travel. In the latter case the observed phase shift provides a direct measurement of the real part of the relative index of refraction. This is in contrast to transmission measurements, which, as we saw in Sec. 3.2, provide a direct measurement of the imaginary part of the index of refraction. In neutron interferometry the phase shifts produced by magnetic and gravitational fields are also observable in a number of experiments of fundamental importance in physics.

As in ordinary optics, it is useful to distinguish between two-beam interferometry, as discussed above, and multiple-beam interference. For neutrons the latter occurs in various thin-film devices that will be described in Sec. 7.5.

7.1. Two-beam interferometers

Types of interferometer

There are two general methods of dividing a wave into coherent parts and this leads to two general types of interferometer. The first method is by *division of wave front,* the classic example of this being the Young double-slit interferometer. The second method is by *division of amplitude* as in, for example, the Michelson interferometer.

The first neutron interferometer was constructed by Maier–Leibnitz and Springer (1962). In this instrument the spatial separation of the incident wave was achieved by the division of wave front technique using an arrangement similar to a Fresnel biprism (Born and Wolf, 1975). Since the angle of deviation produced by a prism is very small for thermal neutrons, typically of the order of a few seconds of arc as we saw in Sec. 3.1, it was only possible to obtain a useful beam separation of 60 μm for a flight path of 10 m. The usefulness of this instrument was further impared by its poor coherence properties. Only the zeroth- and first-order interference fringes were observed and the fringe contrast was greatly reduced when phase-shifting material was inserted into the beams.

Subsequent neutron interferometers based on the division of wave front technique include an instrument similar to a Billet split lens, in which a cylindrical Fresnel zone plate is used instead of a refractive lens (Klein et al., 1981a), and a Young double-slit interferometer (Klein et al., 1981c). The latter instrument was used in a demonstration of the Fizeau effect for neutrons, which will be discussed in Sec. 7.3.

In ordinary optics, interferometers based on the division of amplitude technique usually use a partially reflecting mirror to divide the incident wave into two coherent parts, the reflected and transmitted components. This method could, in principle, also be used for neutrons. The problem is that for thermal neutrons the reflectivity of a mirror is appreciable only for glancing incidence (see Sec. 3.1), so that such an interferometer would have the same drawbacks as those based on the division of wavefront, namely, very small beam separations and very long flight paths.

The wide beam separations that are necessary to provide easy access for material phase-shifting devices or magnetic fields can be obtained by means of Bragg reflection by crystals. However, this method requires the use of large, perfect crystals. Mosaic crystals of the type used as neutron monochromators would be quite unsuitable because the mosaic structure would destroy the spatial coherence of the transmitted and Bragg-reflected beams and, hence, wash out the interference fringes.

In the 1960s, when large, essentially perfect silicon crystals became available at moderate cost from the semiconductor industry, an interferometer based on division of amplitude by Bragg reflection was developed for X-rays by Bonse and Hart (1965) and later for neutrons by Rauch et al. (1974). The development of this perfect-crystal neutron interferometer more than anything else has led to the great resurgence of activity in neutron optics in the last 10 years. This instrument has proved to be extremely useful, not only for the accurate measurement of neutron scattering lengths, but also for fundamental studies in quantum mechanics, gravitation, and other fields.

The perfect-crystal neutron interferometer will be described in Sec. 7.2 and its applications in Sec. 7.3. In the remainder of the present section we shall give a brief formal discussion of neutron interferometry in general.

Model interferometer

Figure 7.1a is a schematic representation of a two-beam interferometer that operates on the division of amplitude principle. Here an incident beam O is divided at P into two coherent parts 1 and 2 that travel different paths and subsequently come together at the point P' where they are superimposed to produce two exit beams labeled + and −.

The wave function of the incident beam at P can be expressed as

$$\psi_0 = A_0 \exp(iS_0/\hbar) \tag{7.1.1}$$

where A_0 and S_0 are real quantities that characterize the amplitude and phase of the wave function. The wave functions of the exit beams are assumed to be of the form

$$\psi_\pm = \psi_1 \pm \psi_2 \tag{7.1.2}$$

in which the contributions from paths 1 and 2 are

$$\psi_1 = A_1 \exp(iS_1/\hbar), \qquad \psi_2 = A_2 \exp(iS_2/\hbar) \tag{7.1.3}$$

For simplicity, we shall also assume that the model interferometer is such that

$$A_1 = A_2 = \tfrac{1}{2}A_0 \tag{7.1.4}$$

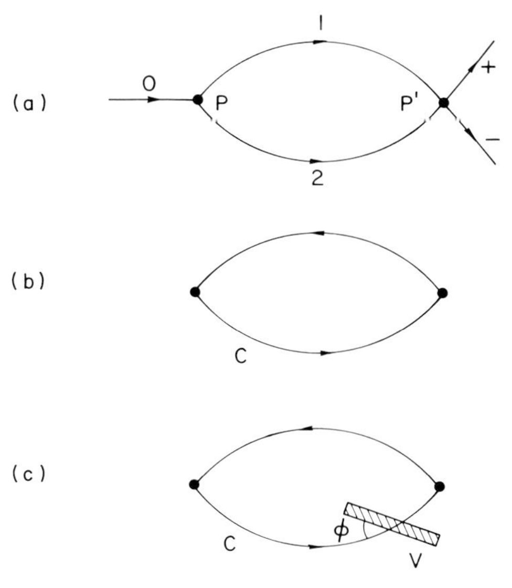

Fig. 7.1 Model interferometer.

Then

$$\psi_{\pm} = \frac{A_0}{2}[\exp(iS_1/\hbar) \pm \exp(iS_2/\hbar)] \tag{7.1.5}$$

Let

$$I_{\pm} = \left|\frac{\psi_{\pm}}{\psi_0}\right|^2 \tag{7.1.6}$$

so that $I_{\pm}$ is the fraction of the incident-neutron intensity that appears in the exit beams $\pm$. Then,

$$\begin{aligned} I_{\pm} &= \tfrac{1}{4}\,|1 \pm \exp(i\chi)|^2 \\ &= \tfrac{1}{2}(1 \pm \cos\chi) \end{aligned} \tag{7.1.7}$$

where χ is the relative phase, or *phase shift,* of the two components,

$$\chi = \frac{S_2 - S_1}{\hbar} \tag{7.1.8}$$

The sinusoidal variation of $I_{\pm}$ with χ represents the interference fringes and $\chi/2\pi$ is called the *fringe order number.* It will be noted that

$$I_+ + I_- = 1 \tag{7.1.9}$$

as required for neutron conservation. Also, the interference fringes vanish in the sum (Eq. 7.1.9) but are enhanced in the difference,

$$I_+ - I_- = \cos\chi \tag{7.1.10}$$

In practice the intensity is usually expressed in the form

$$I = \bar{I}[1 + C\cos(\chi + \gamma)] \tag{7.1.11}$$

where $\bar{I}$ is the average intensity, γ is the *intrinsic phase shift,* and C is the *fringe contrast,*.

$$C = \frac{I_{\max} - I_{\min}}{I_{\max} + I_{\min}} \tag{7.1.12}$$

so that $0 \le C \le 1$. For the model interferometer discussed here

$$\begin{aligned} \bar{I}_{\pm} &= \tfrac{1}{2}, \qquad \gamma_+ = 0 \\ C_{\pm} &= 1, \qquad \gamma_- = \pi \end{aligned} \tag{7.1.13}$$

Phase shift

The phases of the two components at the point P' in Fig. 7.1a are given by

$$\begin{aligned} S_1 &= S_0 + \int_1 \nabla S \cdot \mathbf{dl} \\ S_2 &= S_0 + \int_2 \nabla S \cdot \mathbf{dl} \end{aligned} \tag{7.1.14}$$

where **dl** is a vector element of displacement along paths 1 and 2, and S is the eikonal that was discussed in Sec. 3.3. The phase shift (Eq. 7.1.8) is then given by

$$\chi = \frac{1}{\hbar}\int_2 - \int_1 \nabla S \cdot \mathbf{dl} \tag{7.1.15}$$

or, alternatively,

$$\chi = \frac{1}{\hbar}\int_C \nabla S \cdot \mathbf{dl} \tag{7.1.16}$$

where C denotes the closed contour shown in Fig. 7.1b.

At any point on C the local neutron momentum is given, according to Eq. 3.3.8 by $\hbar\mathbf{K} = \nabla S$. Also, $\mathbf{K} = K\hat{\mathbf{K}}$, where $\hat{\mathbf{K}}$ denotes a unit vector in the direction of $\mathbf{K}$, and $K = nk$, where n is the local index of refraction and k is wave vector *in vacuo*. Hence, the phase shift becomes

$$\chi = kL \tag{7.1.17}$$

where

$$L = \int_C n\hat{\mathbf{K}} \cdot \mathbf{dl} \tag{7.1.18}$$

Since $\hat{\mathbf{K}}$ is parallel to **dl** on path 2 and antiparallel to **dl** on path 1, it is evident that L is the *optical path difference*.

Let us write

$$L = L' + L'' \tag{7.1.19}$$

where

$$L' = \int_C \hat{\mathbf{K}} \cdot \mathbf{dl}, \qquad L'' = \int_C (n-1)\hat{\mathbf{K}} \cdot \mathbf{dl} \tag{7.1.20}$$

Then L' is the actual difference in length between paths 1 and 2, and L'' is the additional contribution from the index of refraction. In some interferometers, such as the conventional Michelson interferometer, the interference fringes are produced by varying L'. However, in the perfect-crystal neutron interferometer, which we are primarily interested in here, L' is constant and the interference fringes are produced by varying L''. In this case the constant component of the phase shift, kL', can be included in the intrinsic phase shift γ in Eq. 7.1.11, whereupon

$$\chi = kL'' = k\int_C (n-1)\hat{\mathbf{K}} \cdot \mathbf{dl} \tag{7.1.21}$$

Suppose the phase shift arises from the presence of a homogeneous material medium in the lower part of C as illustrated in Fig. 7.1c. Then

$$n = \begin{cases} \text{const}, & \text{inside } V \\ 1, & \text{outside } V \end{cases} \tag{7.1.22}$$

and

$$\chi = (n-1)kD \tag{7.1.23}$$

where

$$D = \int \hat{\mathbf{K}} \cdot \mathbf{dl} \tag{7.1.24}$$

and the integration is now restricted to the segment of C inside V. In other words, D is the path length inside V. Thus, for example, if the medium is a plane slab of thickness d, then

$$D = d \operatorname{cosec} \phi \tag{7.1.25}$$

where ϕ is the angle that the path makes with the surface of the slab. The preceding expressions for χ and D agree with our earlier results, Eqs. 3.2.6 and 3.2.7.

Lambda thickness

Since $k = 2\pi/\lambda$, where λ is the incident-neutron wavelength, the phase shift (Eq. 7.1.23) can also be expressed as

$$\chi = -2\pi\left(\frac{D}{D_\lambda}\right) \tag{7.1.26}$$

where D_λ is called the *λ-thickness* of the material and is given by

$$D_\lambda = \frac{\lambda}{1-n} \tag{7.1.27}$$

According to Eq. 3.1.4, $n^2 = 1 - \xi$, so that to first order

$$D_\lambda = \frac{2\lambda}{\xi} \tag{7.1.28}$$

The value of ξ is typically 2×10^{-6} when $\lambda = 2$ Å so that $D_\lambda \simeq 200\ \mu\text{m} = 0.2$ mm.

Interference fringes are produced by measuring I_+ or I_- as a function of ϕ. The separation of neighboring fringes corresponds to a phase change $\delta\chi = 2\pi$ or

$$\delta D = -(D \cot \phi)\, \delta\phi = -D_\lambda \tag{7.1.29}$$

so that

$$\delta\phi = \frac{D_\lambda}{D} \tan \phi \tag{7.1.30}$$

Thus, for example, with $\phi = 45$ degrees, $d = 5.0$ mm, and $D_\lambda = 0.2$ mm we get $\delta\phi = 0.028$ radians $= 1.6$ degrees. This is then the angle through which the slab must be rotated to produce a single interference fringe. Conversely, a measurement of the fringe separation $\delta\phi$ can be used to determine the λ-thickness D_λ and, hence, the index of refraction n.

The accurate determination of D_λ from neutron interferometry measurements requires that the faces of the sample be accurately parallel, so that the thickness d is constant. In particular, any variation δd across the width of the neutron beam must be such that $\delta d \ll D_\lambda$ in order that the corresponding variation $\delta\chi \ll 2\pi$. For example, if $\delta d = 1\ \mu$m, $D_\lambda = 200\ \mu$m, and $\phi = 45$ degrees, then $\delta\chi = 0.044$ radians $= 2.5$ degrees.

Optical smoothness

Random variations in the sample thickness d due to surface roughness will tend to reduce the transverse coherence of the wave function and, hence, the fringe contrast. Again, this effect will be negligible as long as the variation $\delta d \ll D_\lambda$, so that $\delta\chi \ll 2\pi$. In this case the surface can be said to be *optically smooth.*

There is a myth that a surface must be polished smooth to within a small fraction of a wavelength (i.e., $\delta d \ll \lambda$) in order to be optically smooth. It is easy to see how this myth arose, because for the refraction of visible light by glass $n \simeq 1.5$ when $\lambda = 5000$ Å $= 0.5\ \mu$m and, hence, $D_\lambda \simeq 2\lambda = 1\ \mu$m. In this case the criterion for optical smoothness is indeed $\delta d \ll \lambda$. However, this is not true in general and, in particular, it is not true for X-rays or neutrons.

It will be noted that if $\delta d \simeq 1\ \mu$m then the surface is not optically smooth for light but it is for thermal neutrons. Thus, the requirements for surface smoothness are, in fact, less severe for neutrons than for light in spite of the fact that the neutron wavelength is three orders of magnitude smaller.

External fields

In neutron interferometry, gravitational and magnetic fields can also give significant contributions to the phase shift χ. In general, if v denotes the total potential energy of the neutron, including the contribution from external fields (see Eq. 3.3.1), then it follows from Eq. 3.3.21 that

$$n = 1 - \frac{v}{2E} \tag{7.1.31}$$

where $E = (\hbar k)^2/2m$ is the incident-neutron energy. Since v and, hence, n may vary continuously along the contour C, the phase shift must be calculated from Eq. 7.1.21 and we then find that

$$\chi = -\frac{m}{\hbar^2 k}\int_C v\hat{\mathbf{K}} \cdot \mathbf{dl} \tag{7.1.32}$$

Examples of the effect of external fields in neutron interferometry will be discussed in Sec. 7.3.

7.2. Perfect-crystal neutron interferometer

In its conventional form (Fig. 7.2) a perfect-crystal neutron interferometer consists of three parallel equally spaced slabs that are produced by

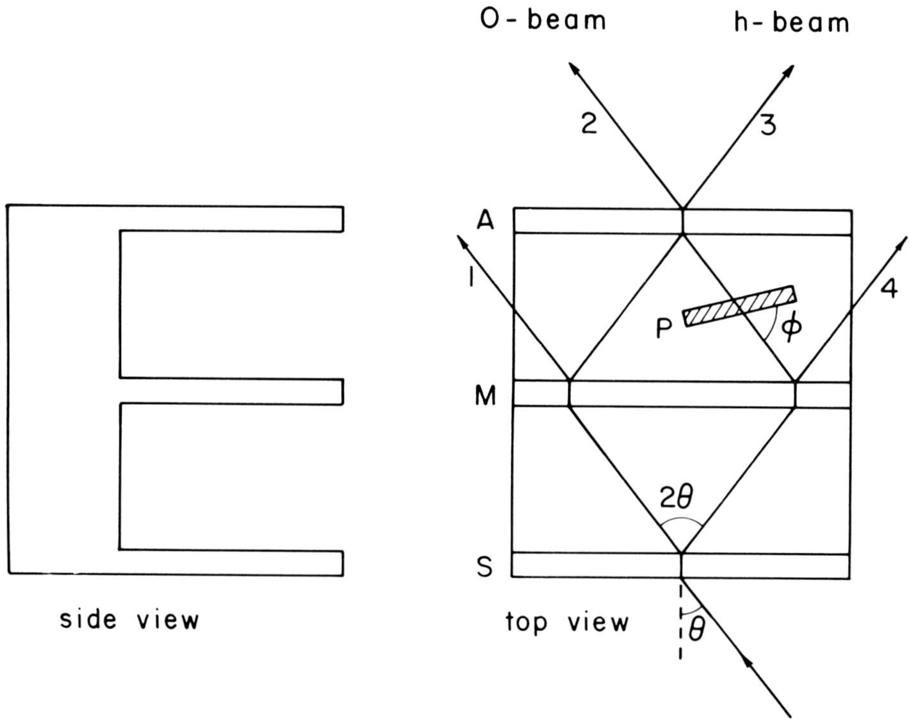

Fig. 7.2 Triple Laue-case perfect-crystal neutron interferometer.

cutting two wide grooves in a large, perfect silicon crystal. The crystal is cut such that the Bragg planes of interest, usually (220), are perpendicular to the faces of the slabs. The parallel alignment of the Bragg planes in the three slabs is then automatically ensured by the fact that the device is cut from a single crystal.

A collimated incident-neutron beam is split into two coherent parts by a symmetric Laue-case reflection in the slab S. The slab M then acts as a mirror that brings the two beams back together at the analyser A. The interference fringes produced by rotating the phase shifter P can then be observed by means of a detector in either the 0-beam or the h-beam.

Since this instrument employs Laue-case reflections in each of the three slabs it is called a *triple Laue-case* (*or LLL*) *interferometer*. It is topologically equivalent to a Mach–Zehnder interferometer in ordinary optics (Born and Wolf, 1975). There are numerous other designs (Bonse and Graeff, 1977; Bonse and Rauch, 1979) some of which employ Bragg-case reflections while others employ both Bragg and Laue-case reflections. Some designs have only two slabs, while others have four or even five slabs. However, the LLL interferometer shown in Fig. 7.2 is generally a good, if not the best, configuration for most applications and in what follows we shall confine our attention to this instrument.

To avoid defocusing effects and loss of contrast, the distances between the slabs in Fig. 7.2 must be equal within a tolerance of about 1 μm. The thicknesses of the three slabs must also be equal to within a similar accuracy. The relative dimensions in this figure correspond to those of the original LLL neutron interferometer of Rauch et al. (1974), for which the slab separation was 27.2936(9) mm and the slab thickness 4.3954(8) mm.

Beam splitting by Bragg reflection

To calculate the intensities of the exit beams from the interferometer in Fig. 7.2 we must use the solutions of the dynamical diffraction equations for each of the three slabs. We begin by considering the first crystal S, which splits the incident wave into transmitted and Bragg-reflected components. The corresponding wave functions are:

$$\psi = \begin{cases} \exp(i\mathbf{k}\cdot\mathbf{r}), & \text{incident} \\ t\exp(i\mathbf{k}\cdot\mathbf{r}), & \text{transmitted} \\ r\exp(i\mathbf{k}'\cdot\mathbf{r}), & \text{Bragg-reflected} \end{cases} \tag{7.2.1}$$

Here we have normalized the incident wave to unit amplitude so that, in the notation of Sec. 6.2,

$$t = \frac{a''}{a}, \qquad r = \frac{a'}{a}, \qquad \mathbf{k}' = \mathbf{k} - \mathbf{K}_h - \mathbf{q}_1 \tag{7.2.2}$$

The present problem is complicated by the fact that, in general, the various beams in Fig. 7.2 make slightly different angles with the Bragg planes and, as a result, the values of r and t are also different. However, if the incident beam satisfies Bragg's law exactly ($\theta = \theta_h$) then $\mathbf{q}_1 = 0$ and Bragg's law is satisfied exactly for all the Bragg reflections in Fig. 7.2. For the (220) reflection the quantities r and t then have the same values for each Bragg reflection too.

For simplicity we shall also treat the crystal as a rigid lattice. Then $\hat{F}_h = 1$ for the (220) reflection, and Eqs. 6.3.12 and 6.3.13 are again valid. It then follows from Eq. 6.2.5 that

$$r = \tfrac{1}{2}[\exp(-2iy) - 1], \qquad t = \tfrac{1}{2}[\exp(-2iy) + 1] \tag{7.2.3}$$

in which $y = \pi d/\Delta_h$ as before (see Eq. 6.2.17), where d is the thickness of the crystal slab and Δ_h is the extinction length.

The reflectivity and transmissivity, which give the intensities of the Bragg-reflected and transmitted beams relative to the incident beam, are given according to Eq. 6.2.10 by

$$R = |r|^2 = \sin^2 y, \qquad T = |t|^2 = \cos^2 y \tag{7.2.4}$$

in agreement with Eq. 6.3.8. Here we have used the fact that y is real when absorption is negligible, as it is in silicon. We note again that

$$R + T = 1 \tag{7.2.5}$$

as required for neutron conservation (cf. Eq. 6.2.13).

It follows from Eq. 7.2.3 that

$$r^*t = -rt^* = \frac{i}{2}\sin(2y) \tag{7.2.6}$$

which is purely imaginary. Hence,

$$\arg(r) = \arg(t) \pm \frac{\pi}{2} \tag{7.2.7}$$

In other words, the Bragg-reflected and transmitted waves are 90 degrees out of phase. Although this is true for the special case under consideration here, it is not, as Zeilinger (1981) has emphasized, true in general.

Exit-beam intensities

Applying the relations (7.2.1) to each Bragg reflection in Fig. 7.2 we see that the wave functions for the exit beams are

$$\begin{aligned}
\psi_1 &= (tt)\exp(i\mathbf{k}\cdot\mathbf{r})\\
\psi_2 &= (rrt + tsrr)\exp(i\mathbf{k}\cdot\mathbf{r})\\
\psi_3 &= (trt + rsrr)\exp(i\mathbf{k}'\cdot\mathbf{r})\\
\psi_4 &= (tr)\exp(i\mathbf{k}'\cdot\mathbf{r})
\end{aligned} \tag{7.2.8}$$

Here $s = \exp(i\chi)$ is the phase shift produced by the sample P.

Since $I_j \equiv |\psi_j|^2$ is the fraction of the incident-neutron intensity that appears in the jth exit beam, it then follows with the help of Eqs. 7.2.4 and 7.2.6 that

$$\begin{aligned}
I_1 &= T^2\\
I_2 &= 2R^2T(1+\cos\chi)\\
I_3 &= R(R^2+T^2) - 2R^2T\cos\chi\\
I_4 &= RT
\end{aligned} \tag{7.2.9}$$

It will be noted that the interference fringes occur only in the beams 2 and 3, i.e. the 0-beam and the h-beam, and not in beams 1 and 4. Also,

$$I_2 + I_3 = R, \qquad I_1 + I_4 = T \tag{7.2.10}$$

so that

$$I_1 + I_2 + I_3 + I_4 = 1 \tag{7.2.11}$$

as required for neutron conservation.

When the intensities are expressed in the conventional form (Eq. 7.1.11) we see that for the 0-beam,

$$\bar{I}_2 = 2R^2T, \qquad C_2 = 1, \qquad \gamma_2 = 0 \tag{7.2.12}$$

and for the h-beam,

$$\bar{I}_3 = R(R^2 + T^2), \qquad C_3 = \frac{2RT}{R^2 + T^2}, \qquad \gamma_3 = \pi \qquad (7.2.13)$$

Note, in particular, that the fringe contrast $C_2 = 1$ always, but $C_3 = 1$ only if $R = T = \frac{1}{2}$.

General case

The expressions (7.2.9) for the exit-beam intensities assume an ideally collimated monoenergetic incident beam for which Bragg's law is satisfied exactly ($\theta = \theta_h$). In practice, the neutron wave vectors $\mathbf{k}$ in the incident beam have a finite distribution of magnitudes and directions, so that the observed intensities depend on the complete intrinsic rocking curve of the interferometer. This rocking curve can be calculated by applying the general solutions (Eq. 6.2.5) of the dynamical diffraction equations to each of the three slabs in Fig. 7.2. Such calculations have been discussed in detail by many authors (Rauch and Suda, 1974; Petrascheck and Folk, 1976; Bonse and Graeff, 1977; Bonse and Rauch, 1979; Staudenmann et al., 1980). It is found that the exit-beam intensities are again of the form

$$I_2 = A(1 + \cos\chi), \qquad I_3 = B - A\cos\chi \qquad (7.2.14)$$

but the values of the coefficients A and B are different from those in Eq. 7.2.9. The fringe contrasts are given by

$$C_2 = 1, \qquad C_3 = \frac{A}{B} \qquad (7.2.15)$$

and the value of C_3 is typically 40 percent.

These results apply only to an ideal interferometer. In reality, the fringe contrast is significantly reduced as a result of slight imperfections in the construction of the interferometer and residual strains in the crystal itself. In practice one usually finds that $C_2 \simeq 30$ percent and $C_3 \simeq 10$ percent.

The quantities A and B are intrinsic properties of the interferometer itself, while χ is a property of the phase-shifting device that is inserted inside the interferometer to produce the interference fringes. Figure 7.3 shows the interference fringes for both the 0-beam (I_2) and the h-beam (I_3) that were observed by Rauch et al. (1974) in the original paper on the perfect-crystal neutron interferometer. In this work an aluminum slab was used as the phase shifter P in Fig. 7.2. It will be noted that the interference fringes for the 0 and h-beams are 180 degrees out of phase, as they should be according to Eq. 7.2.14.

The phase shift χ determines the period of the interference fringes and is the fundamental quantitiy of interest as far as the applications of neutron interferometry are concerned. In Sec. 7.3 we shall describe some

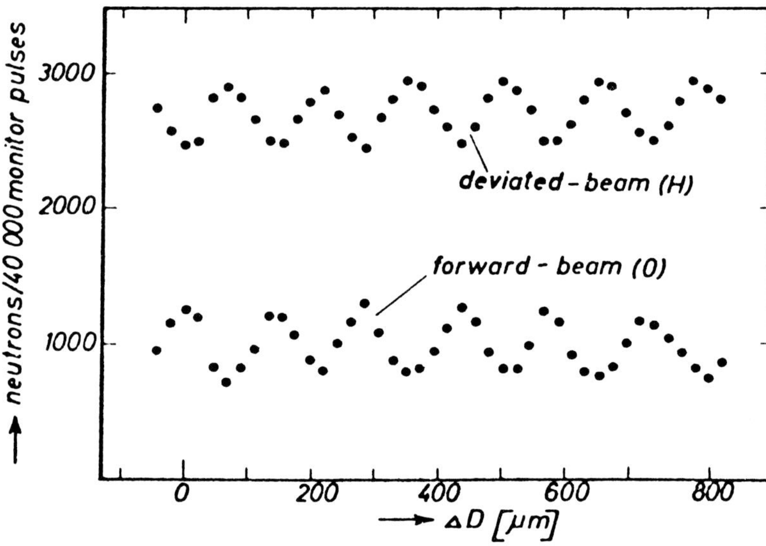

Fig. 7.3 Interference fringes observed by Rauch et al. (1974) in the original paper on the perfect-crystal neutron interferometer.

of the more important of these applications. Before we do this, however, there is one final general subject that must be addressed.

Complex phase shifts

Up to this point we have assumed that the phase shift χ is real. In fact, the index of refraction and, hence, the phase shift are complex quantities (see Sec. 3.2),

$$n = n' + in'', \qquad \chi = \chi' + i\chi'' \tag{7.2.16}$$

where, according to Eq. 7.1.23,

$$\chi' = (n' - 1)kD, \qquad \chi'' = n''kD \tag{7.2.17}$$

Alternatively, $\chi'' = \mu D/2$ where $\mu = 2kn''$ is the absorption coefficient (Eq. 3.2.11).

The intensity of the 0-beam is given in general by

$$\begin{aligned} I_2 &= \frac{A}{2}|1 + \exp(i\chi)|^2 \\ &= \frac{A}{2}[1 + \exp(-2\chi'') + 2\exp(-\chi'')\cos\chi'] \end{aligned} \tag{7.2.18}$$

If $\chi'' \ll 1$ this reduces to Eq. 7.2.14 as it should. Expressing this result in the form of Eq. 7.1.11, we get

$$I_2 = \bar{I}_2[1 + C_2 \cos \chi'] \tag{7.2.19}$$

where the average intensity is

$$\bar{I}_2 = \frac{A}{2}[1 + \exp(-2\chi'')] \tag{7.2.20}$$

and the fringe contrast is

$$C_2 = \operatorname{sech} \chi'' \tag{7.2.21}$$

Thus, the real part of the index of refraction n' determines χ' and, hence, the period of the interference fringes. The imaginary part of the index of refraction n'' determines χ'' and, hence, the average intensity and the fringe contrast.

If $\chi'' \ll 1$ then

$$\bar{I}_2 = A(1 - \chi''), \qquad C_2 = 1 - \frac{\chi''^2}{2} \tag{7.2.22}$$

Thus, for example, if $\chi'' = 0.1$ then $\bar{I}_2 = 0.9A$ and $C_2 = 0.995$. The primary effect of absorption is, therefore, to reduce the average intensity $\bar{I}_2$. The reduction in the fringe contrast C_2 is a second-order effect that is appreciable only if $\chi'' \gtrsim 1$, that is, only for very strong absorption or very thick samples.

In practice, n'' is more conveniently and accurately measureable in a conventional transmission experiment (see Sec. 3.2). Thus, neutron interferometry is primarily of interest as a method for the accurate measurement of n'.

7.3. Applications of neutron interferometry

Coherent scattering length measurement

The main application of neutron interferometry is in the accurate measurement of coherent scattering lengths. We note first that the λ-thickness (Eq. 7.1.28) can be expressed with the help of Eq. 3.1.7 in the form

$$D_\lambda = \frac{2\pi}{\lambda \rho b} \tag{7.3.1}$$

Thus, if the neutron wavelength λ and the number density of the sample ρ are known, the value of D_λ obtained from interferometry measurements can be used to determine the value of the bound coherent scattering length b. This was first done by Bauspiess et al. (1974) who measured b for aluminum and bismuth. In subsequent years values of b for many of the elements in the periodic table have been measured by this method. A selection of these results is given in Table 3.3. We refer the reader to the

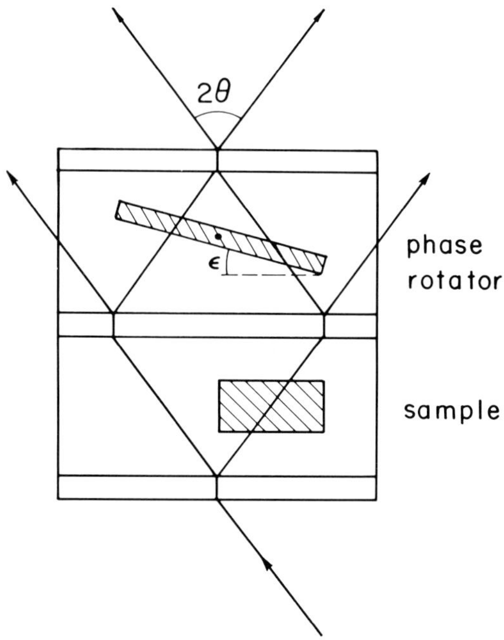

Fig. 7.4 Typical experimental setup for scattering-length measurements by neutron interferometry.

review article by Bauspiess et al., (1978) for detailed technical information on this application and on neutron interferometry in general.

Figure 7.4 illustrates a typical experimental setup for scattering-length measurements by neutron interferometry. The first slot in the interferometer contains the sample, which is held in a fixed position in one beam. The second slot contains a plane slab of material (usually aluminum) that intersects both beams and can be rotated about a central axis perpendicular to the plane of the figure. This slab is referred to variously as a *phase rotator, phase shifter,* or *phase flag* and its orientation is defined by the angle ε. The phase shift is then given by

$$\chi = \begin{cases} \chi_r + \chi_s, & \text{sample in} \\ \chi_r, & \text{sample out} \end{cases} \tag{7.3.2}$$

where χ_r is the contribution from the phase rotator and χ_s that from the sample. It follows from Eqs. 7.1.25 and 7.1.26 that

$$\chi_s = -2\pi \frac{d_s}{D_{\lambda s}} \sec\theta \tag{7.3.3}$$

and

$$\chi_r = -2\pi \frac{d_r}{D_{\lambda r}} [\sec(\theta + \varepsilon) - \sec(\theta - \varepsilon)] \tag{7.3.4}$$

Here θ is the Bragg angle, d_s and d_r are the thicknesses of the sample and the phase rotator respectively, and $D_{\lambda s}$ and $D_{\lambda r}$ are the corresponding λ-thicknesses.

Interference fringes are produced by varying ε, first with the sample in and then with the sample out. The presence of the sample produces a measureable shift χ_s in the fringe pattern from which one can obtain $D_{\lambda s}$ and, hence, the scattering length of the sample b_s. Figure 7.5 shows some typical results obtained by Kaiser et al. (1986) for ^{235}U. The decrease in the fringe contrast when the sample is in is due mainly to the large absorption cross section of ^{235}U (see the discussion at the end of Sec 7.2).

It is evident from Eq. 7.2.19 that, strictly speaking, the interference fringes are determined by the real part of the phase shift χ' and, hence, that the quantity that is obtained from the measured value of χ_s' is the real part of the scattering length b_s'. Since the interference fringes are independent of the sign of χ', the interferometry measurements only determine the sign of χ_s' relative to χ_r' and, hence, the sign of b_s' relative to b_r'. The phase rotator is usually aluminum, for which it is known from mirror reflection measurements that b_r' is positive. This enables one to determine the absolute sign of b_s'.

For gaseous samples the interference fringes are usually produced by varying the pressure in the gas vessel and, hence, the number density ρ. In this case b_s' can be obtained (except for its sign) directly from the measured fringe spacing. The sign of b_s' can be determined by noting the

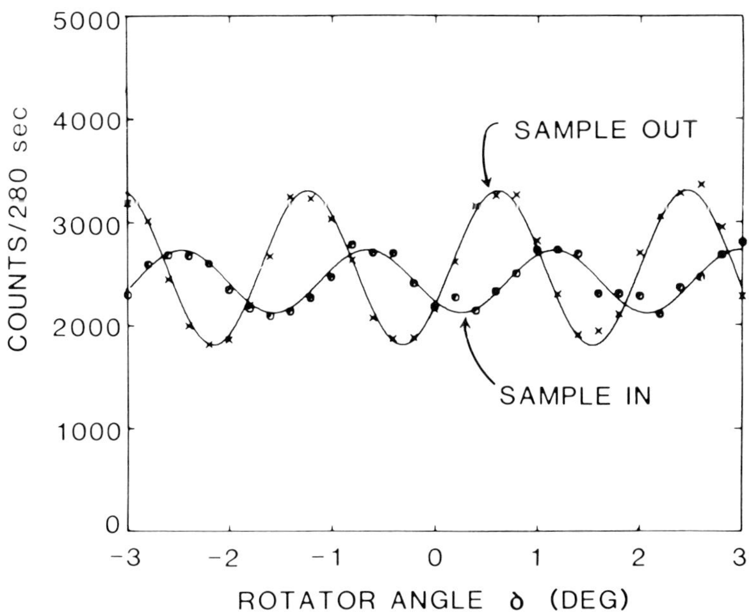

Fig. 7.5 Interference fringes measured by Kaiser et al. (1986) for ^{235}U as described in the text.

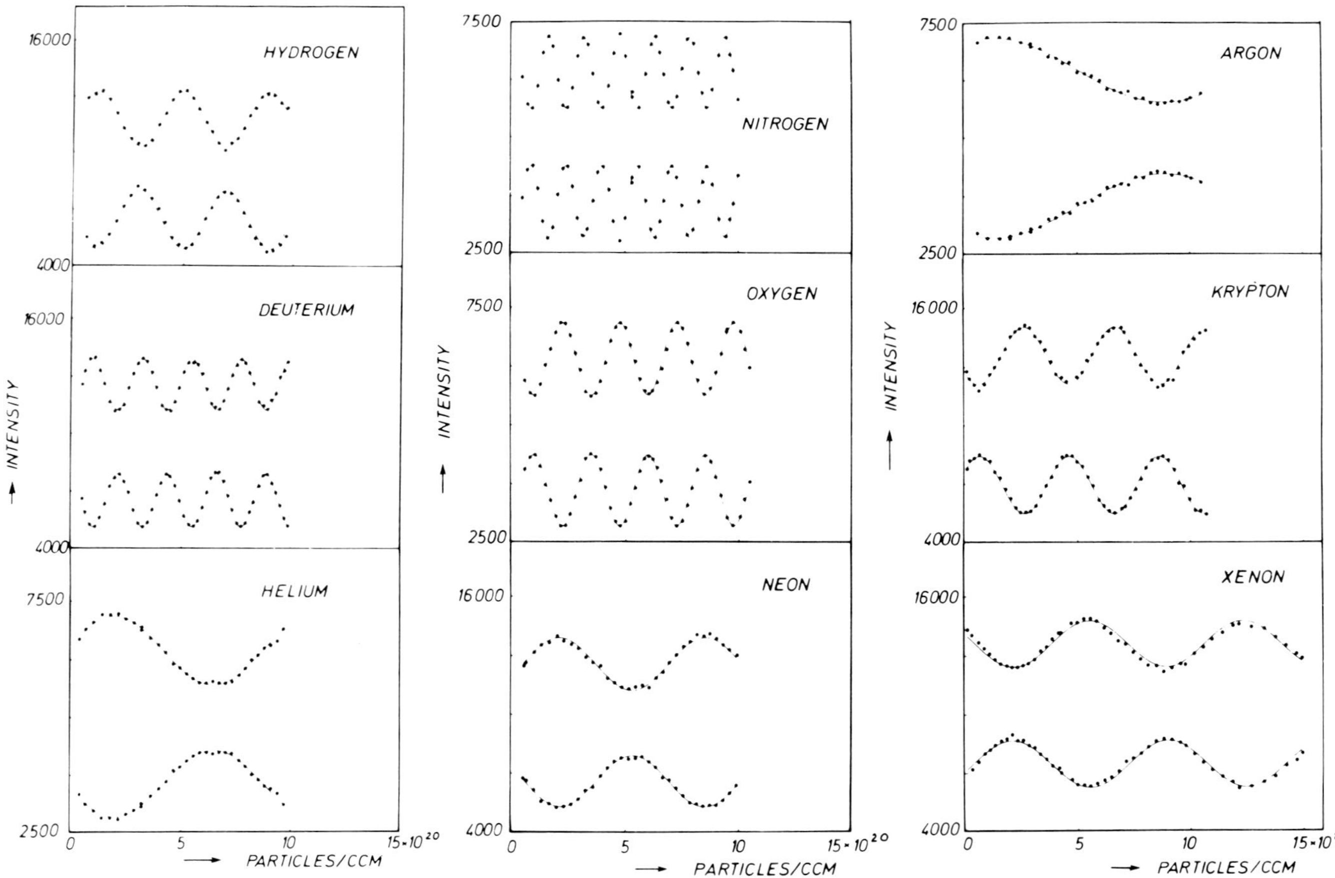

Fig. 7.6 Interference fringes measured by Kaiser et al. (1979) for various gases.

direction in which the fringe pattern shifts when the phase flag is rotated through a small angle ε. The interference patterns measured in this way by Kaiser et al. (1979) for a number of gaseous elements are shown in Fig. 7.6.

Spinor rotation

Before we discuss spinor rotation it is useful to recall some fundamental properties of angular momentum in quantum mechanics. Let us consider, for example, a particle with spin $\mathbf{s}$. Since $\mathbf{s}^2$ and s_z commute, these operators possess a common set of eigenstates,

$$\mathbf{s}^2 |sm\rangle = s(s+1) |sm\rangle, \qquad s_z |sm\rangle = m |sm\rangle \tag{7.3.5}$$

where $m = s, s-1, \ldots -s$. If the system is rotated through an angle α about an axis defined by the unit vector $\hat{\mathbf{n}}$, then the corresponding rotated spin state, or spinor, is given by

$$|sm\rangle' = \exp(-i\boldsymbol{\alpha} \cdot \mathbf{s}) |sm\rangle \tag{7.3.6}$$

where $\boldsymbol{\alpha} = \alpha\hat{\mathbf{n}}$. In particular, for a rotation through 2π about the z-axis,

$$\begin{aligned} |sm\rangle' &= \exp(-2\pi i s_z) |sm\rangle \\ &= \exp(-2\pi i m) |sm\rangle \\ &= (-1)^{2m} |sm\rangle \end{aligned} \tag{7.3.7}$$

Hence,

$$|sm\rangle' = \begin{cases} |sm\rangle, & s = 0, 1, 2, \ldots \\ -|sm\rangle, & s = \frac{1}{2}, \frac{3}{2}, \frac{5}{2}, \ldots \end{cases} \tag{7.3.8}$$

When a classical object is rotated through 2π it returns to its original state. This is true for quantum-mechanical spin states only if s is an integer. If s is a half-integer, the rotation introduces a phase factor -1 and it is necessary to rotate the system through 4π before $|sm\rangle' = |sm\rangle$.

It was first pointed out by Bernstein (1967) that, since the factor -1 represents the relative phase between the rotated and unrotated spinors, it could in principle be observed for a spin-$\frac{1}{2}$ particle such as the neutron by using a magnetic field to rotate the spinor in one branch of a two-beam interferometer.

To demonstrate this explicitly we begin by noting that the energy of interaction of a neutron magnetic dipole momentum $\boldsymbol{\mu}$ with a magnetic field $\mathbf{B}$ is given by $v = -\boldsymbol{\mu} \cdot \mathbf{B}$, so that the resulting phase shift (Eq. 7.1.32) becomes

$$\chi = \frac{m\boldsymbol{\mu}}{\hbar^2 k} \cdot \int_C \mathbf{B}\hat{\mathbf{K}} \cdot \mathbf{dl} \tag{7.3.9}$$

Suppose one beam in the interferometer passes between the poles of an electromagnet, so that

$$\mathbf{B} = \begin{cases} \mathbf{B}_0, & \text{between the poles} \\ 0, & \text{outside} \end{cases} \tag{7.3.10}$$

Then

$$\chi = \left(\frac{mD}{\hbar^2 k}\right)\boldsymbol{\mu}\cdot\mathbf{B}_0 \tag{7.3.11}$$

where D is the path length in the pole region,

$$D = \int \hat{\mathbf{K}}\cdot d\mathbf{l} \tag{7.3.12}$$

In reality, the magnetic field does not vanish identically outside the pole region so that Eq. 7.3.10 is an oversimplification and the quantity D in Eq. 7.3.11 must be regarded as an effective path length.

Since $\hbar k/m$ is the neutron velocity, it follows that the quantity

$$t = \frac{mD}{\hbar k} \tag{7.3.13}$$

is the transit time through the pole region. Hence,

$$\chi = \left(\frac{t}{\hbar}\right)\boldsymbol{\mu}\cdot\mathbf{B}_0 = \frac{-Ht}{\hbar} \tag{7.3.14}$$

where $H = -\boldsymbol{\mu}\cdot\mathbf{B}_0$ is the spin Hamiltonian. The phase factor is therefore

$$\exp(i\chi) = \exp(-iHt/\hbar) \tag{7.3.15}$$

This is the unitary operator that propagates the neutron spinor through the transit time t.

The neutron magnetic dipole moment is given, according to Eq. 3.5.9, by $\boldsymbol{\mu} = 2\gamma\beta_N\mathbf{s}$, where β_N is the nuclear magneton and $\gamma = -1.913$. Hence, Eq. 7.3.14 becomes

$$\chi = -\boldsymbol{\alpha}\cdot\mathbf{s} \tag{7.3.16}$$

where

$$\boldsymbol{\alpha} = \boldsymbol{\omega}_0 t \tag{7.3.17}$$

and

$$\boldsymbol{\omega}_0 = -\frac{2\gamma\beta_N}{\hbar}\mathbf{B}_0 \tag{7.3.18}$$

Since γ is negative $\boldsymbol{\omega}_0$ and $\boldsymbol{\alpha}$ are in the same direction as $\mathbf{B}_0$. The phase factor (Eq. 7.3.15) now becomes

$$\exp(i\chi) = \exp(-i\boldsymbol{\alpha}\cdot\mathbf{s}) \tag{7.3.19}$$

which, according to Eq. 7.3.6, is the unitary operator that rotates the neutron spinor through an angle α about the direction of the magnetic field. Since this rotation occurs in the time interval t, we see that ω_0 is the angular frequency with which the neutron spinor precesses about $\mathbf{B}_0$. In other words, ω_0 is the Larmor frequency.

To calculate the intensities of the exit beams, the expressions (7.2.14) must be averaged over the neutron spin distribution in the incident beam. Thus, for example, if the incident neutrons are unpolarized, the intensity of the 0-beam is given by

$$I_2 = \frac{1}{2s+1} \sum_m \langle sm | A(1+\cos\chi) | sm \rangle \tag{7.3.20}$$

With the z-axis in the direction of $\mathbf{B}_0$, we then get

$$I_2 = A\left[1 + \frac{1}{2s+1} \sum_m \cos(m\alpha)\right] \tag{7.3.21}$$

For the neutron, $s = \frac{1}{2}$ and $m = \pm\frac{1}{2}$ so that

$$I_2 = A[1 + \cos(\alpha/2)] \tag{7.3.22}$$

Interference fringes can be produced by varying B_0 and, hence, α and the period is $\alpha = 4\pi$. On the other hand, for a particle with $s = 1$,

$$I_2 = \tfrac{2}{3}A(2 + \cos\alpha) \tag{7.3.23}$$

and the period of the interference fringes is $\alpha = 2\pi$.

When Bernstein (1967) first proposed this type of experiment the only existing neutron interferometer was the Fresnel biprism of Maier-Leibnitz and Springer (1962). The beam separation in this instrument ($\simeq 60\ \mu$m) was too small to permit a practical demonstration of the effect. It was only after the development of the perfect-crystal neutron interferometer that such a demonstration became feasible. Results were then reported simultaneously by Rauch et al. (1975) and Werner et al. (1975). Both groups found the expected 4π-result that is shown in Fig. 7.7.

At about the same time, Klein and Opat (1975) argued that this property of spinor rotation could also be demonstrated by measuring the Fresnel diffraction of neutrons by a ferromagnetic domain boundary (see Sec. 6.5), and the results of a successful experiment were reported the following year (Klein and Opat, 1976).

The phase shift produced by the simultaneous presence of a magnetic field and a material medium has been investigated both theoretically and experimentally by Badurek et al. (1976a,b). An interesting feature of their results is that the 0 and h-beams become partially polarized. In more recent years other static and dynamic properties of quantum spin-state superposition have been studied by, for example, Summhammer et al. (1982, 1983) and Badurek et al. (1983, 1986) using the perfect-crystal neutron interferometer.

Gravitationally induced quantum interference

A neutron with mass m and position $\mathbf{r}$ has an energy of interaction with the earth's gravitational field that is given by $v = -m\mathbf{g}\cdot\mathbf{r}$, where $\mathbf{g}$ is the gravitational acceleration. This produces a gravitational contribution to the phase shift χ that is given by Eq. 7.1.32. If we choose a system of Cartesian coordinate axes such that the z-axis is vertical, then $\mathbf{g}$ is in the

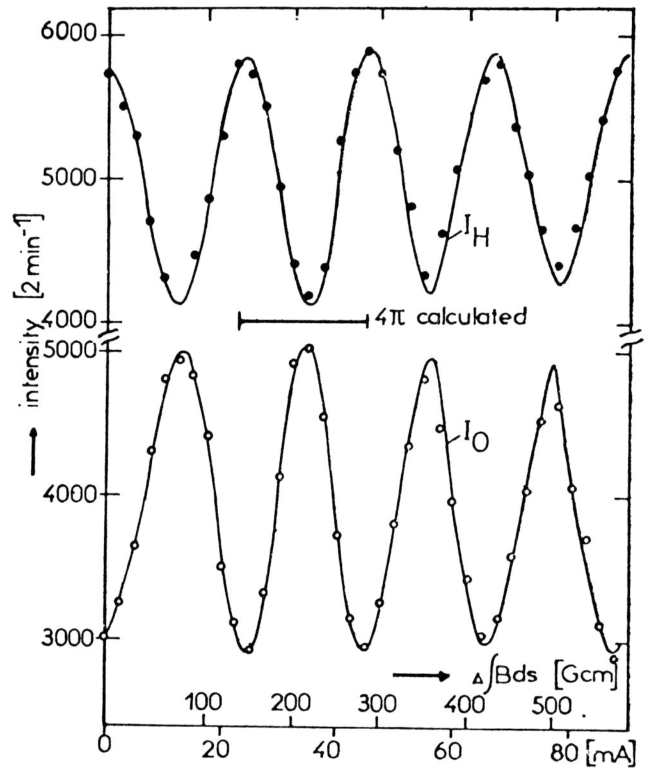

Fig. 7.7 Measurements of spinor rotation by neutron interferometry (Rauch et al., 1975).

$-z$ direction and the gravitational phase shift is

$$\chi = -\frac{m^2 g}{\hbar^2 k} \int_C z\hat{\mathbf{K}} \cdot \mathbf{dl} \tag{7.3.24}$$

In most experiments the interferometer has a fixed orientation such that the contour C is in the xy plane. Then $z = 0$ everywhere on C and $\chi = 0$. In such experiments we can therefore ignore gravity as we have, in fact, done up to this point.

It was first pointed out by Overhauser and Colella (1974) that if the interferometer were rotated about a horizontal axis through an angle ϕ then χ would no longer vanish, and the variation of χ with ϕ would produce measurable interference fringes. Since χ depends on both g and $\hbar$, Overhauser and Colella called this effect *gravitationally induced quantum interference*.

If a perfect-crystal neutron interferometer is rotated about the incident beam through an angle ϕ then the contour integral in Eq. 7.3.24 can easily be evaluated and one finds (Overhauser and Colella, 1974) that

$$\chi = \left(\frac{m^2 g}{\hbar^2 k}\right) A \sin \phi \tag{7.3.25}$$

where A is the area of the contour C,

$$A = 2L^2 \tan \theta \tag{7.3.26}$$

Here θ is the Bragg angle, L is the distance between the crystal slabs, and we have neglected the finite thickness of these slabs for simplicity. We have also neglected the curvature of the neutron beams due to the free fall of the neutron within the interferometer. Overhauser and Colella (1974) showed that this effect gives no contribution to χ to first order in g.

The phase shift (Eq. 7.3.25) can also be expressed in the form

$$\chi = 2\pi N \sin \phi \tag{7.3.27}$$

where N is the number of interference fringes that are produced when the interferometer is rotated through 90 degrees,

$$N = \left(\frac{m}{h}\right)^2 g\lambda A \tag{7.3.28}$$

Suppose, for example, that $\lambda = 1.5$ Å. Then $\theta = 23.0$ degrees for the (220) reflection in silicon. If $L = 3.5$ cm we then get $A = 10.4$ cm^2. With $g = 980$ cm s^{-2}, Eq. 7.3.28 gives $N = 9.8$. In this case we would expect to observe approximately 10 interference fringes when the interferometer is rotated through 90 degrees.

The first measurements of gravitationally induced quantum interference were made by Colella, Overhauser, and Werner (1975) in what is now called the COW experiment. They found a slight discrepancy between the observed and calculated values of N even after correcting for the finite thickness of the crystal slabs. They were able to show that this discrepancy is, in fact, due to the bending of the interferometer under its own weight when it is rotated out of the horizontal plane. An example of the interference fringes observed in later experiments of Staudenmann et al. (1980) with $\lambda = 1.060$ Å when ϕ varies from -40 degrees to $+40$ degrees is shown in Fig. 7.8.

In the foregoing discussion we have ignored any possible effect of centrifugal and Coriolis forces due to the rotation of the earth. The centrifugal force simply modifies the magnitude and direction of the effective gravitational acceleration $\mathbf{g}$. The centrifugal force is, therefore, already included implicitly in the measured local value of $\mathbf{g}$, so that it needs no further consideration. The Coriolis force, on the other hand, depends on the neutron velocity and, hence, must be dealt with explicitly.

The effect of the Coriolis force on the phase shift in neutron

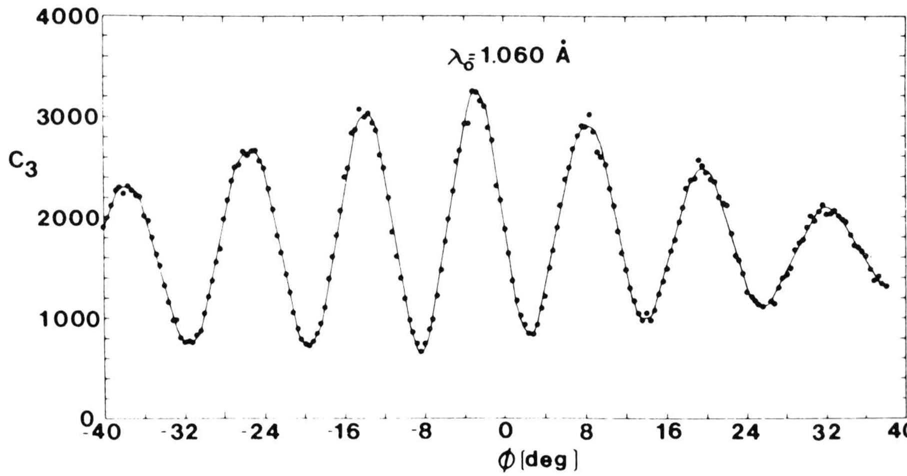

Fig. 7.8 Gravitationally-induced quantum interference measured by Staudenmann et al. (1980).

interferometry was first discussed by Page (1975), who showed that it gives a contribution to χ proportional to $\boldsymbol{\omega} \cdot \mathbf{A}$, where $\boldsymbol{\omega}$ is the angular velocity of the earth's rotation and $\mathbf{A}$ is the normal area of the contour C. This effect is analogous to the Sagnac effect in ordinary optical interferometry and is small, typically 2 percent of the gravitational phase shift (Eq. 7.3.27). The neutron Sagnac effect was first observed by Werner et al. (1979) by measuring the interference fringes that are produced when the interferometer is rotated about a vertical axis. In this arrangement the gravitational phase shift (Eq. 7.3.27) is constant and, hence, gives no contribution to the frequency of the interference fringes. Later, the neutron Sagnac effect was also observed by Atwood et al. (1984) with a two-crystal (LL) interferometer.

A detailed discussion of gravitationally induced quantum interference and the Sagnac effect together with additional experimental results has been given by Staudenmann et al. (1980). These authors also show that the measurements provide experimental confirmation of the equivalence of gravitational and inertial mass, and hence of the universality of free fall, in the quantum limit. A complete test of the equivalence principle in general relativity requires that experiments also be performed with an accelerated interferometer. Such measurements were later made by Bonse and Wroblewski (1983, 1984).

Fizeau effect

Up to this point we have assumed that the phase-shifting medium P in Fig. 7.2 is at rest relative to the interferometer. If the medium is in motion, however, the phase shift χ will change and, hence, produce interference phenomena analogous to the Fizeau effect in ordinary optics. The theory of the neutron Fizeau effect was first discussed by Horne and Zeilinger (1979) and later, in more detail, by Horne et al. (1983) and Peres (1983).

Let us suppose that relative to the interferometer the neutron has velocity $\mathbf{u}$ and the phase-shifting medium has velocity $\mathbf{w}$. Then the velocity of the neutron relative to the medium is

$$\mathbf{u}' = \mathbf{u} - \mathbf{w} \tag{7.3.29}$$

This nonrelativistic (or Galilean) transformation is appropriate in the present case since $(u/c)^2 \simeq 10^{-10}$ for thermal neutrons. Since $m\mathbf{u} = \hbar\mathbf{k}$ we then get

$$\mathbf{k}' = \mathbf{k} - \frac{m}{\hbar}\mathbf{w} \tag{7.3.30}$$

The important point to recognize now is that the phase shift χ is a scalar and, hence, is invariant under a Galilean transformation. Thus, when the medium is in motion with velocity $\mathbf{w}$ relative to the interferometer, the phase shift has the same value $\chi(\mathbf{k}')$ in both reference systems. If the medium is at rest relative to the interferometer, the phase shift has the value $\chi(\mathbf{k})$. Hence, the phase shift due to the motion of the medium, i.e., the Fizeau effect, is

$$\delta\chi = \chi(\mathbf{k}') - \chi(\mathbf{k}) \tag{7.3.31}$$

Combining Eqs. 7.1.23, 7.1.25, and 7.1.31, we see that, if the medium is a plane slab of thickness d, then

$$\chi(\mathbf{k}) = -\frac{mdv}{\hbar^2 k_z} \tag{7.3.32}$$

where v is the optical potential and $k_z = k \sin\phi$ is the component of $\mathbf{k}$ perpendicular to the face of the slab. Hence, to first order,

$$\frac{\delta\chi}{\chi} = \frac{\delta v}{v} - \frac{\delta k_z}{k_z} \tag{7.3.33}$$

Let us first assume that $\mathbf{w}$ is in the x direction and, hence, is parallel to the face of the slab. In this case the medium is in motion but its boundaries are at rest relative to the interferometer. Since $\delta k_z = 0$ we get

$$\frac{\delta\chi}{\chi} = \frac{\delta v}{v} = \frac{1}{v}\left(\frac{dv}{dE}\right)\delta E \tag{7.3.34}$$

Since $\delta E = -\hbar k_x w$ to first order and $k_x/k_z = \cot\phi$, it follows that

$$\delta\chi = \left(\frac{md}{\hbar}\right)\left(\frac{dv}{dE}\right) w \cot\phi \tag{7.3.35}$$

In the elementary theory of dispersion the optical potential v is given by Eq. 2.8.43, which means that it is constant. In this case, $dv/dE = 0$, so that $\delta\chi = 0$. In other words, the neutron Fizeau effect should vanish when the medium is moving parallel to its boundaries. Arif et al. (1985) have performed such an experiment in which a rapidly rotating quartz disk was inserted in a perfect-crystal neutron interferometer. No Fizeau effect was observed and an upper limit $|dv/dE| \leq 2.1 \times 10^{-8}$ was obtained for quartz at the incident-neutron wavelength $\lambda = 1.268$ Å.

It has been suggested (Klein et al. 1981c; Horne et al., 1983) that a nonvanishing Fizeau effect will occur near an (n, γ) resonance where the scattering amplitude and, hence, v have a strong energy dependence. Sears (1985b) has pointed out that v also becomes energy dependent when local-field effects are taken into account (see Chap. 4). He showed that for quartz $|dv/dE| \simeq 10^{-8}$ when $\lambda \gtrsim 13$ Å, but that this quantity is of the order of 10^{-12} for the wavelength at which the experiments of Arif et al. (1985) were performed. For most practical purposes, therefore, δv is effectively zero for thermal neutrons and Eq. 7.3.33 reduces to

$$\frac{\delta\chi}{\chi} = -\frac{\delta k_z}{k_z} \tag{7.3.36}$$

Let us now assume that $\mathbf{w}$ is in the z direction and, hence, is perpendicular to the face of the slab. In this case the medium and its boundaries have the same velocity, $\mathbf{w}$. Here, $\delta k_z = -mw/\hbar$ so that

$$\frac{\delta\chi}{\chi} = \frac{w}{u} \operatorname{cosec} \phi \tag{7.3.37}$$

The relative change in χ due to the motion of the slab is therefore proportional to the velocity ratio. This expression was first verified by Klein et al. (1981c) to an accuracy of about 9 percent using a Young double-slit interferometer and later, with higher accuracy ($\simeq$1.5 percent), the Bonse and Rumpf (1986) using a perfect-crystal neutron interferometer.

Quaternion quantum mechanics

An important class of applications of neutron interferometry arises from its ability to provide direct tests of many of the fundamental assumptions of quantum mechanics. For example, we have already seen that neutron interferometry provides a simple demonstration of the superposition principle for quantum-mechanical states.

Another fundamental assumption of quantum mechanics is that the scalar product $\langle \phi \mid \psi \rangle$ of any two state vectors in Hilbert space is a complex number. An immediate consequence of this assumption is that all wave functions, matrix elements, etc., are also complex numbers. However, there is no a priori reason why $\langle \phi \mid \psi \rangle$ must necessarily belong to the complex field. If, for example, one requires that the quantities $\langle \phi \mid \psi \rangle$ be numbers that can be added and multiplied in a unique way, then there are essentially three possibilities: $\langle \phi \mid \psi \rangle$ is either real, complex, or quaternion. In all three number fields the basic arithmetic operations are (with one exception) associative, distributive, and commutative. The exception is in the case of quaternions for which multiplication is noncommutative.

Although quantum mechanics is, in practice, based on the complex field, it can also be developed in a logically consistent way using real numbers or quaternions instead. It is found that real quantum mechanics is essentially equivalent to complex quantum mechanics, but quaternion quantum mechanics is distinctly different because multiplication is then noncommutative (Kaneno, 1960; Finkelstein et al., 1962).

The choice between complex and quaternion quantum mechanics is one that can only be made on the basis of experiment. Peres (1979) has devised a number of experiments that could, in principle, be used to detect the presence of quaternions in quantum mechanics. One of these, which will be described, is based on neutron interferometry.

In the perfect-crystal neutron interferometer, the phase shifter P in Fig. 7.2 introduces a phase factor $s = \exp(i\chi)$ in the neutron wave function (7.2.8). If absorption is negligible, then, in conventional complex quantum mechanics, χ is real and s is a unimodular complex number, $|s| = 1$. In quaternion quantum mechanics, s is a unimodular quaternion.

Suppose the phase shifter consists of two slabs of different materials a and b. Then the phase factor is $s_b s_a$ if a is in front of b and $s_a s_b$ if a is behind b. In complex quantum mechanics, $s_b s_a = s_a s_b$ so that the interference fringe pattern is independent of whether a is in front of b or behind it. In quaternion quantum mechanics, on the other hand, $s_b s_a \neq s_a s_b$ in general so that one might observe a change in the interference fringe pattern when the order of the two slabs is reversed. Such an experiment has been performed by Kaiser et al. (1984) using slabs of titanium and aluminum. They found that the fringe pattern is, in fact, invariant under an interchange of the two slabs to an accuracy of 1 part in 30,000. Thus, there is no indication of any significant quaternion component in the phase factor.

Nonlinear quantum mechanics

Yet another fundamental assumption of quantum mechanics is that observables are represented by linear operators in Hilbert space. Thus, for example, the Hamiltonian H is a linear operator and the corresponding Schrödinger equation, $i\hbar\dot{\psi} = H\psi$, is a linear equation. This is somewhat surprising when one considers that in all other branches of physics (classical mechanics, electromagnetism, hydrodynamics, general relativity, etc.) the fundamental equations are always nonlinear.

There is, of course, no a priori reason why the Schrödinger equation need necessarily be linear. Any nonlinear variant of the Schrödinger equation is formally equivalent to adding a ψ-dependent term v_{NL} to the potential energy. This problem has been considered by Bialynicki-Birula and Mycielski (1976), who showed that as long as v_{NL} is of the form $F(|\psi|^2)$ then, regardless of the functional form of F, most of the usual properties of the Schrödinger equation remain intact. For example, ψ still has all the properties required for the probability interpretation of quantum mechanics, the Schrödinger equation remains Galilean invariant, etc. In fact, the only obvious difficulty is that the nonlinear term will in general produce correlations between particles even if there is no interaction between them. This violates the separability principle according to which systems that are dynamically independent should also be statistically independent. In other words, if a system consists of two noninteracting parts a and b then its wave function should be of the form $\psi = \psi_a \psi_b$.

It is easily verified that the separability principle will be satisfied if

$$v_{NL} = -b \ln(a^3 |\psi|^2), \tag{7.3.38}$$

where a and b are constants. Bialynicki–Birula and Mycielski (1976) showed that, within the previously mentioned class of nonlinear potentials, Eq. 7.3.38 is the only one that satisfies the separability principle. They pointed out that, while b is a universal constant, the constant a is arbitrary since changing its value is simply equivalent to adding a constant to v_{NL}. They obtained an upper limit $b < 4 \times 10^{-10}$ eV from the measured Lamb shift in hydrogen.

Shimony (1979) later showed that neutron interferometry could also be used to look for a nonlinear term of the type of Eq. 7.3.38 in the Schrödinger equation. In particular he showed, using the WKB approximation, that the effect of the nonlinear term (Eq. 7.3.38) is that the phase shift χ now depends on the position of the phase shifter P in Fig. 7.2 and not simply on its orientation. Thus, for example, if the phase shifter is repositioned downstream in the beam, a shift in the interference fringe pattern will occur if $b \neq 0$. He estimated that with then-existing technology the previous upper limit on b might be reduced by two orders of magnitude. The experiment was soon performed with a double-crystal (LL) interferometer by Shull et al. (1980), who found no measurable effect and were, in fact, able to reduce the upper limit by three orders of magnitude to $b < 3.4 \times 10^{-13}$ eV.

The presence of a nonlinear term in the Schrödinger equation would, of course, have an effect on all neutron-optical phenomena. For example, Gähler et al. (1981) showed that in the case of Fresnel diffraction by a straight edge the nonlinear term (Eq. 7.3.38) would produce a shift in the positions of the diffraction maxima. No measurable shift was observed (see Fig. 6.18) and these authors were able to obtain an even smaller limit, $b < 3.3 \times 10^{-15}$ eV, on the magnitude of the nonlinear coupling constant. In summary, therefore, there is no indication of any significant nonlinearity in the Schrödinger equation.

There have been many other suggestions of how neutron interferometry might be used to test the fundamental assumptions and predictions of quantum mechanics. For example, we have already discussed the case of spinor rotation. In addition, measurements with a two-crystal interferometer have been used to verify the expected absence of the Aharonov–Bohm effect for neutrons (Greenberger et al., 1981). For a detailed discussion of this and other possible applications, see Greenberger and Overhauser (1979) and Greenberger (1983).

Materials science

Little work has been done so far on materials science applications of neutron interferometry. Nevertheless, test measurements suggest that neutron interferometry does, in fact, have several potential applications in the area of nondestructive testing and these will undoubtedly be exploited in the future.

One such application is the measurement of hydrogen concentration in metals. In such a binary system the scattering length density ρb in the expression (Eq. 7.3.1) for D_λ is replaced by $\rho_M b_M + \rho_H b_H$ where M refers to the metal atoms and H to hydrogen (see Eq. 3.1.8). Since the bound coherent scattering lengths b_M and b_H are already known, and the number density ρ_M can be measured independently, the value of D_λ obtained from neutron interferometry measurements can be used to determine ρ_H. Rauch et al. (1978) have used this method to measure the concentration of hydrogen and deuterium in vanadium and niobium, and have demonstrated that the sensitivity of the technique is about 0.03 to 0.06 atom percent or 6 to 12 ppm by weight.

Rauch et al. (1978) also noted that when the hydrogen concentration increased to the point at which precipitation occurred, there was a marked decrease in the fringe contrast that they attributed to the effect of small-angle scattering by the precipitates. In more recent work, Rauch and Seidl (1987) have shown that the same effect can be used to obtain information on precipitates in alloys and on ferromagnetic domains that is complementary to the type of information that is obtained from a direct measurement of the small-angle scattering. They also demonstrated that macroscopic inhomogeneities in materials can be studied directly by a photographic method that they called *phase topography*.

7.4. Coherence and interferometry

Up to this point we have assumed that the incident-neutron beam is described by a single plane wave with wave vector $\mathbf{k}$. Such a beam is said to be *totally coherent*. In reality, the thermal fission reactors that are used as sources in neutron optics experiments are analogous to black-body sources in ordinary optics and produce neutron beams that contain an *incoherent* admixture of all wave vectors with a roughly Maxwellian distribution corresponding to the temperature of the moderator. The interferometer filters out all wave vectors that do not satisfy Bragg's law to within the width of the rocking curve of the instrument. Even though this width is, or course, very small, it is still not zero. Consequently, the exit beams contain a distribution of plane waves with wave vectors that have almost the same magnitude and direction but with essentially random phases. The exit beams are then said to be *partially coherent*, for reasons that will become clear later.

The purpose of the present section is to discuss the coherence properties of thermal neutron beams and the way in which the degree of coherence affects the fringe contrast in neutron interferometry experiments. For more details, see Hamilton et al. (1983) and Klein and Opat (1984).

Wave packets

The wave function ψ for a neutron in free space satisfies the Schrödinger equation

$$L\psi = 0 \tag{7.4.1}$$

where

$$L = \frac{\hbar^2}{2m}\Delta + i\hbar\frac{\partial}{\partial t} \tag{7.4.2}$$

In ordinary optics the components of the electromagnetic field also satisfy a homogeneous linear equation of the type (7.4.1) in which

$$L = \Delta - \frac{1}{c^2}\frac{\partial^2}{\partial t^2} \tag{7.4.3}$$

where c is the velocity of light.

In both cases the wave equation (7.4.1) has a plane-wave solution

$$\psi(\mathbf{r}, t) = a\exp[i(\mathbf{k}\cdot\mathbf{r} - \omega_{\mathbf{k}}t)] \tag{7.4.4}$$

in which

$$\omega_{\mathbf{k}} = \begin{cases} \hbar k^2/2m, & \text{neutrons} \\ ck, & \text{light} \end{cases} \tag{7.4.5}$$

The general solution of Eq. 7.4.1 can therefore be expressed as a coherent superposition of plane waves, that is, as a *wave packet*,

$$\psi(\mathbf{r}, t) = \int a(\mathbf{k})\exp[i(\mathbf{k}\cdot\mathbf{r} - \omega_{\mathbf{k}}t)]\,d\mathbf{k} \tag{7.4.6}$$

The wave packets that we are primarily interested in are ones that have been prepared by Bragg reflection by a perfect crystal. The crystal has the effect of modifying the values of the amplitudes $a(\mathbf{k})$ (see Eqs. 6.2.5 and 6.2.6). In particular, it filters out all components in the incident-neutron wave packet that do not satisfy Bragg's law to within the width of the reflectivity curve. Thus, if $\mathbf{k}$ henceforth denotes the central wave vector in the reflected-neutron wave packet then $a(\mathbf{k}') \simeq 0$ unless $\mathbf{k}' \simeq \mathbf{k}$.

Let us suppose for simplicity that $a(\mathbf{k}')$ does not depend on $\mathbf{k}$ and $\mathbf{k}'$ separately but only on the difference $\mathbf{q} = \mathbf{k}' - \mathbf{k}$. Then Eq. 7.4.6 becomes

$$\psi(\mathbf{r}, t) = G(\mathbf{r}, t) \exp[i(\mathbf{k} \cdot \mathbf{r} - \omega_{\mathbf{k}} t)] \tag{7.4.7}$$

where

$$G(\mathbf{r}, t) = \int a(\mathbf{q}) \exp\{i[\mathbf{q} \cdot \mathbf{r} - (\omega_{\mathbf{k}+\mathbf{q}} - \omega_{\mathbf{k}})t]\}\, d\mathbf{q} \tag{7.4.8}$$

The reflected-neutron wave packet can therefore be thought of as a plane wave with wave vector $\mathbf{k}$ that is modulated by the envelope function $G(\mathbf{r}, t)$.

We now make a Taylor series expansion,

$$\omega_{\mathbf{k}+\mathbf{q}} = \omega_{\mathbf{k}} + \mathbf{q} \cdot \mathbf{v} + O(q^2) \tag{7.4.9}$$

where $\mathbf{v}$ is the *group velocity*,

$$\mathbf{v} = \nabla_{\mathbf{k}} \omega_{\mathbf{k}} = \begin{cases} \hbar\mathbf{k}/m, & \text{neutrons} \\ c\hat{\mathbf{k}} & \text{light} \end{cases} \tag{7.4.10}$$

and $\hat{\mathbf{k}}$ denotes a unit vector in the direction of $\mathbf{k}$. Then

$$G(\mathbf{r}, t) = \int a(\mathbf{q}) \exp\{i[\mathbf{q} \cdot (\mathbf{r} - \mathbf{v}t) - O(q^2 t)]\}\, d\mathbf{q} \tag{7.4.11}$$

Since $a(\mathbf{q}) \simeq 0$ unless $\mathbf{q} \simeq 0$, we can evidently neglect the $q^2 t$ term for small t. In this case

$$G(\mathbf{r}, t) = G(\mathbf{r} - \mathbf{v}t) \tag{7.4.12}$$

where

$$G(\mathbf{r}) \equiv G(\mathbf{r}, 0) = \int a(\mathbf{q}) \exp(i\mathbf{q} \cdot \mathbf{r})\, d\mathbf{q} \tag{7.4.13}$$

The wave packet of Eq. 7.4.7 then becomes

$$\psi(\mathbf{r}, t) = G(\mathbf{r} - \mathbf{v}t) \exp[i\mathbf{k} \cdot (\mathbf{r} - \mathbf{u}t)] \tag{7.4.14}$$

in which, $\mathbf{u}$ is the *phase velocity*,

$$\mathbf{u} = \frac{\omega_{\mathbf{k}}}{k} \hat{\mathbf{k}} = \begin{cases} \hbar\mathbf{k}/2m, & \text{neutrons} \\ c\hat{\mathbf{k}}, & \text{light} \end{cases} \tag{7.4.15}$$

Thus, the envelope moves like a rigid body with the group velocity $\mathbf{v}$. Within the envelope $\psi(\mathbf{r}, t)$ is a plane wave with wave vector $\mathbf{k}$ that moves with the phase velocity $\mathbf{u}$.

For light, $\mathbf{u} = \mathbf{v}$, so that the plane-wave factor in Eq. 7.4.14 is at rest relative to the envelope. In other words, the entire wave packet moves like a rigid body. For neutrons, on the other hand, $\mathbf{u} = \mathbf{v}/2$, so that the plane-wave factor moves backward with velocity $-\mathbf{u}$ relative to the envelope.

The $q^2 t$ term in Eq. 7.4.11, which we have neglected, vanishes identically for a light wave is one dimension. In all other cases this term is nonvanishing and produces a spreading or *dispersion* of the wave packet at large t.

Thermal-neutron beams

A system that is described by a single wave function ψ is said to be in a *pure state*. Otherwise, it is in a *mixed state* and an appropriate ensemble of different wave functions is required for its description.

A collimated neutron beam from a thermal fission reactor is an example of a system that is in a mixed state. The neutron number density in the beam is sufficiently small that neutron–neutron collisions are neglibible and the beam can, therefore, be regarded as a system of free neutrons. The pure states of a single free neutron are wave packets of the type of Eq. 7.4.6, so that the beam can be represented by an appropriate ensemble of such wave packets.

It is important to keep in mind that the wave packets in the ensemble are not intended to correspond to the individual neutrons in the beam. If this were so then each neutron in the beam would be in a pure state. In fact, each neutron is in a mixed state that is represented by the ensemble of wave packets. Since all the neutrons are statistically equivalent, we need only consider a single ensemble of wave packets. However, there is clearly not a one-to-one correspondence between the neutrons and the wave packets.

Since the neutron beam is prepared by random collisions with the atoms in the moderator, we may assume that the coefficients $a(\mathbf{k})$ in the corresponding ensemble are random variables subject only to the constraint that the momentum distribution for the ensemble is the Maxwell–Boltzmann distribution at the temperature of the moderator.

Coherence

The macroscopic properties of a thermal-neutron beam are described by the density matrix

$$\Gamma(\mathbf{r}, t) = \langle \psi(0, 0)^* \psi(\mathbf{r}, t) \rangle \tag{7.4.16}$$

where $\psi(\mathbf{r}, t)$ is the wave packet of Eq. 7.4.6 and the brackets $\langle \cdots \rangle$ denote the ensemble average. Thus, $\Gamma(\mathbf{r}, t)$ is the autocorrelation function of $\psi(\mathbf{r}, t)$. Since $L\psi = 0$ according to Eq. 7.4.1, it follows that $L\Gamma = 0$ too. In other words, $\Gamma(\mathbf{r}, t)$ satisfies the same wave equation as does $\psi(\mathbf{r}, t)$.

The density matrix $\Gamma(\mathbf{r}, t)$ is said to describe the *coherence* of the neutron beam. In particular, $\Gamma(\mathbf{r}, 0)$ describes its *spatial coherence* and $\Gamma(0, t)$ its *temporal coherence*. In optics, the density matrix $\Gamma(\mathbf{r}, t)$ is called the *mutual coherence function* (Mandel and Wolf, 1965; Born and Wolf, 1975; Hamilton et al., 1983).

Since the amplitudes $a(\mathbf{k})$ in Eq. 7.4.6 are generated by random processes, we may assume that

$$\langle a(\mathbf{k}) \rangle = 0 \tag{7.4.17}$$

and that $a(\mathbf{k})$ and $a(\mathbf{k}')$ are statistically independent if $\mathbf{k} \neq \mathbf{k}'$. Hence,

$$\langle a(\mathbf{k})^* a(\mathbf{k}') \rangle = \rho(\mathbf{k})\, \delta(\mathbf{k} - \mathbf{k}') \tag{7.4.18}$$

in which

$$\rho(\mathbf{k})^* = \rho(\mathbf{k}) \geq 0 \tag{7.4.19}$$

Hence, the density matrix (Eq. 7.4.16) becomes

$$\Gamma(\mathbf{r}, t) = \int \rho(\mathbf{k}) \exp[i(\mathbf{k} \cdot \mathbf{r} - \omega_{\mathbf{k}} t)]\, d\mathbf{k} \tag{7.4.20}$$

One can also easily verify that

$$\langle \psi(\mathbf{r}, t)^* \psi(\mathbf{r}', t') \rangle = \Gamma(\mathbf{r}' - \mathbf{r}, t' - t) \tag{7.4.21}$$

so that the neutron beam is marcosopically homogeneous in space and stationary in time.

The average neutron number density is given according to Eq. 2.4.3 by

$$\rho = \langle |\psi(\mathbf{r}, t)|^2 \rangle = \Gamma(0, 0) = \int \rho(\mathbf{k})\, d\mathbf{k} \tag{7.4.22}$$

Thus, $\rho(\mathbf{k})\, d\mathbf{k}$ is the average number of neutrons per unit volume in $d\mathbf{k}$. The average neutron current density is

$$\begin{aligned} \mathbf{J} &= \left\langle \operatorname{Re}\left[\psi(\mathbf{r}, t)^* \frac{\hbar}{im} \nabla \psi(\mathbf{r}, t)\right] \right\rangle \\ &= \int \rho(\mathbf{k}) \frac{\hbar \mathbf{k}}{m}\, d\mathbf{k} \end{aligned} \tag{7.4.23}$$

We see that ρ and $\mathbf{J}$ are independent of $\mathbf{r}$ and t as they should be.

Coherence effects in interferometry

The wave function (7.2.8) for the 0-beam from a perfect-crystal neutron interferometer can be expressed as

$$\psi_2 = a(1 + s) \exp(i\mathbf{k} \cdot \mathbf{r}) \tag{7.2.24}$$

where $a = r^2 t$ and $s = \exp(i\chi)$. The phase shift is given, according to Eq. 7.1.17, by $\chi = kL$, where L is the optical path difference. Alternatively, we can write $\chi = \mathbf{k} \cdot \mathbf{L}$ where $\mathbf{L} = L\hat{\mathbf{k}}$ and $\hat{\mathbf{k}}$ is a unit vector in the direction of $\mathbf{k}$ as before.

Then, including the time factor,

$$\psi_2(\mathbf{r}, t) = \psi(\mathbf{r}, t) + \psi(\mathbf{r} + \mathbf{L}, t) \tag{7.4.25}$$

where

$$\psi(\mathbf{r}, t) = a \exp[i(\mathbf{k} \cdot \mathbf{r} - \omega_{\mathbf{k}} t)] \tag{7.4.26}$$

It has been assumed that the incident beam is represented by a plane wave with wave vector **k**. In reality, however, the incident beam is a thermal-neutron beam from a reactor and is represented by a statistical ensemble of wave packets, as discussed earlier. The interferometer filters out all components in each incident-neutron wave packet that do not satisfy Bragg's law to within the width of the rocking curve of the instrument. Thus, the wave function $\psi_2(\mathbf{r}, t)$ for the 0-beam still has the form of Eq. 7.4.25, but $\psi(\mathbf{r}, t)$ is now a wave packet of the type of Eq. 7.4.7. Since the wave vectors in this wave packet differ only slightly in direction, the vector **L** can be defined in terms of the direction of the average wave vector **k**.

The average intensity of the 0-beam is given by

$$\begin{aligned} I_2 &= \langle |\psi_2(\mathbf{r}, t)|^2 \rangle \\ &= 2\Gamma(0, 0) + \Gamma(\mathbf{L}, 0) + \Gamma(-\mathbf{L}, 0) \end{aligned} \tag{7.4.27}$$

where

$$\Gamma(\mathbf{L}, 0) = \int \rho(\mathbf{k}) \exp(i\mathbf{k} \cdot \mathbf{L})\, d\mathbf{k} \tag{7.4.28}$$

Since $\rho(\mathbf{k})$ is real it follows that $\Gamma(-\mathbf{L}, 0) = \Gamma(\mathbf{L}, 0)^*$ and, hence, that

$$I_2 = 2[\Gamma(0, 0) + \operatorname{Re} \Gamma(\mathbf{L}, 0)] \tag{7.4.29}$$

It will be noted that I_2 is independent of **r** and t. This means, for example, that I_2 is independent of where the detector is located in the 0-beam. It also means that I_2 is independent of the spreading of the individual wave packets in the beam (Klein et al., 1983). We stated earlier that, in general, $\Gamma(\mathbf{r}, 0)$ describes the spatial coherence of the neutron beam. Since **L** is in the direction of **k**, it follows that $\Gamma(\mathbf{L}, 0)$ and, hence, I_2 describe the *longitudinal spatial coherence* of the 0-beam.

As emphasized by Comsa (1983), $\Gamma(\mathbf{L}, 0)$ is the Fourier transform of the wave-vector distribution function. If, as before, we let $\mathbf{q} = \mathbf{k}' - \mathbf{k}$, where $\mathbf{k}'$ denotes an arbitrary wave vector in the beam and **k** the central wave vector, then

$$\Gamma(\mathbf{L}, 0) = \rho C(\mathbf{L}) \exp(i\chi) \tag{7.4.30}$$

where

$$C(\mathbf{L}) = \frac{1}{\rho} \int \rho(\mathbf{k} + \mathbf{q}) \exp(i\mathbf{q} \cdot \mathbf{L})\, d\mathbf{q} \tag{7.4.31}$$

and has the normalization property $C(0) = 1$. Since $\Gamma(0, 0) = \rho$ we get

$$I_2 = 2\rho\{1 + \operatorname{Re}[C(\mathbf{L}) \exp(i\chi)]\} \tag{7.4.32}$$

Let us assume for simplicity that $\rho(\mathbf{k} + \mathbf{q})$ is an even function of **q**.

Then $C(\mathbf{L})$ is real and

$$I_2 = \bar{I}_2[1 + C(\mathbf{L}) \cos \chi] \tag{7.4.33}$$

where $\bar{I}_2 = 2\rho$ and $C(\mathbf{L})$ is the fringe contrast (see Eq. 7.1.11). If $\psi(\mathbf{r}, \mathrm{t})$ were a plane wave then we would have $\rho(\mathbf{k} + \mathbf{q}) = \rho\, \delta(\mathbf{q})$ and $C(\mathbf{L}) = 1$ as before (see Eq. 7.2.12). The finite distribution of wave vectors that is always present in reality therefore leads to a reduction in the fringe contrast.

In general, $\psi(\mathbf{r}, t)$ is a wave packet of the type of Eq. 7.4.7. Thus,

$$\psi(\mathbf{L}, 0) = G(\mathbf{L}) \exp(i\chi), \qquad \psi(0, 0) = G(0) \tag{7.4.34}$$

where $G(\mathbf{L}) \equiv G(\mathbf{L}, 0)$ is the envelope of the wave packet at $t = 0$. Hence,

$$\begin{aligned} \Gamma(\mathbf{L}, 0) &= \langle \psi(0, 0)^* \psi(\mathbf{L}, 0) \rangle \\ &= \rho C(\mathbf{L}) \exp(i\chi) \end{aligned} \tag{7.4.35}$$

where

$$C(\mathbf{L}) = \frac{1}{\rho} \langle G(0)^* G(\mathbf{L}) \rangle \tag{7.4.36}$$

Thus, the fringe contrast $C(\mathbf{L})$ is given, equivalently, by the autocorrelation function of the wave-packet envelope $G(\mathbf{L})$.

Earlier, we discussed a number of effects that reduce the fringe contrast in neutron interferometry. These included imperfections in the interferometer, absorption in the sample, etc. Kaiser et al. (1983) have measured the fringe contrast as a function of L using a series of aluminum slabs and have demonstrated that the above effects account for only 60 percent of the measured fringe contrast. they attributed the remaining 40 percent to the finite longitudinal coherence of the neutron beam as represented by Eq. 7.4.36. The measured $C(\mathbf{L})$ curve was found to be approximately Gaussian with a full width at half maximum (FWHM) of 94 Å. This then represents the *longitudinal coherence length* of the neutron wave packets. This does not, of course, mean that the individual wave packets are themselves Gaussian. In fact, $G(\mathbf{L})$ is an essentially random function of $\mathbf{L}$. On the other hand, it does mean that the wave-vector distribution function $\rho(\mathbf{k} + \mathbf{q})$ is an approximately Gaussian function of $\mathbf{q}$ with a FWHM that is inversely proportional to the longitudinal coherence length.

Fourth-order coherence

The effects just discussed are referred to as *second-order coherence effects,* since they depend on correlations in the wave function at two space-time points (see Eq. 7.4.21). In general, one can define a whole hierarchy of density matrices in which the nth member describes the correlations in the wave function at n space-time points. Of particular interest is the fourth-order density matrix $\langle |\psi(\mathbf{r}, t)|^2\, |\psi(\mathbf{r}', t')|^2 \rangle$, which represents intensity correlations at two space-time points. The most notable example of this type of *fourth-order coherence* effect is the Hanbury–Brown and Twiss experiment in ordinary optics (Mandel and Wolf, 1965).

Goldberger et al. (1963) pointed out that an analogous experiment is also possible with particles. In particular, they showed that it should in principle be possible to observe intensity correlations with particle beams from two incoherent sources that are scattered from the same target and observed with two correlated detectors. However, Klein et al. (1983) assert that such an experiment is still far beyond experimental capabilities for thermal neutrons.

7.5. Multiple-beam interference

In neutron optics the most important examples of multiple-beam interference occur in thin-film devices such as multilayer monochromators and supermirrors. In this final section we present a brief discussion of the theory and application of such devices. More information can be found in the articles by Saxena and Schoenborn (1977), Sears (1983b), Majkrzak (1984), and Majkrzak and Passell (1985).

Multilayer monochromators

Figure 7.9 represents an idealized multilayer monochromator consisting of alternating layers of two different materials A and B that have been

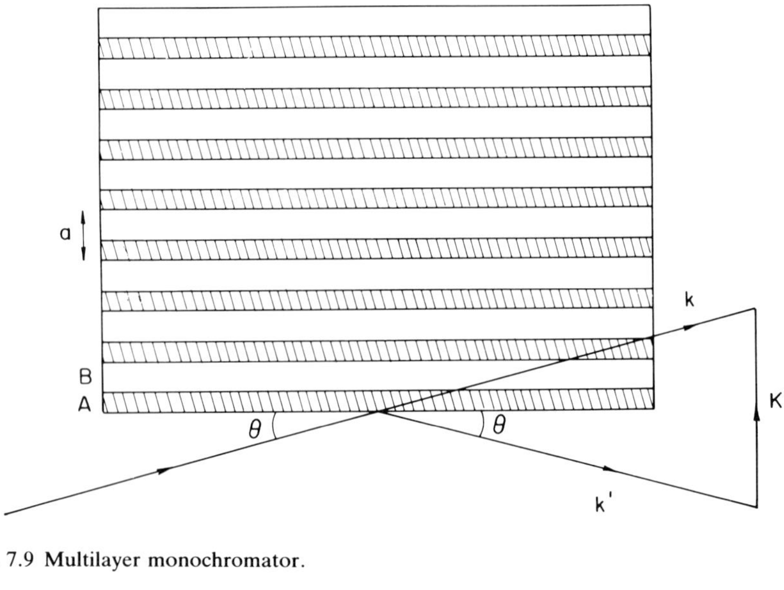

Fig. 7.9 Multilayer monochromator.

deposited by vacuum evaporation or sputtering on a suitable substrate such as an optical flat. The total thickness of the multilayers is then $d = Na$ where a is the unit bilayer thickness and N is the total number of bilayers.

Suppose a collimated monoenergetic beam of neutrons with wave vector $k = 2\pi/\lambda$ is incident on the device at an angle θ, as shown in Fig. 7.9. The incident beam will then be partially reflected and refracted at each interface, and the various reflected waves will interfere with one another such that for some values of θ the net reflected intensity will be small while for other values θ it will be large.

A convenient way of understanding the behavior of such a multilayer device is to think of it as a one-dimensional crystal in which a is the lattice constant. The results derived in Chaps. 5 and 6 can then easily be modified to obtain expressions for the reflecting properties of the multilayer. Thus, for example, the reflected intensity will be a maximum when λ and θ satisfy Bragg's law (Eq. 5.2.24),

$$n\lambda = 2a \sin \theta \tag{7.5.1}$$

where $n = 1, 2, 3, \ldots$ is the order of the reflection.

The reflected intensities for two Ge–Mn multilayers measured by Schoenborn et al. (1974) in the original paper on this subject are shown in Fig. 7.10. The data, shown on a logarithmic scale, are for multilayers

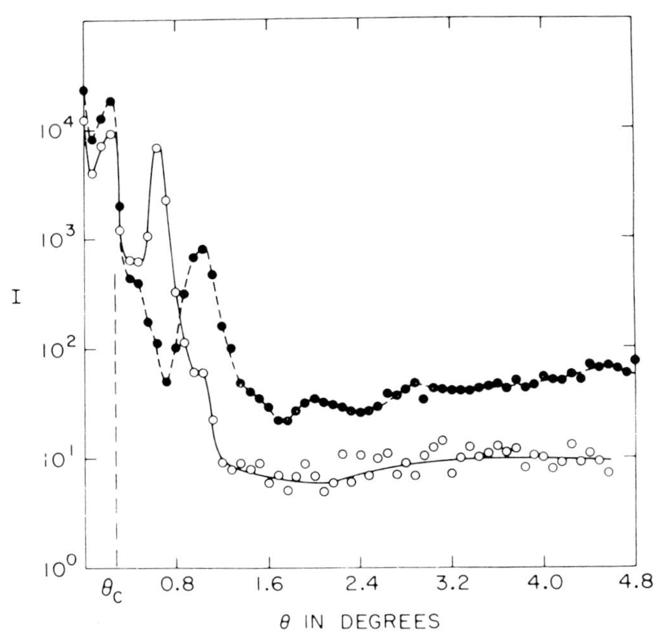

Fig. 7.10 Reflected intensities for two Ge–Mn multilayers measured by Schoenborn et al. (1974).

with $a = 120$ Å (solid circles) and $a = 191$ Å (open circles). In both cases the total number of bilayers is $N = 10$ and the measurements were performed with neutrons of wavelength $\lambda = 4.2$ Å. The most prominent features in Fig. 7.10 are the first-order Bragg peaks that occur at $\theta = 1.00$ degree for the 120 Å multilayer and at $\theta = 0.63$ degree for the 191 Å multilayer. The higher-order peaks with $n \geq 2$ are almost entirely suppressed, for reasons that will be discussed later. The mirror reflection of neutrons at $\theta < \theta_c$, which was discussed in Chap. 3, is also present in Fig. 7.10.

Structure factor

The integrated intensities of the Bragg peaks in a crystal are determined by the unit-cell structure factor (Eq. 5.2.17). The corresponding quantity for a multilayer is

$$F = \int b(\mathbf{r}) \exp(i\mathbf{K} \cdot \mathbf{r})\, d\mathbf{r} \tag{7.5.2}$$

where $b(\mathbf{r})$ is the scattering length density, $\mathbf{K}$ is the reciprocal lattice vector, and the integration extends over one unit bilayer. The reciprocal lattice vector $\mathbf{K}$ is normal to the surface of the multilayer and has a magnitude $K = 2\pi n/a$.

Let us choose a system of Cartesian coordinate axes such that $\mathbf{K}$ is in the z direction. Then the characteristic property of a multilayer is that $b(\mathbf{r})$ is a periodic function of z and is independent of x and y. Hence,

$$G \equiv \frac{F}{V_0} = \frac{1}{a} \int_0^a b(z) \exp(iKz)\, dz, \tag{7.5.3}$$

where $V_0 = Aa$ is the volume of a unit bilayer, the area of which is A.

For an ideal bilayer,

$$b(z) = \begin{cases} \rho_A b_A \equiv f_A, & 0 < z < sa \\ \rho_B b_B \equiv f_B, & sa < z < a \end{cases} \tag{7.5.4}$$

where ρ_A and ρ_B are the number densities of the materials A and B, b_A and b_B are the corresponding bound coherent scattering lengths, and s is the fractional thickness of the A-layer. Hence,

$$G = (f_A - f_B) \frac{\exp(2ns\pi i) - 1}{2n\pi i} \tag{7.5.5}$$

and

$$|G| = \left| (f_A - f_B) \frac{\sin(ns\pi)}{n\pi} \right| \tag{7.5.6}$$

In particular, if the A and B layers are equally thick, so that $s = 1/2$, then

$$|G| = \begin{cases} |f_A - f_B|/n\pi, & n = 1, 3, \ldots \\ 0, & n = 2, 4, \ldots \end{cases} \tag{7.5.7}$$

Thus, the even-order reflections are absent when $s = \frac{1}{2}$.

In practice, the A and B materials will tend to interdiffuse so that the boundary between these two components will not be sharply defined. In the extreme case in which the scattering length density varies sinusoidally,

$$b(z) = \frac{1}{2}\left[(f_A + f_B) + (f_A - f_B)\sin\frac{2\pi z}{a}\right] \tag{7.5.8}$$

we get

$$G = \frac{i}{4}(f_A - f_B)\,\delta_{n,1} \tag{7.5.9}$$

so that

$$|G| = \begin{cases} |f_A - f_B|/4, & n = 1 \\ 0, & n = 2, 3, 4, \ldots \end{cases} \tag{7.5.10}$$

In this case the higher-order reflections are all absent and only the first-order reflection occurs. This situation is apparently realized to a good approximation for the Ge–Mn multilayers in Fig. 7.10.

Optimum number of bilayers

The reflected-neutron flux is proportional to the integrated reflectivity (Sears, 1983b). Thus, an important consideration in the design of a multilayer monochromator is the number of bilayers that are needed to saturate the integrated reflectivity. According to Eq. 6.3.2, the integrated reflectivity of a Bragg peak is proportional to $\tanh y$, where (see Eq. 5.2.85),

$$y = \frac{\pi d}{\Delta} \tag{7.5.11}$$

Here $d = Na$ is the total thickness of the multilayer when the number of bilayers is N, and Δ is the extinction length,

$$\Delta = \frac{\pi \sin\theta}{\lambda\,|G|} \tag{7.5.12}$$

Thus, for the first-order reflection, we can put

$$y = \frac{N}{N_0} \tag{7.5.13}$$

where

$$N_0 = \frac{1}{2a^2\,|G|} \tag{7.5.14}$$

In particular, for an ideal bilayer, where $|G|$ is given by Eq. 7.5.7,

$$N_0 = \frac{\pi}{2a^2\,|f_A - f_B|} \tag{7.5.15}$$

The proportionality factor $\tanh y$ has the value 0.76 when $N = N_0$, and

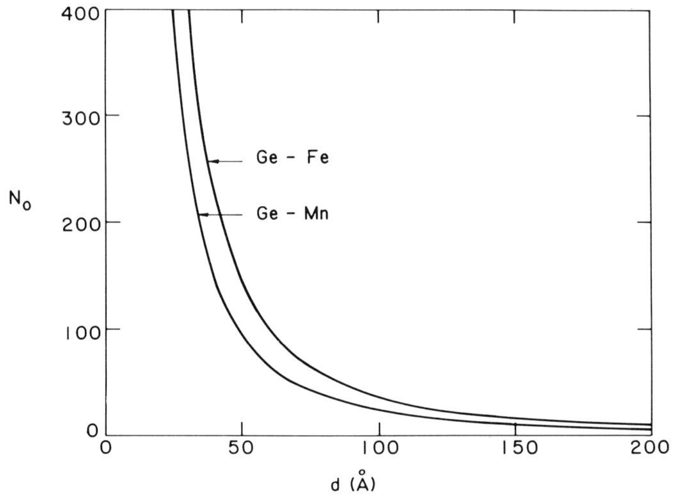

Fig. 7.11 The quantity N_0 that determines the optimum number of bilayers as a function of the bilayer thickness a for two different bilayers (from Sears, 1983b).

0.96 when $N = 2N_0$. Thus, the reflected-neutron flux will be significantly less than its maximum value if $N < N_0$, and it will be essentially saturated if $N > 2N_0$. For an ideal bilayer it is therefore desirable that N be somewhere between N_0 and $2N_0$. In reality, variations in the uniformity and thickness of the bilayers will reduce the reflectivity so that the optimum value of N may be somewhat larger than $2N_0$.

Figure 7.11 shows N_0 as a function of a as calculated by Sears (1983b) from Eq. 7.5.15 for Ge–Mn and Ge–Fe multilayers. To reduce the length of multilayer needed to produce a reflected beam of a specified width, it is desirable that the Bragg angle θ be as large as possible (see Eq. 3.1.35). This, in turn, requires that a be as small as possible. It is difficult to produce uniformly thick layers with $a < 30$ Å. If one takes $a \simeq 50$ Å as a suitable compromise, then it is evident from Fig. 7.11 that N_0 will be typically 100 to 200 and, hence, that values of N of the order of 200 to 400 will be needed to saturate the integrated reflectivity of an ideal multilayer.

It is evident from Eq. 7.5.15 that N_0, and hence the optimum number of bilayers, is independent of λ. On the other hand, the Bragg angle θ and, hence, the width of the reflected beam increase with increasing λ. Thus, it is clear that multilayer monochromators are most useful at long wavelengths where $\lambda \gtrsim 4$ Å, say. In contrast, crystal monochromators are most useful at short wavelengths where $\lambda \lesssim 4$ Å (see Fig. 5.15).

Polarizing monochromators

If one of the materials in the unit bilayer is ferromagnetic then, as we shall see below, the Bragg-reflected beam will be polarized. This effect was first demonstrated by Lynn et al. (1976) and such devices are becoming of increasing interest for use in polarized-beam spectrometers.

Suppose the A-layer is nonmagnetic while the B-layer is ferromagnetic. Then $f_A = \rho_A b_A$ as before but $f_B = \rho_B(b_B \pm b_M)$. Here b_M is the magnetic scattering length (Eq. 3.5.13) and the $\pm$ refers to neutrons with spin parallel (+) or antiparallel (−) to the magnetization. Hence, Eq. 7.5.15 becomes

$$N_{\pm} = \frac{\pi}{2a^2\,|\Delta f \mp f_M|} \qquad (7.5.16)$$

where $\Delta f = \rho_A b_A - \rho_B b_B$ and $f_M = \rho_B b_M$. Suppose the materials A and B are chosen such that $f_M \simeq \Delta f$. In this case $N_+ \gg N_-$, so that if $N \simeq 2N_-$ the reflectivity will be very large for the − spins and practically zero for the + spins and, hence, the reflected beam will be almost completely polarized in the minus direction. This situation occurs for Ge–Fe multilayers and polarizations $\gtrsim 98$ percent can be obtained in practice.

Figure 7.12 shows the reflectivity of a Ge–Fe multilayer measured by

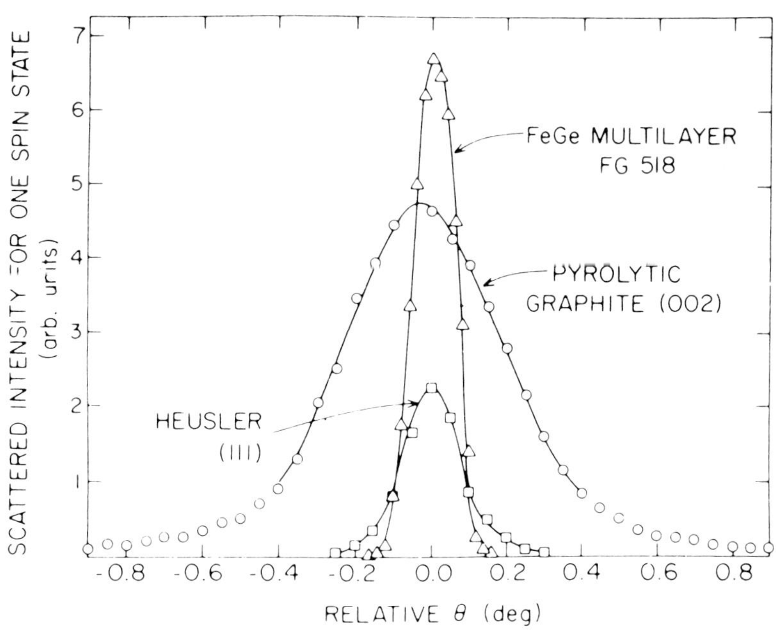

Fig. 7.12 Comparison of the measured reflectivities for one spin state of a Ge–Fe multilayer, a Heusler crystal, and a pyrolytic graphite crystal (from Majkrzak and Passell, 1985).

Majkrzak and Passell (1985) with a polarized incident beam of 2.4 Å neutrons. The unit bilayer thickness is $a = 55$ Å and the total number of bilayers is $N = 1500$. Figure 7.12 also shows the corresponding measurements for the (111) reflection of a Heusler crystal and the (002) reflections of pyrolytic graphite. As discussed in Sec. 5.6, Heusler crystals are the current preferred choice for polarizing monochromators, while pyrolytic graphite is, in many respects, the best monochromator crystal for unpolarized neutrons.

Relative to the Ge–Fe multilayer, the integrated reflectivity is a factor of 4 larger for pyrolytic graphite and a factor of 2 smaller for the Heusler crystal. It is therefore evident that, in this respect, the multilayer is not as good as pyrolytic graphite for unpolarized neutrons but is clearly superior to the Heusler crystal for polarized neutrons. There are, nevertheless, situations in which the Heusler crystal would be preferable, such as for short wavelengths or when a wide reflected beam is required.

Supermirrors

Consider a stack of multilayers with gradually increasing values of a such that the Bragg peaks of the individual multilayers overlap. Then the reflectivity will be large for all values of θ up to some maximum value θ_m determined by the smallest value of a in the stack, and the reflectivity will be small when $\theta > \theta_m$. The same effect can clearly be produced in a single multilayer that contains a suitable gradient in the unit bilayer thickness a. In practice, the cut-off angle θ_m can be a factor of 3 or more larger than the ordinary critical angle for total reflection (Eq. 3.1.34). Such a device, called a *supermirror,* was first proposed by Mezei (1976) and subsequently demonstrated by Mezei and Dagleish (1977).

Detailed calculations of the properties of supermirrors have been given by Schelten and Mika (1979) and Ebisawa et al. (1979). The primary application of supermirrors is for neutron polarization. Nonpolarizing supermirrors are, however, finding increasing use in neutron guide tubes (see Sec. 3.1). A review of current supermirror applications has been given by Schärpf (1982).

8

TRANSPORT EFFECTS IN NEUTRON DIFFRACTION

The single crystals that are used as samples in neutron crystallography studies or as monochromators in neutron spectrometers are mosaic crystals. The nature of multiple Bragg reflection is completely different in a mosaic crystal from that in a perfect crystal. In the latter case it is described by the dynamical theory of neutron diffraction given in Chap. 6 and appears as a coherent transfer of intensity back and forth between the transmitted and the Bragg-reflected waves. This, in turn, gives rise to characteristic dynamical diffraction effects such as Pendellösung interference.

In a mosaic crystal, on the other hand, the spatial coherence of the neutron wave function is destroyed because the mosaic blocks scatter incoherently (see Sec. 5.3). The net result is that, while the cross sections for individual collisions must still be computed quantum-mechanically, the neutrons can otherwise be regarded as classical particles. Thus, if a beam of neutrons is incident on a mosaic crystal with a wave vector that satisfies Bragg's law for some set of reflection planes, the neutrons will be reflected back and forth between these planes like classical particles until they leave the crystal. As illustrated in Fig. 5.9, those neutrons that are Bragg reflected $0, 2, 4, \ldots$ times will then contribute to the external transmitted beam, while those that are Bragg reflected $1, 3, 5, \ldots$ times will contribute to the external Bragg-reflected beam.

When the neutrons are treated as classical particles, the scattered-neutron distribution is determined by a transport equation similar to the Boltzmann equation rather than by the Schrödinger equation. Neutron transport theory was originally developed to describe the thermalization and diffusion of neutrons in bulk matter and now forms the theoretical basis for reactor physics and engineering (Davison and Sykes, 1957). It is also used to calculate multiple-scattering corrections for thermal-neutron scattering experiments on condensed matter (Vineyard, 1954; Sears, 1975) in addition to the present subject of multiple Bragg reflection in mosaic crystals (Sears, 1977).

The use of neutron transport theory in the present context requires that multiple Bragg reflection within an individual mosaic block be negligible. This, in turn, requires that the linear dimensions of the mosaic blocks be small in comparison with the extinction length as discussed in Sec. 5.3. In what follows we shall assume that this condition is well satisfied so that the effect of primary extinction can be neglected.

8.1. Neutron transport theory

Neutron transport equation

The fundamental quantity of interest in neutron transport theory is the *phase-space distribution function* $f(\mathbf{r}, \mathbf{k}, t)$, which is defined such that $f(\mathbf{r}, \mathbf{k}, t)\, d\mathbf{r}\, d\mathbf{k}$ is the average number of neutrons in the volume element $d\mathbf{r}$ with wave vector in $d\mathbf{k}$ at time t. The quantity $\mathbf{k}$ is now not to be thought of as a wave vector per se but simply as the classical velocity in units of $\hbar/m$ so that $\hbar\mathbf{k} \equiv m\mathbf{v}$. In general,

$$f(\mathbf{r}, \mathbf{k}, t) = \left\langle \sum_j \delta(\mathbf{r} - \mathbf{r}_j)\, \delta(\mathbf{k} - \mathbf{k}_j) \right\rangle \tag{8.1.1}$$

where $\mathbf{r}_j$ is the position and $\mathbf{k}_j$ the wave vector of the jth neutron at time t, and the brackets $\langle \cdots \rangle$ denote an appropriate ensemble average.

The variation of $f(\mathbf{r}, \mathbf{k}, t)$ with time is due partly to the variation of $\mathbf{r}_j$ (the streaming effect) and partly to the variation of $\mathbf{k}_j$ (the collision effect). Thus,

$$\begin{aligned}\frac{\partial f}{\partial t} &= \left\langle \sum_j \left\{ \dot{\mathbf{r}}_j \cdot \frac{\partial}{\partial \mathbf{r}_j} + \dot{\mathbf{k}}_j \cdot \frac{\partial}{\partial \mathbf{k}_j} \right\} \delta(\mathbf{r} - \mathbf{r}_j)\, \delta(\mathbf{k} - \mathbf{k}_j) \right\rangle \\ &\equiv \left(\frac{\partial f}{\partial t}\right)_s + \left(\frac{\partial f}{\partial t}\right)_c\end{aligned} \tag{8.1.2}$$

The streaming term can be expressed as

$$\left(\frac{\partial f}{\partial t}\right)_s = -\mathbf{v} \cdot \nabla f \tag{8.1.3}$$

where $\nabla = \partial/\partial\mathbf{r}$ is the gradient operator, and the collision term as

$$\left(\frac{\partial f}{\partial t}\right)_c = \left\langle \sum_j \frac{\mathbf{F}_j}{\hbar} \cdot \frac{\partial}{\partial \mathbf{k}_j} \delta(\mathbf{r} - \mathbf{r}_j)\, \delta(\mathbf{k} - \mathbf{k}_j) \right\rangle \tag{8.1.4}$$

where $\mathbf{F}_j = \hbar\dot{\mathbf{k}}_j = m\dot{\mathbf{v}}_j$ is the force on the jth neutron. For simplicity we shall ignore the contribution to $\mathbf{F}_j$ from external forces such as gravitational or magnetic fields since these are unimportant for the applications of interest to us here. In this case $\mathbf{F}_j$ can be assumed to represent the short-range neutron–atom interactions. The contribution to $\mathbf{F}_j$ from neutron–neutron interactions is always completely negligible because, in practice, the neutron number density is many orders of magnitude smaller than the atomic number density in condensed matter.

From a phenomenological point of view, the collision term can be expressed alternatively as

$$\left(\frac{\partial f}{\partial t}\right)_c = \int [p_s(\mathbf{k}', \mathbf{k}) f(\mathbf{r}, \mathbf{k}', t) - p_s(\mathbf{k}, \mathbf{k}') f(\mathbf{r}, \mathbf{k}, t)]\, d\mathbf{k}' - p_a(\mathbf{k}) f(\mathbf{r}, \mathbf{k}, t) \tag{8.1.5}$$

where $p_s(\mathbf{k}, \mathbf{k}')\,d\mathbf{k}'$ is the probability per unit time that a neutron with wave vector $\mathbf{k}$ will be scattered into $d\mathbf{k}'$, and $p_a(\mathbf{k})$ is the probability per unit time that this neutron will be absorbed. The neutron transport equation then becomes

$$\left\{\frac{\partial}{\partial t} + \mathbf{v} \cdot \nabla\right\} f(\mathbf{r}, \mathbf{k}, t) =$$

$$\int [p_s(\mathbf{k}', \mathbf{k}) f(\mathbf{r}, \mathbf{k}', t) - p_s(\mathbf{k}, \mathbf{k}') f(\mathbf{r}, \mathbf{k}, t)]\, d\mathbf{k}' - p_a(\mathbf{k}) f(\mathbf{r}, \mathbf{k}, t) \quad (8.1.6)$$

The scattering and absorption probabilities also depend on $\mathbf{r}$, although we have not indicated this explicitly. Thus, in free space, these quantities vanish, while in a homogeneous medium they have constant values that are characteristic of the medium. These quantities are, moreover, independent of $f(\mathbf{r}, \mathbf{k}, t)$, so that the neutron transport equation is linear. This is in contrast to the kinetic theory of gases in which the collision term describes the effect of binary collisions between the atoms and is quadratic in $f(\mathbf{r}, \mathbf{k}, t)$. The corresponding transport equation, i.e., the Boltzmann equation, is then nonlinear.

Equation of continuity

When the neutron transport equation (8.1.6) is integrated over all $\mathbf{k}$, the scattering term vanishes and we obtain the equation of continuity (cf. Eq. 2.4.5),

$$\frac{\partial n}{\partial t} + \nabla \cdot \mathbf{J} = -s \quad (8.1.7)$$

where n is the neutron number density,

$$n(\mathbf{r}, t) = \int f(\mathbf{r}, \mathbf{k}, t)\, d\mathbf{k} \quad (8.1.8)$$

$\mathbf{J}$ is the neutron current density,

$$\mathbf{J}(\mathbf{r}, t) = \int \mathbf{v} f(\mathbf{r}, \mathbf{k}, t)\, d\mathbf{k} \quad (8.1.9)$$

and s is the neutron sink density,

$$s(\mathbf{r}, t) = \int p_a(\mathbf{k}) f(\mathbf{r}, \mathbf{k}, t)\, d\mathbf{k} \quad (8.1.10)$$

A particular solution of the neutron transport equation in free space is

$$f(\mathbf{r}, \mathbf{k}, t) = n_0\, \delta(\mathbf{k} - \mathbf{k}_0) \quad (8.1.11)$$

which represents a collimated monoenergetic beam of neutrons with wave vector $\mathbf{k}_0$. Here, the number density is

$$n(\mathbf{r}, t) = n_0 \quad (8.1.12)$$

and the current density is

$$\mathbf{J}(\mathbf{r}, t) = n_0 \mathbf{v}_0 \tag{8.1.13}$$

where $m\mathbf{v}_0 = \hbar \mathbf{k}_0$.

Another important quantity in neutron transport theory is the flux:

$$\Phi(\mathbf{r}, t) = \int v f(\mathbf{r}, \mathbf{k}, t)\, d\mathbf{k} \tag{8.1.14}$$

In earlier chapters we have used the term *flux* as though it were synonomous with the magnitude of the current density. In fact, this is true for the collimated monoenergetic beam (Eq. 8.1.11) where

$$\Phi(\mathbf{r}, t) = n_0 v_0 \tag{8.1.15}$$

but it is not true in general.

Scattering and absorption probabilities

The scattering and absorption probabilities $p_s(\mathbf{k}, \mathbf{k}')$ and $p_a(\mathbf{k})$ are intrinsic properties of the material medium and are independent of its size and shape. To determine these quantities we can therefore consider a sample of the material whose volume V is small enough that multiple scattering is negligible. In other words, these quantities are given *rigorously* by the kinematical theory of scattering. The effect of multiple scattering is described by the neutron transport equation itself.

The double differential scattering cross section $d^2\sigma$ is defined as the average number of incident neutrons with wave vector $\mathbf{k}$ that are scattered into $d\mathbf{k}'$ per unit time per unit incident flux. Thus,

$$d^2\sigma = \frac{nV}{nv} p_s(\mathbf{k}, \mathbf{k}')\, d\mathbf{k}' \tag{8.1.16}$$

where n is the incident-neutron number density, nv is the incident-neutron flux, and nV is the average number of incident neutrons inside the sample. It then follows from the expression (2.5.18) for $d\mathbf{k}'$ that

$$\frac{d^2\sigma}{d\Omega\, d\varepsilon'} = V\left(\frac{m}{\hbar}\right)^2 \frac{k'}{k} p_s(\mathbf{k}, \mathbf{k}') \tag{8.1.17}$$

The total scattering cross section per unit volume is given by

$$\Sigma_s(\mathbf{k}) = \frac{1}{V}\iint d^2\sigma = \frac{1}{v} p_s(\mathbf{k}) \tag{8.1.18}$$

where $\mathbf{p}_s(\mathbf{k})$ is the probability per unit time that a neutron with wave vector $\mathbf{k}$ will be scattered,

$$p_s(\mathbf{k}) = \int p_s(\mathbf{k}, \mathbf{k}')\, d\mathbf{k}' \tag{8.1.19}$$

Similarly, the absorption cross section per unit volume is given by

$$\Sigma_a(\mathbf{k}) = \frac{1}{v} p_a(\mathbf{k}) \tag{8.1.20}$$

and the total collision cross section per unit volume is

$$\Sigma(\mathbf{k}) = \Sigma_s(\mathbf{k}) + \Sigma_a(\mathbf{k}) \tag{8.1.21}$$

The neutron transport equation (8.1.6) can therefore be expressed as

$$\left\{\frac{1}{v}\frac{\partial}{\partial t} + \hat{\mathbf{k}} \cdot \nabla + \Sigma(\mathbf{k})\right\} f(\mathbf{r}, \mathbf{k}, t) = \int \Gamma(\mathbf{k}', \mathbf{k}) f(\mathbf{r}, \mathbf{k}', t)\, d\mathbf{k}' \tag{8.1.22}$$

where $\hat{\mathbf{k}}$ is a unit vector in the direction of $\mathbf{k}$ and $\Gamma(\mathbf{k}', \mathbf{k})$ is called the *transference function* and is given by

$$\Gamma(\mathbf{k}', \mathbf{k}) = \frac{1}{v} p_s(\mathbf{k}', \mathbf{k}) \tag{8.1.23}$$

For steady-state conditions the distribution function is independent of t and the neutron transport equation reduces to

$$\{\hat{\mathbf{k}} \cdot \nabla + \Sigma(\mathbf{k})\} f(\mathbf{r}, \mathbf{k}) = \int \Gamma(\mathbf{k}', \mathbf{k}) f(\mathbf{r}, \mathbf{k}')\, d\mathbf{k}' \tag{8.1.24}$$

which is the basic equation for the work to be described.

8.2. Neutron transport in mosaic crystals

Mosaic crystals

The neutron transport equation (8.1.24) applies to any material medium that is macroscopically homogeneous, and the coefficients $\Sigma(\mathbf{k})$ and $\Gamma(\mathbf{k}', \mathbf{k})$ in this equation are characteristic of the structure of the material. In particular, for a mosaic crystal the scattering cross section contains contributions from both Bragg scattering and diffuse scattering. Thus, $\Sigma_s(\mathbf{k})$ is of the form (cf. Eq. 5.2.62)

$$\Sigma_s(\mathbf{k}) = \sum_{h \neq 0} \Sigma_h(\mathbf{k}) + \Sigma_d(\mathbf{k}) \tag{8.2.1}$$

where $\Sigma_h(\mathbf{k})$ is the cross section per unit volume for Bragg scattering via the reciprocal lattice vector $\mathbf{K}_h$, and $\Sigma_d(\mathbf{k})$ is the cross section per unit volume for diffuse scattering. The quantity $\Sigma_h(\mathbf{k})$ has been discussed in detail in Chap. 5. Its most important property is that it is appreciably different from zero only if $\mathbf{k}$ satisfies Bragg's law, that is, only if $\mathbf{k}$ lies on or near the Brillouin zone boundary that bisects $\mathbf{K}_h$. Thus,

$$\Sigma_h(\mathbf{k}) \simeq 0 \qquad \text{unless } \mathbf{k} \cdot \mathbf{K}_h \simeq \tfrac{1}{2} K_h^2 \tag{8.2.2}$$

Similarly, from Eq. 8.1.21,

$$\Sigma(\mathbf{k}) = \sum_{h\neq 0} \Sigma_h(\mathbf{k}) + \Sigma'(\mathbf{k}) \tag{8.2.3}$$

where $\Sigma'(\mathbf{k})$ is the cross section per unit volume for all collisions other than Bragg scattering,

$$\Sigma'(\mathbf{k}) = \Sigma_d(\mathbf{k}) + \Sigma_a(\mathbf{k}) \tag{8.2.4}$$

The transference function can be approximated by

$$\Gamma(\mathbf{k}', \mathbf{k}) = \sum_{h\neq 0} \Sigma_h(\mathbf{k}')\, \delta(\mathbf{k}' - \mathbf{k} - \mathbf{K}_h) + \Gamma_d(\mathbf{k}', \mathbf{k}) \tag{8.2.5}$$

in which the first term is the contribution from Bragg scattering and the second is that from diffuse scattering. The effect of the mosaic spread is contained in $\Sigma_h(\mathbf{k}')$, as discussed in Sec. 5.3. The double differential cross section for diffuse scattering is given according to Eqs. 8.1.17 and 8.1.23 by

$$\frac{d^2\sigma_d}{d\Omega\, d\varepsilon'} = V\left(\frac{m}{\hbar}\right)\frac{k'^2}{k}\Gamma_d(\mathbf{k}, \mathbf{k}') \tag{8.2.6}$$

Hence, it follows from Eq. 2.9.2 that

$$\Gamma_d(\mathbf{k}, \mathbf{k}') = \frac{\hbar\rho}{mk'} S(\mathbf{q}, \omega) \tag{8.2.7}$$

where $\rho = N/V$ is the atomic number density and $S(\mathbf{q}, \omega)$ is the scattering function.

The coefficients $\Sigma(\mathbf{k})$ and $\Gamma(\mathbf{k}', \mathbf{k})$ in the neutron transport equation (8.1.24) are now completely specified, so that we can proceed to solve this equation subject to appropriate boundary conditions.

Generalized Darwin equations

Consider a mosaic crystal of arbitrary size and shape that is completely bathed in a collimated monoenergetic beam of incident neutrons, each with wave vector $\mathbf{k}_0$. We want to find the corresponding solution of the transport equation (8.1.24) at all points $\mathbf{r}$ inside the crystal and for all wave vectors $\mathbf{k}$. We consider a trial solution of the form

$$f(\mathbf{r}, \mathbf{k}) = \sum_h n_h(\mathbf{r})\, \delta(\mathbf{k} - \mathbf{k}_h) + f_d(\mathbf{r}, \mathbf{k}) \tag{8.2.8}$$

in which $n_0(\mathbf{r})$ is the nuetron number density in the transmitted beam with wave vector $\mathbf{k}_0$ and, for $h \neq 0$, $n_h(\mathbf{r})$ is the neutron number density in the beam that is Bragg reflected via the reciprocal lattice vector $\mathbf{K}_h$ and has wave vector $\mathbf{k}_h = \mathbf{k}_0 - \mathbf{K}_h$. Finally, $f_d(\mathbf{r}, \mathbf{k})$ is the phase-space distribution function for the diffusely scattered neutrons. The number of Bragg-reflected beams with nonvanishing number density $n_h(\mathbf{r})$ will, of course, depend on the incident-neutron wave vector $\mathbf{k}_0$, as will be discussed later.

Substituting the trial solution (Eq. 8.2.8) into the transport equation (8.1.24) with $\Sigma(\mathbf{k})$ of the form of Eq. 8.2.3 and $\Gamma(\mathbf{k}', \mathbf{k})$ of the form of Eq. 8.2.5 we find that

$$\sum_h \left\{[\hat{\mathbf{k}}_h \cdot \nabla + \Sigma(\mathbf{k}_h)]n_h(\mathbf{r}) - \sum_{h'(\neq h)} \Sigma_{h-h'}(\mathbf{k}_{h'})n_{h'}(\mathbf{r})\right\} \delta(\mathbf{k} - \mathbf{k}_h)$$

$$+ \left\{[\hat{\mathbf{k}} \cdot \nabla + \Sigma'(\mathbf{k})]f_d(\mathbf{r}, \mathbf{k}) - \sum_h \Gamma_d(\mathbf{k}_h, \mathbf{k})n_h(\mathbf{r}) - \int \Gamma_d(\mathbf{k}', \mathbf{k})f_d(\mathbf{r}, \mathbf{k}')\, d\mathbf{k}'\right\}$$

$$+ \sum_{h\neq 0} \{\Sigma_h(\mathbf{k})f_d(\mathbf{r}, \mathbf{k}) - \Sigma_{-h}(\mathbf{k} - \mathbf{K}_h)f_d(\mathbf{r}, \mathbf{k} - \mathbf{K}_h)\} = 0 \quad (8.2.9)$$

in which $h - h'$ refers to the reciprocal lattice vector $\mathbf{K}_{h-h'} \equiv \mathbf{K}_h - \mathbf{K}_{h'}$. The third group of terms in Eq. 8.2.9 describes the effect of the Bragg reflection of neutrons that have already been diffusely scattered. This group of terms is different from zero only if $\mathbf{k}$ is in the neighborhood of a Brillouin zone boundary and, except in special experimental arrangements, is negligible in comparison with the second group of terms. In other words, there is a negligible probability that a diffusely scattered neutron will have a wave vector $\mathbf{k}$ that satisfies the condition for Bragg reflection. By neglecting the third group of terms we therefore assume that, once a neutron has been diffusely scattered, then any subsequent scattering collisions are also diffuse scatterings and not Bragg reflections. In this case diffuse scattering is equivalent to absorption as far as its effect on the attenuation of the transmitted and Bragg-reflected beams is concerned.

Thus, after omitting the third group of terms in Eq. 8.2.9, it follows that this equation can hold for an arbitrary value of $\mathbf{k}$ only if

$$[\hat{\mathbf{k}}_h \cdot \nabla + \Sigma(\mathbf{k}_h)]n_h(\mathbf{r}) = \sum_{h'(\neq h)} \Sigma_{h-h'}(\mathbf{k}_{h'})n_{h'}(\mathbf{r}) \quad (8.2.10)$$

and

$$[\hat{\mathbf{k}} \cdot \nabla + \Sigma'(\mathbf{k})]f_d(\mathbf{r}, \mathbf{k}) = \sum_h \Gamma_d(\mathbf{k}_h, \mathbf{k})n_h(\mathbf{r}) + \int \Gamma_d(\mathbf{k}', \mathbf{k})f_d(\mathbf{r}, \mathbf{k}')\, d\mathbf{k}' \quad (8.2.11)$$

The Eqs. 8.2.10 are called the *generalized Darwin equations*. These are a system of homogeneous linear equations that can be solved, subject to boundary conditions given below, to determine the neutron number densities $n_h(\mathbf{r})$ in the transmitted and Bragg-reflected beams. Equation 8.2.11 is an inhomogeneous linear equation for the phase-space distribution function of the diffusely scattered neutrons. The inhomogeneous term in Eq. 8.2.11 is determined by the solution of Eq. 8.2.10 and its presence indicates that the transmitted and Bragg-reflected beams are each a source of diffuse scattering.

Integral form of the transport equation

We shall assume that the shape of the crystal is such that its surface is nowhere concave. This restriction is convenient to avoid the possibility that a neutron could re-enter the crystal after having passed out through the surface. Let $L(\mathbf{r}, \hat{\mathbf{k}})$ denote the distance from an arbitrary point $\mathbf{r}$ inside the crystal to the surface in the direction $-\hat{\mathbf{k}}$, as illustrated in Fig. 8.1. Since the surface is nowhere concave, the only neutrons that can enter the crystal from the outside are the incident neutrons. Thus, Eq. 8.2.10 must be solved subject to the boundary conditions

$$n_h(\mathbf{r} - L(\mathbf{r}, \hat{\mathbf{k}}_h)\hat{\mathbf{k}}_h) = n\,\delta_{h0} \tag{8.2.12}$$

where n is the incident-neutron number density. The boundary condition for Eq. 8.2.11 is

$$f_d(\mathbf{r} - L(\mathbf{r}, \hat{\mathbf{k}})\hat{\mathbf{k}}, \mathbf{k}) = 0 \tag{8.2.13}$$

Let $\Lambda_{\mathbf{k}}$ denote the following linear integral operator:

$$\Lambda_{\mathbf{k}}\{g(\mathbf{r})\} \equiv \int_0^{L(\mathbf{r},\hat{\mathbf{k}})} \exp[-\Sigma(\mathbf{k})\xi]g(\mathbf{r} - \xi\hat{\mathbf{k}})\,d\xi \tag{8.2.14}$$

where $g(\mathbf{r})$ is an arbitrary function of $\mathbf{r}$, and let $\Lambda'_{\mathbf{k}}$ denote the corresponding operator when $\Sigma(\mathbf{k})$ is replaced by $\Sigma'(\mathbf{k})$. Then one can easily verify that the system of differential equations (8.2.10) and the boundary conditions (Eq. 8.2.12) are together equivalent to the following system of integral equations,

$$n_h(\mathbf{r}) = \delta_{h0} n \exp[-\Sigma(\mathbf{k}_h)L(\mathbf{r}, \hat{\mathbf{k}}_h)] + \sum_{h'(\neq h)} \Sigma_{h-h'}(\mathbf{k}_{h'})\Lambda_{\mathbf{k}_h}\{n_{h'}(\mathbf{r})\} \tag{8.2.15}$$

Similarly, the inhomogeneous integrodifferential equation (8.2.11) and the boundary condition (Eq. 8.2.13) are together equivalent to the

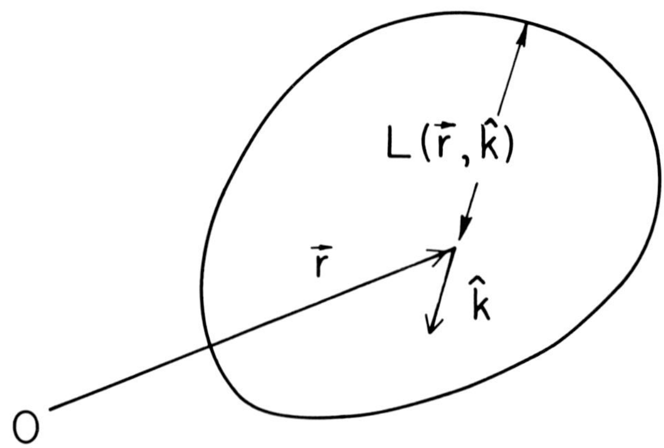

Fig. 8.1 The quantity $L(\mathbf{r}, \hat{\mathbf{k}})$.

following inhomogeneous Fredholm equation,

$$f_d(\mathbf{r}, \mathbf{k}) = \sum_h \Gamma_d(\mathbf{k}_h, \mathbf{k})\Lambda'_{\mathbf{k}}\{n_h(\mathbf{r})\} + \int \Gamma_d(\mathbf{k}', \mathbf{k})\Lambda'_{\mathbf{k}}\{f_d(\mathbf{r}, \mathbf{k}')\}\, d\mathbf{k}' \quad (8.2.16)$$

The iterated solutions of Eqs. 8.2.15 and 8.2.16 generate multiple-scattering expansions in which the successive terms represent contributions from single, double, triple, . . . scattering (Sears 1975, 1977). In neutron crystallography and other condensed-matter applications the samples are always chosen small enough that such expansions are rapidly convergent. In optical devices such as monochromator crystals, on the other hand, such multiple-scattering expansions are not rapidly convergent and a direct solution of Eqs. 8.2.10 or 8.2.15 is required.

Transmissivity and reflectivity

The quantities we are primarily interested in are the transmissivity, $T(\mathbf{k}_0)$, which is defined as the fraction of incident neutrons that are transmitted by the crystal when the incident-neutron wave vector is $\mathbf{k}_0$, and the reflectivity, $R_h(\mathbf{k}_0)$, which is defined as the fraction of incident neutrons that are Bragg-reflected via $\mathbf{K}_h$ when the incident-neutron wave vector is $\mathbf{k}_0$. Equivalently, $T(\mathbf{k}_0)$ is the ratio of the transmitted current to the incident current, and $R_h(\mathbf{k}_0)$ is the ratio of the reflected current to the incident current.

When the phase-space distribution function is of the form of Eq. 8.2.8, the current density (Eq. 8.1.9) becomes

$$\mathbf{J}(\mathbf{r}) = \sum_h n_h(\mathbf{r})\mathbf{v}_h + \mathbf{J}_d(\mathbf{r}) \quad (8.2.17)$$

in which $\mathbf{v}_h = (\hbar/m)\mathbf{k}_h$ is the neutron velocity and $n_h(\mathbf{r})\mathbf{v}_h$ is the current density in the hth beam, while $\mathbf{J}_d(\mathbf{r})$ is the current density of the diffusely-scattered neutrons,

$$\mathbf{J}_d(\mathbf{r}) = \int \mathbf{v} f_d(\mathbf{r}, \mathbf{k})\, d\mathbf{k} \quad (8.2.18)$$

The transmitted current is

$$\int_{S(\hat{\mathbf{k}}_0)} n_0(\mathbf{r})\mathbf{v}_0 \cdot \hat{\mathbf{e}}\, dS \quad (8.2.19)$$

in which dS denotes an element of area at a point $\mathbf{r}$ on the surface of the crystal, $\hat{\mathbf{e}}$ is a unit vector in the direction of the outward normal to the surface at that point, and $S(\hat{\mathbf{k}}_0)$ is that part of the surface of the crystal which is visible from the direction $\hat{\mathbf{k}}_0$. Since the surface is nowhere concave $S(\hat{\mathbf{k}}_0)$ is, more precisely, the subset of surface points for which $\mathbf{v}_0 \cdot \hat{\mathbf{e}} > 0$. The incident current is $nv_0A(\hat{\mathbf{k}}_0)$, where $A(\hat{\mathbf{k}}_0)$ is the cross-sectional area of the crystal perpendicular to the direction $\hat{\mathbf{k}}_0$,

$$A(\hat{\mathbf{k}}_0) = \int_{S(\hat{\mathbf{k}}_0)} \hat{\mathbf{k}}_0 \cdot \hat{\mathbf{e}}\, dS \quad (8.2.20)$$

The transmissivity is therefore given by

$$T(\mathbf{k}_0) = \frac{1}{nA(\hat{\mathbf{k}}_0)} \int_{S(\hat{\mathbf{k}}_0)} n_0(\mathbf{r})\hat{\mathbf{k}}_0 \cdot \hat{\mathbf{e}}\, dS \tag{8.2.21}$$

and, similarly, the reflectivity is given for $h \neq 0$ by

$$R_h(\mathbf{k}_0) = \frac{1}{nA(\hat{\mathbf{k}}_0)} \int_{S(\hat{\mathbf{k}}_h)} n_h(\mathbf{r})\hat{\mathbf{k}}_h \cdot \hat{\mathbf{e}}\, dS \tag{8.2.22}$$

where we have used the fact that $v_h = v_0$ since the Bragg scattering is elastic.

The basic problem is, therefore, to solve the generalized Darwin equations (8.2.10) or the equivalent integral equations (8.2.15) to determine $n_h(\mathbf{r})$ and, hence, to compute the transmissivity (Eq. 8.2.21) and the reflectivity (Eq. 8.2.22). The solution of the generalized Darwin equations depends on the number of reciprocal lattice sites that lie on or near the surface of the Ewald sphere corresponding to $\mathbf{k}_0$. For the m-beam case, in which there are m such points, Eq. 8.2.10 reduces to a system of m coupled equations for the transmitted beam and the $m-1$ Bragg-reflected beams.

Thus, in the one-beam case in which there are no Bragg-reflected beams, $n_0(\mathbf{r})$ is simply attenuated exponentially by absorption and diffuse scattering in accordance with Lambert's law (Eq. 3.2.10). In the two-beam case, Eq. 8.2.10 reduces to the ordinary Darwin equations (Darwin, 1922) for the transmitted beam and the single Bragg-reflected beam. For $m \geq 3$ the generalized Darwin equations provide a basis for the discussion of *parasitic reflections* and the *Renninger effect* in mosaic crystals. The three-beam case has been discussed by Moon and Shull (1964) for the special case of a plane slab. In what follows we shall, for simplicity, confine our attention to the two-beam case.

Two-beam case

Suppose the incident-neutron wave vector $\mathbf{k}_0$ satisfies the condition for Bragg reflection via one particular reciprocal lattice vector $\mathbf{K}_h$. In this case Eq. 8.2.10 reduces to the ordinary Darwin equations (Darwin, 1922),

$$\begin{aligned} [\hat{\mathbf{k}}_0 \cdot \nabla + \Sigma(\mathbf{k}_0)]n_0(\mathbf{r}) &= \Sigma_{-h}(\mathbf{k}_h)n_h(\mathbf{r}) \\ [\hat{\mathbf{k}}_h \cdot \nabla + \Sigma(\mathbf{k}_h)]n_h(\mathbf{r}) &= \Sigma_h(\mathbf{k}_0)n_0(\mathbf{r}) \end{aligned} \tag{8.2.23}$$

in which

$$\begin{aligned} \Sigma(\mathbf{k}_0) &= \Sigma_h(\mathbf{k}_0) + \Sigma'(\mathbf{k}_0) \\ \Sigma(\mathbf{k}_h) &= \Sigma_{-h}(\mathbf{k}_h) + \Sigma'(\mathbf{k}_h) \end{aligned} \tag{8.2.24}$$

and $n_{h'}(\mathbf{r}) = 0$ if $h' \neq 0$, h. The corresponding integral equations (8.2.15) become

$$\begin{aligned} n_0(\mathbf{r}) &= n \exp[-\Sigma(\mathbf{k}_0)L(\mathbf{r}, \hat{\mathbf{k}}_0)] + \Sigma_{-h}(\mathbf{k}_h)\Lambda_{\mathbf{k}_0}\{n_h(\mathbf{r})\} \\ n_h(\mathbf{r}) &= \Sigma_h(\mathbf{k}_0)\Lambda_{\mathbf{k}_h}\{n_0(\mathbf{r})\} \end{aligned} \tag{8.2.25}$$

In what follows we shall adopt the abbreviated notation

$$\begin{aligned} T &= T(\mathbf{k}_0), & R &= R_h(\mathbf{k}_0) \\ \Sigma_h &= \Sigma_h(\mathbf{k}_0), & \Sigma_{-h} &= \Sigma_{-h}(\mathbf{k}_h) \\ \Lambda_0 &= \Lambda_{\mathbf{k}_0}, & \Lambda_h &= \Lambda_{\mathbf{k}_h} \end{aligned} \tag{8.2.26}$$

The coupled equations (8.2.25) can then be combined to give

$$n_0(\mathbf{r}) = n \exp[-\Sigma(\mathbf{k}_0)L(\mathbf{r}, \hat{\mathbf{k}}_0)] + \Sigma_{-h}\Sigma_h\Lambda_0\Lambda_h\{n_0(\mathbf{r})\} \tag{8.2.27}$$

Multiple-scattering expansion

The iterated solution of Eq. 8.2.27 gives the multiple-scattering expansion

$$n_0(\mathbf{r}) = \sum_{j=0}^{\infty} n_0^{(2j)}(\mathbf{r}), \tag{8.2.28}$$

in which $n_0^{(2j)}(\mathbf{r})$ is the contribution to the transmitted beam from neutrons that have been Bragg reflected $2j$ times and is given by the recursion relation

$$n_0^{(2j)}(\mathbf{r}) = \begin{cases} n \exp[-\Sigma(\mathbf{k}_0)L(\mathbf{r}, \hat{\mathbf{k}}_0)], & j = 0 \\ \Sigma_{-h}\Sigma_h\Lambda_0\Lambda_h\{n_0^{(2j-2)}(\mathbf{r})\}, & j \geq 1 \end{cases} \tag{8.2.29}$$

Also, from Eq. 8.2.25,

$$n_h(\mathbf{r}) = \sum_{j=0}^{\infty} n_h^{(2j+1)}(\mathbf{r}) \tag{8.2.30}$$

in which $n_h^{(2j+1)}(\mathbf{r})$ is the contribution to the Bragg-reflected beam from neutrons that have been Bragg reflected $2j+1$ times and is given by

$$n_h^{(2j+1)}(\mathbf{r}) = \Sigma_h\Lambda_h\{n_0^{(2j)}(\mathbf{r})\} \tag{8.2.31}$$

The transmissivity (Eq. 8.2.21) can therefore be expressed as

$$T = \sum_{j=0}^{\infty} T^{(2j)} \tag{8.2.32}$$

in which

$$T^{(2j)} = \frac{V}{A(\hat{\mathbf{k}}_0)}(\Sigma_{-h}\Sigma_h)^j H_{2j} \tag{8.2.33}$$

where V is the volume of the crystal and

$$H_{2j} = \frac{1}{V}\int_{S(\hat{\mathbf{k}}_0)} (\Lambda_0\Lambda_h)^j\{\exp[-\Sigma(\mathbf{k}_0)L(\mathbf{r}, \hat{\mathbf{k}}_0)]\}\hat{\mathbf{k}}_0 \cdot \hat{\mathbf{e}}\, dS \tag{8.2.34}$$

The reflectivity (Eq. 8.2.22) can likewise be expressed as

$$R = \sum_{j=0}^{\infty} R^{(2j+1)} \tag{8.2.35}$$

in which

$$R^{(2j+1)} = \frac{V}{A(\hat{\mathbf{k}}_0)}\Sigma_{-h}^j\Sigma_h^{j+1}H_{2j+1} \tag{8.2.36}$$

and

$$H_{2j+1} = \frac{1}{V}\int_{S(\hat{\mathbf{k}}_h)} \Lambda_h(\Lambda_0\Lambda_h)^j\{\exp[-\Sigma(\mathbf{k}_0)L(\mathbf{r}, \hat{\mathbf{k}}_0)]\}\hat{\mathbf{k}}_h \cdot \hat{\mathbf{e}}\, dS \quad (8.2.37)$$

The problem of calculating the transmissivity and reflectivity in the two-beam case is therefore reduced to evaluating the multidimensional integrals of Eqs. 8.2.34 and 8.2.37 for the quantities H_j. The evaluation of these quantities is facilitated by the general theorem

$$\int_{S(\hat{\mathbf{k}})} \Lambda_{\mathbf{k}}\{g(\mathbf{r})\}\hat{\mathbf{k}} \cdot \hat{\mathbf{e}}\, dS = \int_V \exp[-\Sigma(\mathbf{k})L(\mathbf{r}, -\hat{\mathbf{k}})]g(\mathbf{r})\, d\mathbf{r} \quad (8.2.38)$$

from which it follows, for example, that

$$H_1 = \frac{1}{V}\int_V \exp[-\Sigma(\mathbf{k}_0)L(\mathbf{r}, \hat{\mathbf{k}}_0) - \Sigma(\mathbf{k}_h)L(\mathbf{r}, -\hat{\mathbf{k}}_h)]\, d\mathbf{r} \quad (8.2.39)$$

The approximate calculation of the preceding integrals has been discussed by, for example, Sears (1977, 1984a).

Let us suppose for the moment that the incident-neutron wave vector $\mathbf{k}_0$ moves away from the Bragg position so that $\Sigma_h = \Sigma_{-h} = 0$ according to Eq. 8.2.2. Then Eq. 8.2.25 reduces to

$$\begin{aligned} n_0(\mathbf{r}) &= n \exp[-\Sigma'(\mathbf{k}_0)L(\mathbf{r}, \hat{\mathbf{k}}_0)] \\ n_h(\mathbf{r}) &= 0 \end{aligned} \quad (8.2.40)$$

In this case the reflectivity (Eq. 8.2.22) vanishes and the transmissivity (Eq. 8.2.21) reduces to Lambert's law (Eq. 3.2.10) in which the absorption coefficient is

$$\mu = \Sigma'(\mathbf{k}_0) \quad (8.2.41)$$

as it should according to Eqs. 3.2.14 and 8.2.14. Thus, neutron transport theory and dynamical diffraction theory give the same result when Bragg's law is not satisfied.

Let us now return to the two-beam case in which $\mathbf{k}_0$ satisfies the condition for Bragg reflection via $\mathbf{K}_h$. We shall assume for simplicity that Friedel's law is satisfied (see Sec. 5.2) so that $\Sigma_{-h} = \Sigma_h$. Then, from Eq. 8.2.24,

$$\Sigma(\mathbf{k}_0) = \Sigma(\mathbf{k}_h) = \mu + \Sigma_h \quad (8.2.42)$$

and, hence,

$$H_1 = A(\mu + \Sigma_h) \quad (8.2.43)$$

where $A(\mu)$ is the *absorption factor,*

$$A(\mu) = \frac{1}{V}\int_V \exp\{-\mu[L(\mathbf{r}, \hat{\mathbf{k}}_0) + L(\mathbf{r}, -\hat{\mathbf{k}}_h)]\}\, d\mathbf{r} \quad (8.2.44)$$

In the kinematical theory of neutron diffraction the reflectivity is given, according to Eq. 5.2.65, by

$$R_k = \Sigma_h t \quad (8.2.45)$$

Table 8.1 The average crystal thickness in the direction of the incident beam, t, and the average path length of a singly reflected neutron through the crystal, l, for crystals of various shapes. The angles ϕ and ϕ' are defined in Fig. 5.7. The values for the cylinder are for equatorial reflections where the plane of scattering is perpendicular to the axis

Shape	Dimensions	t	l
Plane slab	Thickness d	$d \operatorname{cosec} \phi$	$\frac{d}{2}(\operatorname{cosec} \phi + \operatorname{cosec} \phi')$
Cylinder	Radius r	$\frac{\pi}{2} r$	$\frac{16}{3\pi} r$
Sphere	Radius r	$\frac{4}{3} r$	$\frac{3}{2} r$

in which $t \equiv V/A(\hat{\mathbf{k}}_0)$ is the average crystal thickness in the direction of the incident beam (see Table 8.1). Hence, from Eq. 8.2.36,

$$R^{(2j+1)} = R_k \Sigma_h^{2j} H_{2j+1} \tag{8.2.46}$$

In particular, the contribution from single reflections is given by

$$R^{(1)} = R_k A(\mu + \Sigma_h) \tag{8.2.47}$$

Extinction factor

According to Eq. 5.2.70,

$$\Sigma_h = QW(\theta_1) \tag{8.2.48}$$

where $W(\theta_1)$ is the normalized rocking curve. Here $\theta_1 = \theta - \theta_h$ where θ is the angle of incidence on the Bragg planes and θ_h is the Bragg angle. The kinematical reflectivity (Eq. 8.2.45) can therefore be expressed in the torm

$$R_k = \rho_k W(\theta_1) \tag{8.2.49}$$

in which ρ_k is the integrated reflectivity in the kinematical theory,

$$\rho_k = \int R_k \, d\theta_1 = Qt \tag{8.2.50}$$

Hence, it follows from Eq. 8.2.46 that

$$R^{(2j+1)} = \rho_k Q^{2j} H_{2j+1} W(\theta_1)^{2j+1} \tag{8.2.51}$$

in which H_{2j+1} is to be evaluated with

$$\Sigma(\mathbf{k}_0) = \Sigma(\mathbf{k}_h) = \mu + QW(\theta_1) \tag{8.2.52}$$

The integrated reflectivity,

$$\rho = \int R \, d\theta_1 \tag{8.2.53}$$

can therefore be expressed in the form

$$\rho = \rho_k E \tag{8.2.54}$$

in which the factor E has a multiple-scattering expansion of the form

$$E = \sum_{j=0}^{\infty} E^{(2j+1)} \tag{8.2.55}$$

where $E^{(2j+1)}$ is the contribution from neutrons that have been Bragg-reflected $2j+1$ times and is given by

$$E^{(2j+1)} = Q^{2j} \int H_{2j+1} W(\theta_1)^{2j+1}\, d\theta_1 \tag{8.2.56}$$

In particular, the single-scattering contribution is given, according to Eq. 8.2.43, by

$$E^{(1)} = \int A(\mu + QW(\theta_1)) W(\theta_1)\, d\theta_1 \tag{8.2.57}$$

The quantity E will be called the *extinction factor* since it describes the reduction in the integrated reflectivity due to the attenuation of the transmitted and Bragg-reflected beams inside the crystal that is neglected in the kinematical theory. This attenuation contains contributions, not only from absorption and diffuse scattering, but also a self-consistent contribution from multiple Bragg reflection. The latter contribution is called *secondary extinction.*

First-order extinction correction

For a small crystal the absorption factor (Eq. 8.2.44) can be evaluated by expanding the exponential, and one finds that

$$A(\mu) = 1 - \mu l + O(l^2) \tag{8.2.58}$$

where l is the average path length of a singly reflected neutron through the crystal (see Table 8.1). Hence, Eq. 8.2.57 becomes

$$E^{(1)} = 1 - (\mu + gQ)l + O(l^2), \tag{8.2.59}$$

where g is the *Darwin g-value,*

$$g = \int W(\theta_1)^2\, d\theta_1 \tag{8.2.60}$$

Thus, for example, for the Gaussian model (Eq. 5.3.32) we get

$$g = \frac{1}{2\sqrt{\pi}\,\eta} \tag{8.2.61}$$

where η is the effective mosaic spread. Since $E^{(2j+1)} = O(l^{2j})$ for $j \geq 1$, it follows that likewise

$$E = 1 - (\mu + gQ)l + O(l^2) \tag{8.2.62}$$

Thus, we see that absorption and diffuse scattering are negligible if $\mu l \ll 1$ and secondary extinction is negligible if $gQl \ll 1$. The latter condition is equivalent to our earlier result (Eq. 5.3.35).

For the small mosaic crystals that are used as samples in neutron crystallography, the first-order approximation (Eq. 8.2.62) for the extinction correction will often be adequate. Approximations for E beyond the linear region have been given by, for example, Sears (1977) and in other references cited therein.

8.3. Monochromator crystals

For the large mosaic crystals that are used as monochromators in neutron spectrometers, the multiple-scattering expansions, Eqs. 8.2.32 and 8.2.35, are not rapidly convergent and it is necessary to solve the Darwin equations directly. Fortunately, monochromator crystals are usually in the form of plane slabs, in which case a direct solution of the Darwin equations is relatively straightforward.

Darwin equations for a plane slab

Let us therefore consider a mosaic crystal in the form of a plane slab of thickness d and area S, and assume that the lateral dimensions of the slab are sufficiently large that edge effects can be neglected. If we choose a system of Cartesian coordinate axes such that the faces of the slab are the planes $z = 0$ and $z = d$, it then follows that $n_h(\mathbf{r}) = n_h(z)$, independent of x and y. The neutron currents are then given by

$$\begin{aligned} I_0(z) &= n_0(z)v_0 S \sin\phi \\ I_h(z) &= n_h(z)v_h S \sin\phi' \end{aligned} \tag{8.3.1}$$

where ϕ and ϕ' are the angles that the incident and Bragg-reflected beams make with the surface of the slab (Fig. 5.7) and $v_h = v_0$ since the Bragg reflection is elastic. Thus, $I_0(0)$ is the incident current, $I_0(d)$ is the transmitted current, $I_h(0)$ is the Bragg-reflected current in the Bragg case (Fig. 5.7a), and $I_h(d)$ is the Bragg-reflected current in the Laue case (Fig. 5.7b).

The coupled differential equations (8.2.23) now reduce to the form given originally by Darwin (1922),

$$\begin{aligned} \left[\frac{d}{dz} + \alpha + \beta\right] I_0(z) &= \beta' I_h(z) \\ \left[\pm\frac{d}{dz} + \alpha' + \beta'\right] I_h(z) &= \beta I_0(z) \end{aligned} \tag{8.3.2}$$

in which the plus sign refers to the Laue case and the minus sign to the

Bragg case. The coefficients are given by

$$\begin{aligned} \alpha &= \mu \operatorname{cosec} \phi, \qquad & \beta &= \Sigma_h \operatorname{cosec} \phi \\ \alpha' &= \mu \operatorname{cosec} \phi', \qquad & \beta' &= \Sigma_{-h} \operatorname{cosec} \phi' \end{aligned} \tag{8.3.3}$$

where μ is the absorption coefficient (Eq. 8.2.41).

The Darwin equations (8.3.2) must be solved for the boundary conditions

$$\begin{aligned} I_h(d) &= 0, \qquad & &\text{(Bragg case)} \\ I_h(0) &= 0, \qquad & &\text{(Laue case)} \end{aligned} \tag{8.3.4}$$

The transmissivity is then given by

$$T = \frac{I_0(d)}{I_0(0)} \tag{8.3.5}$$

and the reflectivity by

$$R = \begin{cases} I_h(0)/I_0(0), & \text{(Bragg case)} \\ I_h(d)/I_0(0), & \text{(Laue case)} \end{cases} \tag{8.3.6}$$

One-beam case

Let us suppose for the moment that the incident-neutron wave vector $\mathbf{k}_0$ moves away from the Bragg position so that $\beta = \beta' = 0$. In this case there is no Bragg-reflected beam, so that $I_h(z) = 0$, and Eq. 8.3.2 reduces to

$$\left[\frac{d}{dz} + \alpha\right] I_0(z) = 0 \tag{8.3.7}$$

The solution of this equation is

$$I_0(z) = I_0(0) \exp(-\alpha z) \tag{8.3.8}$$

so that the transmissivity (Eq. 8.3.5) becomes

$$T = \exp(-\alpha d) = \exp(-\mu D) \tag{8.3.9}$$

in which $D = d \operatorname{cosec} \phi$. This is, again, Lambert's law (Eq. 3.2.10), as it should be.

Two-beam case

Let us now return to the two-beam case where $\mathbf{k}_0$ satisfies the condition for Bragg reflection via $\mathbf{K}_h$. Absorption is usually of minor importance in comparison with secondary extinction and can often be neglected entirely. In the case of zero absorption, where $\alpha = \alpha' = 0$, the Darwin equations (8.3.2) reduce to

$$\frac{d}{dz} I_0(z) = \pm \frac{d}{dz} I_h(z) = \beta' I_h(z) - \beta I_0(z) \tag{8.3.10}$$

in which the plus sign now refers to the Bragg case and the minus sign to the Laue case.

The solution of these equations for the Bragg case is

$$\begin{aligned} I_0(z) &= I_0(0)\left[\frac{\beta \exp[(\beta-\beta')(d-z)]-\beta'}{\beta \exp[(\beta-\beta')d]-\beta'}\right] \\ I_h(z) &= I_0(0)\left[\frac{\beta\{\exp[(\beta-\beta')(d-z)]-1\}}{\beta \exp[(\beta-\beta')d]-\beta'}\right] \end{aligned} \tag{8.3.11}$$

and for the Laue case is

$$\begin{aligned} I_0(z) &= I_0(0)\left[\frac{\beta \exp[-(\beta+\beta')z]+\beta'}{\beta+\beta'}\right] \\ I_h(z) &= I_0(0)\left[\frac{\beta\{1-\exp[-(\beta+\beta')z]\}}{\beta+\beta'}\right] \end{aligned} \tag{8.3.12}$$

In both cases $R+T=1$, as required for neutron conservation, and

$$R = \begin{cases} \dfrac{\beta\{\exp[(\beta-\beta')d]-1\}}{\beta \exp[(\beta-\beta')d]-\beta'}, & \text{(Bragg case)} \\ \dfrac{\beta\{1-\exp[-(\beta+\beta')d]\}}{\beta+\beta'}, & \text{(Laue case)} \end{cases} \tag{8.3.13}$$

For a plane slab the kinematical reflectivity (Eq. 8.2.45) becomes $R_k = \beta d$, so that Eq. 8.3.13 can be expressed equivalently as

$$R = \begin{cases} \dfrac{\exp[(1-\zeta)R_k]-1}{\exp[(1-\zeta)R_k]-\zeta}, & \text{(Bragg case)} \\ \dfrac{1-\exp[-(1+\zeta)R_k]}{1+\zeta}, & \text{(Laue case)} \end{cases} \tag{8.3.14}$$

in which

$$\zeta \equiv \frac{\beta'}{\beta} = \frac{\sin\phi}{\sin\phi'} \tag{8.3.15}$$

For symmetric reflections, where $\phi = \phi'$ and $\zeta = 1$, Eq. 8.3.14 reduces to the expressions given by Bacon and Lowde (1948),

$$R = \begin{cases} \dfrac{R_k}{1+R_k}, & \text{(Bragg case)} \\ \tfrac{1}{2}[1-\exp(-2R_k)], & \text{(Laue case)} \end{cases} \tag{8.3.16}$$

Neutron monochromators

Suppose a collimated neutron beam is incident on a monochromator crystal and makes an angle θ with the Bragg planes of interest. In what follows we shall let λ' denote an arbitrary wavelength in the beam and λ

the wavelength for which Bragg's law is satisfied,

$$\lambda = 2d \sin \theta \tag{8.3.17}$$

where d is the distance between the Bragg planes.

The incident-neutron current can be expressed as

$$I = \int_0^\infty I(\lambda')\, d\lambda' \tag{8.3.18}$$

where $I(\lambda')\, d\lambda'$ is the current in $d\lambda'$. The Bragg-reflected current is then

$$I' = \int_0^\infty R(\lambda')I(\lambda')\, d\lambda' \tag{8.3.19}$$

where $R(\lambda')$ is the reflectivity. This quantity can be expressed as

$$R(\lambda') = \sum_{n=1}^{\infty} R_n(\lambda') \tag{8.3.20}$$

where $R_n(\lambda')$ is the contribution from nth-order reflections and has the property

$$R_n(\lambda') \simeq 0 \qquad \text{unless } \lambda' \simeq \lambda/n \tag{8.3.21}$$

The Bragg-reflected current can be expressed similarly as

$$I' = \sum_{n=1}^{\infty} I'_n \tag{8.3.22}$$

where

$$I'_n = R_n I(\lambda/n) \tag{8.3.23}$$

and

$$R_n = \int_0^\infty R_n(\lambda')\, d\lambda' \tag{8.3.24}$$

Alternatively,

$$R_n = \rho_n \left(\frac{\lambda}{n}\right) \cot \theta \tag{8.3.25}$$

where ρ_n is integrated reflectivity of the rocking curve,

$$\rho_n = \int R_n(\lambda')\, d\theta' \tag{8.3.26}$$

Let us write

$$I' = I'_1 \left(1 + \sum_{n=2}^{\infty} C_n\right) \tag{8.3.27}$$

where the first-order reflected current is

$$I'_1 = \rho_1 \lambda \cot \theta\, I(\lambda) \tag{8.3.28}$$

and C_n measures the nth-order contamination,

$$C_n = \frac{I'_n}{I'_1} = \frac{\rho_n I(\lambda/n)}{n \rho_1 I(\lambda)} \tag{8.3.29}$$

Clearly, one would like ρ_1 and, hence, I_1' to be as large as possible while C_n remains negligibly small for all $n \geq 2$.

For an ideal thermal-neutron source, in which the neutrons have a Maxwell–Boltzmann velocity distribution, one can easily show that the incident-neutron current at the monochromator is of the form

$$I(\lambda) \propto \lambda^{-5} \exp\left[-\left(\frac{\lambda_0}{\lambda}\right)^2\right] \tag{8.3.30}$$

where λ_0 is the wavelength of a neutron with energy $k_B T$,

$$\lambda_0 = h(2mk_B T)^{-1/2} \tag{8.3.31}$$

The function $I(\lambda)$ is a maximum when $\lambda = (2/5)^{1/2}\lambda_0$. A thermal-neutron beam from a heavy-water reactor is usually well represented by Eq. 8.3.30 for neutron energies $\lesssim 5k_B T$, that is, for $\lambda \gtrsim \lambda_0/\sqrt{5}$. The effective neutron temperature in the thermal region is typically 10 to 30 K higher than the temperature of the moderator. In the short wavelength (epithermal) region $I(\lambda) \propto \lambda^{-1}$ and is larger than one would expect from Eq. 8.3.30. If, for example, $T = 308$ K we then get $\lambda_0 = 1.76$ Å and $I(\lambda)$ is a maximum when $\lambda = 1.11$ Å.

When Eq. 8.3.30 is applicable, the order contamination is given by

$$C_n = \frac{n^4 \rho_n}{\rho_1} \exp\left[-(n^2 - 1)\left(\frac{\lambda_0}{\lambda}\right)^2\right] \tag{8.3.32}$$

and is particularly large when $\lambda \gtrsim n\lambda_0$. When order contamination is not negligible it can, in practice, be eliminated by the use of filters, as discussed in Sec. 5.6.

Monochromator reflectivity

Let us assume for simplicity that the normalized rocking curve is given by the Gaussian model (Eq. 5.3.32). Then the kinematical reflectivity (Eq. 8.2.49) becomes

$$R_k = \sqrt{2}\, gQt \exp\left[-\frac{1}{2}\left(\frac{\theta - \theta_h}{\eta}\right)^2\right] \tag{8.3.33}$$

where θ_h is the Bragg angle, η the effective mosaic spread, and $g = 1/(2\sqrt{\pi}\,\eta)$ as before (see Eq. 8.2.61). For a plane slab, $t = d \operatorname{cosec} \phi$ according to Table 8.1, while the quantity Q, which is characteristic of the crystal structure itself, is given by Eq. 5.2.71.

Figure 8.2 shows the reflectivity R as a function of $(\theta - \theta_h)/\eta$ for selected values of gQt assuming a symmetric Bragg-case reflection for which R is given by Eq. 8.3.16. The reflectivity is a maximum at $\theta = \theta_h$, where

$$R_{\max} = \frac{\sqrt{2}\, gQt}{1 + \sqrt{2}\, gQt} \tag{8.3.34}$$

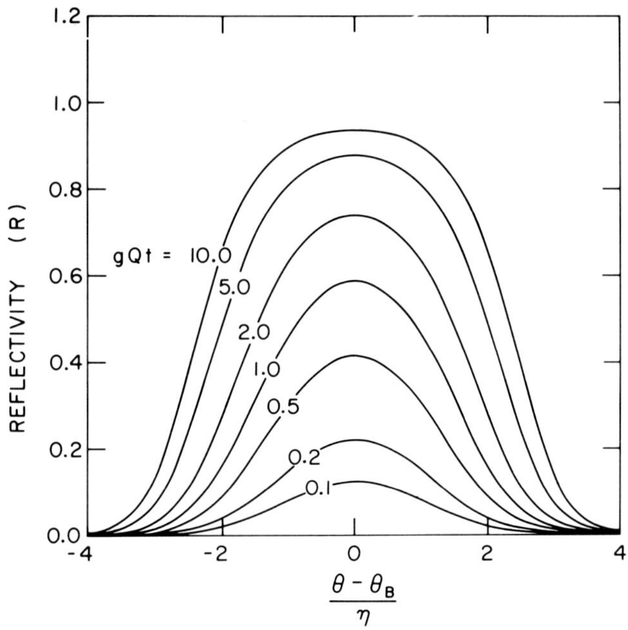

Fig. 8.2 Reflectivity of a plane slab with a Gaussian mosaic in the symmetric Bragg case for various values of gQt.

The full width at half maximum of the reflectivity is given by

$$\Delta\theta = \eta[8 \ln(2 + \sqrt{2}\, gQt)]^{1/2} \tag{8.3.35}$$

The important quantity here is evidently the dimensionless parameter gQt. Since $(gQ)^{-1}$ is the mean free path for Bragg reflection when $\theta = \theta_h$, the quantity gQt is the crystal thickness in the direction of the incident beam in units of the mean free path.

If $gQt \ll 1$, then multiple Bragg reflection is negligible and the reflectivity is given by the kinematical expression (Eq. 8.3.33). Here $R_{\max} \simeq \sqrt{2}\, gQt$ and is inversely proportional to the mosaic spread η. The peak width $\Delta\theta \simeq \eta[8 \ln 2]^{1/2}$, so that the integrated reflectivity $\rho \simeq Qt$, which is independent of η.

If, on the other hand, $gQt \gg 1$, then the peak reflectivity is saturated by multiple Bragg reflection, so that $R_{\max} \simeq 1$, while $\Delta\theta$ and, hence, ρ continue to increase logarithmically with gQt.

Since the reflected-neutron current is proportional to the integrated reflectivity ρ, according to (Eq. 8.3.28), it is desirable that ρ be as large as possible and this, in turn, requires that the crystal be sufficiently thick that $gQt \gtrsim 1$ so as to approach saturation conditions. The values of $(gQ)^{-1}$ for some typical neutron monochromator reflections are shown in Table 8.2 at two different wavelengths and assuming $\eta = 9$ arc minutes.

Table 8.2 The mean free path for Bragg reflection, $(gQ)^{-1}$, for some typical neutron monochromator reflections at two different wavelengths, λ, and assuming an effective mosaic spread $\eta = 9$ arc minutes

	$(gQ)^{-1}$(cm)	
Crystal (*hkl*)	$\lambda = 1$ Å	$\lambda = 3$ Å
Graphite(002)	0.474	0.047
Be(002)	0.546	0.035
Cu(111)	1.04	0.082
Pb(111)	3.33	0.318
Ge(111)	4.29	0.429
Al(111)	9.00	0.783
Si(111)	13.62	1.34

Note that $(gQ)^{-1}$, and hence the crystal thickness needed to saturate the reflectivity depends strongly on both the nature of the crystal and the wavelength.

APPENDIX A

LATTICES

Space lattices

A *space lattice* is a periodic array of points in space of the type shown in Fig. A.1. The points are called *lattice sites* and the position of the lth site is given by the *lattice vector*

$$\mathbf{R}_l = l_1\mathbf{a}_1 + l_2\mathbf{a}_2 + l_3\mathbf{a}_3 \tag{A.1}$$

where $\mathbf{a}_1$, $\mathbf{a}_2$, $\mathbf{a}_3$ are three noncoplanar primitive translation vectors and

$$l_1, l_2, l_3 = 0, \pm 1, \pm 2, \ldots \tag{A.2}$$

Thus, there is a one-to-one correspondence between the lattice sites l and the triads of integers (l_1, l_2, l_3). The vectors $\mathbf{a}_1$, $\mathbf{a}_2$, $\mathbf{a}_3$ span what is called a *primitive unit cell*, the volume of which is

$$V_0 = \mathbf{a}_1 \cdot (\mathbf{a}_2 \wedge \mathbf{a}_3) \tag{A.3}$$

The space in which the preceding lattice is embedded is supposed to be ordinary physical space. We shall call it **r**-*space* in order to distinguish it from wave vector space, which we shall refer to as **k**-*space*.

Reciprocal lattices

To each lattice in **r**-space there corresponds a *reciprocal lattice* in **k**-space. The position of the hth *reciprocal lattice site* is given by the *reciprocal lattice vector*

$$\mathbf{K}_h = h_1\mathbf{b}_1 + h_2\mathbf{b}_2 + h_3\mathbf{b}_3 \tag{A.4}$$

where

$$\begin{aligned} \mathbf{b}_1 &= \frac{2\pi}{V_0}\mathbf{a}_2 \wedge \mathbf{a}_3 \\ \mathbf{b}_2 &= \frac{2\pi}{V_0}\mathbf{a}_3 \wedge \mathbf{a}_1 \\ \mathbf{b}_3 &= \frac{2\pi}{V_0}\mathbf{a}_1 \wedge \mathbf{a}_2 \end{aligned} \tag{A.5}$$

and

$$h_1, h_2, h_3 = 0, \pm 1, \pm 2, \ldots \tag{A.6}$$

Thus, there is a one-to-one correspondence between the reciprocal lattice sites h and the triads of integers (h_1, h_2, h_3). The latter are called *Miller*

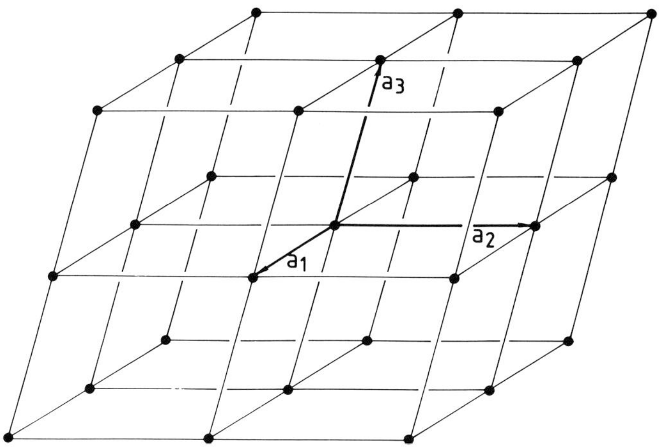

Fig. A.1 Space lattice.

indices. The wave vectors $\mathbf{b}_1, \mathbf{b}_2, \mathbf{b}_3$ span a primitive unit cell in the reciprocal lattice of volume

$$V_0' = \mathbf{b}_1 \cdot (\mathbf{b}_2 \wedge \mathbf{b}_3) = \frac{(2\pi)^3}{V_0} \tag{A.7}$$

It follows from Eq. A.5 that

$$\mathbf{a}_i \cdot \mathbf{b}_j = 2\pi\delta_{ij} \tag{A.8}$$

and, hence, that

$$\mathbf{a}_1 = \frac{2\pi}{V_0'} \mathbf{b}_2 \wedge \mathbf{b}_3$$
$$\mathbf{a}_2 = \frac{2\pi}{V_0'} \mathbf{b}_3 \wedge \mathbf{b}_1 \tag{A.9}$$
$$\mathbf{a}_3 = \frac{2\pi}{V_0'} \mathbf{b}_1 \wedge \mathbf{b}_2$$

The reciprocal of the reciprocal lattice is therefore the original lattice so that the term *reciprocal lattice* is indeed appropriate. The lattice in **r**-space is often called the *direct lattice* in order to distinguish it from the reciprocal lattice in **k**-space.

Finally, it will be noted that

$$\mathbf{K}_h \cdot \mathbf{R}_l = 2\pi(h_1 l_1 + h_2 l_2 + h_3 l_3) = 2\pi \text{ (integer)} \tag{A.10}$$

so that for all h and l

$$\exp(i\mathbf{K}_h \cdot \mathbf{R}_l) = 1 \tag{A.11}$$

Bragg planes

An important consequence of Eq. A.10 is that each reciprocal lattice vector $\mathbf{K}_h$ in **k**-space divides the lattice sites in **r**-space into a set of parallel equally spaced planes called *Bragg planes.* These planes are orthogonal to $\mathbf{K}_h$ and have a separation

$$d_h = \frac{2\pi n}{K_h} \tag{A.12}$$

where n is the largest common factor of h_1, h_2, h_3.

To verify the preceding statement, we note first that the equation $\hat{\mathbf{e}} \cdot \mathbf{r} = d$ represents a plane that is orthogonal to the unit vector $\hat{\mathbf{e}}$ and at a distance d from the origin. Hence, the set of equations $\mathbf{K}_h \cdot \mathbf{r} = 2\pi m$, where $m = 0, \pm 1, \pm 2, \ldots$, represents a family of parallel equally spaced planes that are orthogonal to $\mathbf{K}_h$ and have a separation $2\pi / K_h$. It is evident from Eq. A.10 that each lattice site l lies on one of these planes. However, each plane is not necessarily occupied by lattice sites. Those that are so occupied are the Bragg planes. If h_1, h_2, h_3 have a common factor n, then $\mathbf{K}_h / n$ is also a reciprocal lattice vector and clearly divides the direct lattice into the same set of Bragg planes as $\mathbf{K}_h$ does. Hence, the separation of the Bragg planes is given by Eq. A.12, where n is the largest common factor of h_1, h_2, h_3.

Conversely, each lattice vector $\mathbf{R}_l$ in $\mathbf{r}$-space divides the reciprocal lattice sites in $\mathbf{k}$-space into a set of parallel equally spaced Bragg planes. The planes are orthogonal to $\mathbf{R}_l$ and have a separation given by an expression analogous to Eq. A.12.

Crystallographic conventions

Crystallographers conventionally denote lattice sites simply as $l_1 l_2 l_3$. Here brackets and commas are omitted and negative values are indicated by a bar. Thus, for example, one writes $1\bar{2}3$ rather than $(1, -2, 3)$. Depending on the context, $[l_1 l_2 l_3]$ then denotes either the direction of the line from 000 to $l_1 l_2 l_3$, or a row of direct lattice sites parallel to this direction, or a plane of reciprocal lattice sites perpendicular to this direction.

Similarly, $h_1 h_2 h_3$ denotes a reciprocal lattice site. Depending on the context, $(h_1 h_2 h_3)$ then denotes either the direction of the line from 000 to $h_1 h_2 h_3$, or a row of reciprocal lattice sites parallel to this direction, or a plane of direct lattice sites perpendicular to this direction.

The use of these conventions is illustrated in Fig. A.2 which shows the (001) plane of a simple cubic lattice. The diagonal lines represent the family of (120) planes and are orthogonal to the (120) direction. In a cubic lattice the (120) and [120] directions are the same, but this is not in general true for other lattices.

Another convention is that symmetrically equivalent $[l_1 l_2 l_3]$ values are referred to collectively as $\langle l_1 l_2 l_3 \rangle$. Thus, for example, in a simple cubic lattice $\langle 100 \rangle$ denotes the set [100], $[\bar{1}00]$, [010], $[0\bar{1}0]$, [001], $[00\bar{1}]$. One can therefore say that the $[0\bar{1}0]$ direction is a $\langle 100 \rangle$ direction, and so on. Similarly, symmetrically equivalent $(h_1 h_2 h_3)$ values are referred to

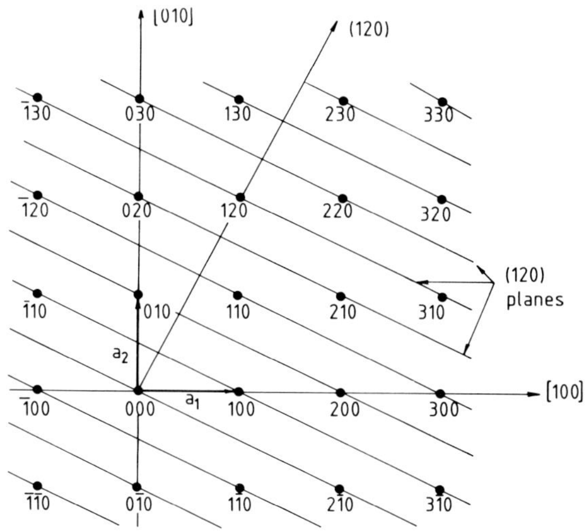

Fig. A.2 The (001) plane of a simple cubic lattice.

collectively as $\{h_1h_2h_3\}$. These latter conventions are, however, often ignored in practice. Thus, one will speak of a (111) plane when one actually means a {111} plane. There is no harm in doing this since it is always clear from the context what is meant.

Lattice functions

A function $f(\mathbf{r})$ is called a *lattice function* if it has the periodicity of a space lattice,

$$f(\mathbf{r}+\mathbf{R}_l)=f(\mathbf{r}), \qquad \text{all } l \tag{A.13}$$

Such lattice functions are often encountered in solid-state physics. For example, the potential energy of an electron in a metal is a lattice function as is also the optical potential of a neutron in a crystal.

It is evident from Eq. A.11 that any function of the form

$$f(\mathbf{r})=\sum_h f_h \exp(-i\mathbf{K}_h\cdot\mathbf{r}) \tag{A.14}$$

is a lattice function. Conversely, any lattice function $f(\mathbf{r})$ can be expressed in the form of Eq. A.14 with

$$f_h=\frac{1}{V_0}\int_{V_0}\exp(i\mathbf{K}_h\cdot\mathbf{r})f(\mathbf{r})\,d\mathbf{r} \tag{A.15}$$

The expansion (A.14) is clearly the lattice analog of a Fourier series.

A function $g(\mathbf{k})$ is called a lattice function if it has the periodicity of a reciprocal lattice,

$$g(\mathbf{k}+\mathbf{K}_h)=g(\mathbf{k}), \qquad \text{all } h \tag{A.16}$$

Again, such lattice functions are often encountered in solid-state physics. For example, electron band energies, phonon and magnon dispersion relations, etc., are all lattice functions of this type.

It is evident from Eq. A.11 that any function of the form

$$g(\mathbf{k}) = \sum_l g_l \exp(i\mathbf{k} \cdot \mathbf{R}_l) \tag{A.17}$$

is a lattice function. Conversely, any lattice function $g(\mathbf{k})$ can be expressed in the form of Eq. A.17 with

$$g_l = \frac{1}{V_0'} \int_{V_0'} \exp(-i\mathbf{k} \cdot \mathbf{R}_l) g(\mathbf{k})\, d\mathbf{k} \tag{A.18}$$

These Fourier expansions can easily be established with the help of the relations

$$\sum_l \exp(\pm i\mathbf{k} \cdot \mathbf{R}_l) = V_0' \sum_h \delta(\mathbf{k} - \mathbf{K}_h) \tag{A.19}$$

and

$$\sum_h \exp(\pm i\mathbf{K}_h \cdot \mathbf{r}) = V_0 \sum_l \delta(\mathbf{r} - \mathbf{R}_l) \tag{A.20}$$

To derive these relations, we first consider the quantity

$$I(\mathbf{k}) = \sum_l \exp(i\mathbf{k} \cdot \mathbf{R}_l) \tag{A.21}$$

The wave vector $\mathbf{k}$ can be expanded in terms of the basis vectors of the reciprocal lattice,

$$\mathbf{k} = k_1\mathbf{b}_1 + k_2\mathbf{b}_2 + k_3\mathbf{b}_3 \tag{A.22}$$

where, according to Eq. A.8,

$$k_j = \mathbf{k} \cdot \mathbf{a}_j / 2\pi \tag{A.23}$$

Then

$$\mathbf{k} \cdot \mathbf{R}_l = 2\pi(k_1 l_1 + k_2 l_2 + k_3 l_3) \tag{A.24}$$

and

$$I(\mathbf{k}) = I(k_1) I(k_2) I(k_3) \tag{A.25}$$

where, for a finite lattice,

$$I(k_j) = \sum_{l_j=-N_j}^{N_j} \exp(2\pi i k_j l_j) \tag{A.26}$$

This geometric series can be summed to give

$$I(k_j) = \frac{\sin[(2N_j + 1)\pi k_j]}{\sin(\pi k_j)} \tag{A.27}$$

from which it is seen that

$$\begin{aligned} &I(k_j + h_j) = I(k_j), && h_j = 0, \pm 1, \pm 2, \ldots \\ &I(-k_j) = I(k_j) && \\ &I(0) = 2N_j + 1 && \\ &I(k_j) = 0 && \text{if } k_j = (2N_j + 1)^{-1} \end{aligned} \tag{A.28}$$

It also follows from Eq. A.27 that

$$\lim_{N_j \to \infty} \int_{-1/2}^{1/2} I(k_j)\, dk_j = 1 \tag{A.29}$$

Hence, we see that for an infinite lattice ($N_j \to \infty$),

$$I(k_j) = \sum_{h_j=-\infty}^{\infty} \delta(k_j - h_j) \tag{A.30}$$

and, therefore,

$$I(\mathbf{k}) = V_0' \sum_h \delta(\mathbf{k} - \mathbf{K}_h) \tag{A.31}$$

where we have used the fact that

$$\delta(\mathbf{k}) = \frac{1}{V_0'} \delta(k_1)\, \delta(k_2)\, \delta(k_3) \tag{A.32}$$

It also follows from Eq. A.28 that $I(-\mathbf{k}) = I(\mathbf{k})$ and, hence, that $I(\mathbf{k})$ remains unchanged if the exponent in Eq. A.21 has the opposite sign. This result is, of course, also evident from the fact that the lattice has inversion symmetry. This completes the proof of Eq. A.19. The relation (A.20) can be obtained in a similar way.

Unit cells

The two-dimensional space lattice shown in Fig. A.3 illustrates a number of important properties of space lattices in general. For example, it is evident that the lattice can be thought of as being generated by the regular repetition in space of a *unit cell.* However, the unit cell is not unique; there are many different kinds of unit cell that will generate the same space lattice.

The *primitive unit cell* in three dimensions is the parallelepiped that is spanned by the vectors $\mathbf{a}_1$, $\mathbf{a}_2$, $\mathbf{a}_3$ and whose volume V_0 is given by Eq. A.3. The primitive unit cell is itself not unique. However, all primitive unit cells for a given lattice have the same volume V_0, as is clear from the fact that there is only one lattice site per primitive unit cell.

The *symmetric unit cell* is defined as those points in space that lie nearer to a given lattice site than to any other lattice site. The symmetric unit cell in three dimensions is, therefore, the convex polyhedron that is produced by constructing the planes that right-bisect the lines joining the given lattice site to all its neighbors. For a given lattice the symmetric unit cell is clearly unique. In the direct lattice it is called the *Wigner–Seitz cell,* and in the reciprocal lattice it is called the *first Brillouin zone.* Since there is only one lattice site per symmetric unit cell, this unit cell has the same volume V_0 as the primitive unit cell. The lattice can also be generated by applying the primitive translations $\mathbf{a}_1$, $\mathbf{a}_2$, $\mathbf{a}_3$ to the symmetric unit cell.

For some lattices it is conventional to use a nonprimitive centered unit

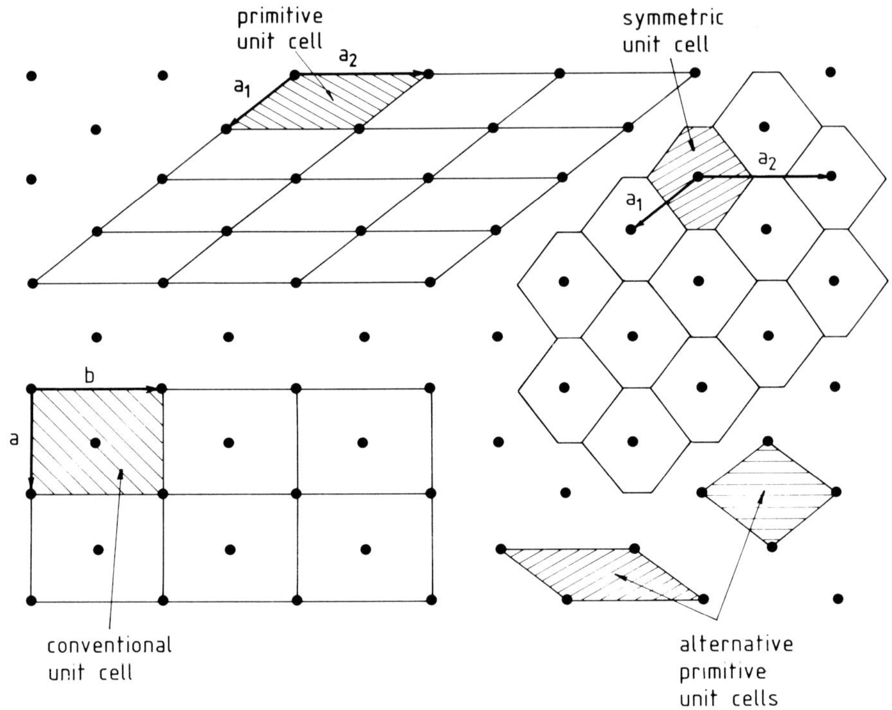

Fig. A.3 Unit cells that generate the same space lattice.

cell that contains more than one lattice site. The *conventional unit cell* in three dimensions is a parallelepiped that is spanned by the vectors **a**, **b**, **c** and has the volume

$$V = \mathbf{a} \cdot (\mathbf{b} \wedge \mathbf{c}) = nV_0 \tag{A.33}$$

where n is the number of lattice sites in the conventional unit cell.

The important point to keep in mind is that the lattice sites l and the corresponding lattice vectors $\mathbf{R}_l$ are completely independent of the choice of the unit cell, this choice being purely a matter of convenience. The advantage of the primitive unit cell is that l_1, l_2, l_3 are independent unrestricted integers, a fact that facilitates the calculation of lattice sums such as that in Eq. A.19. The advantage of the nonprimitive unit cells is that they display more clearly the symmetry and coordination of the lattice. For example, it is not immediately obvious from the primitive unit cell that the lattice in Fig. A.3 has rectangular symmetry or that each lattice site has four nearest neighbors. In theoretical calculations that are intended to apply to any lattice, one always uses a general primitive unit cell. However, in subsequent applications of the theory to a specific lattice it is often advantageous to switch to a suitable nonprimitive unit cell. We might add that crystal structures are always defined in terms of the conventional unit cell, whereas lattice functions are always defined in terms of the symmetric unit cell.

Bravais lattices

The parameters that define a space lattice are the magnitudes and directions of the translation vectors $\mathbf{a}_1$, $\mathbf{a}_2$, $\mathbf{a}_3$ or $\mathbf{a}$, $\mathbf{b}$, $\mathbf{c}$ that characterize the unit cell (see Fig. A.4). Since these parameters can be varied continuously, the number of different lattices is clearly infinite.

Different lattices can be distinguished by their *space-group symmetry,* that is, by the group of translations, rotations, and reflections that map the lattice onto itself. The translational symmetry imposes restrictions on the type of rotational symmetry that can occur. In particular, only 2-, 3-, 4-, and 6-fold axes are compatible with the translational symmetry.

It was first shown by Bravais in 1848 that in three dimensions there are only 14 symmetrically distinct space lattices. These lattices belong to seven different systems, and the lattices within a given system are characterized by a conventional unit cell that is either *primitive* (P), *base-centered* (C), *body-centered* (I), or *face-centered* (F). The conventional unit cells for the 14 *Bravais lattices* are shown in Fig. A.5 and the corresponding lattice parameters are listed in Table A.1. In this table we also give the Bravais lattices in one and two dimensions. For example, the two-dimensional lattice in Fig. A.3 is the centered rectangular net.

When the lattice is defined in terms of the conventional unit cell the lattice vectors are expressed as

$$\mathbf{R}_l = u\mathbf{a} + v\mathbf{b} + w\mathbf{c} \tag{A.34}$$

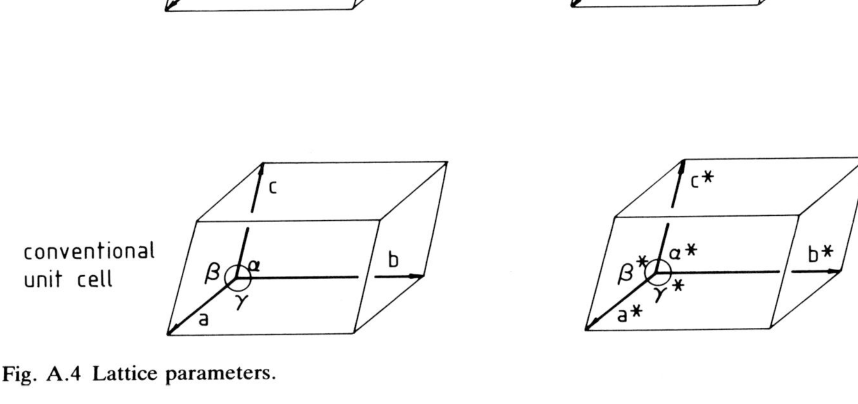

Fig. A.4 Lattice parameters.

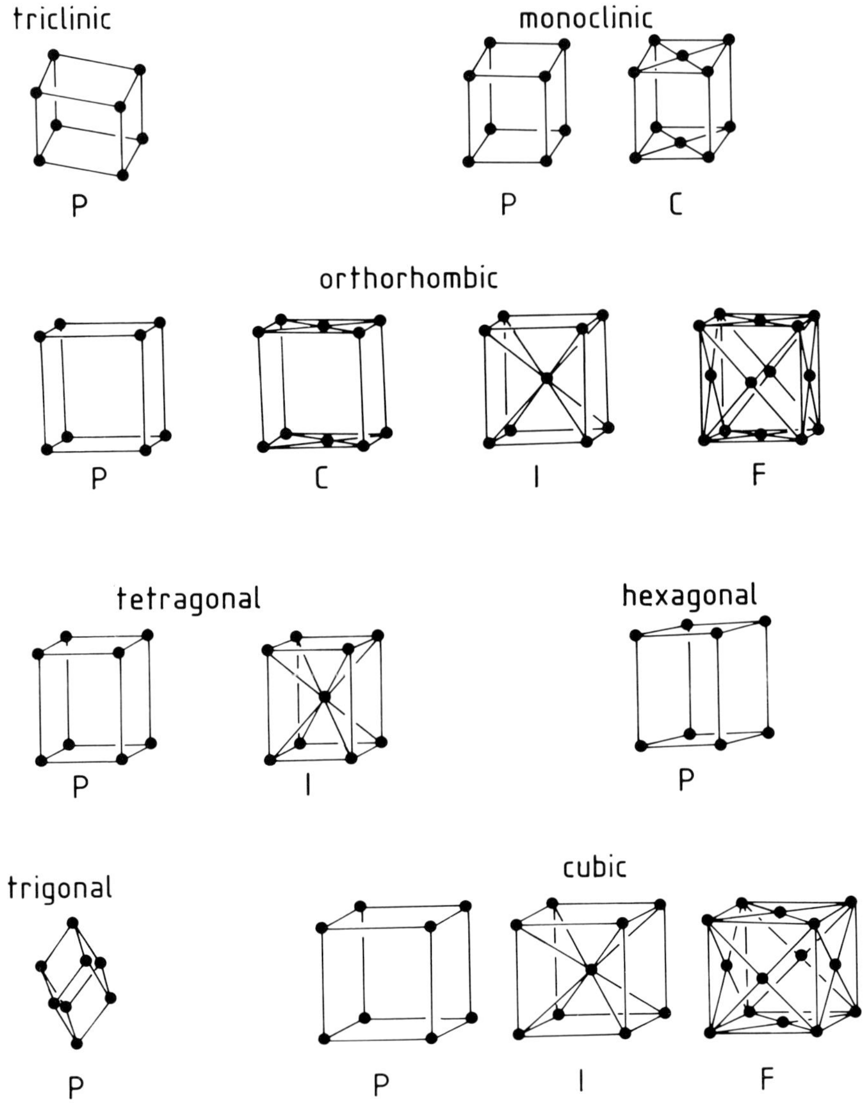

Fig. A.5 Conventional unit cells for the 14 Bravais lattices.

where u, v, w are *rational numbers,* the allowed values of which depend on the type of lattice under consideration. The values of u, v, w for the lattice sites in the conventional unit cell are given in Table A.2. The trigonal lattice is represented in Fig. A.5 by a primitive rhombohedral unit cell. However, this lattice can also be referred to a hexagonal unit cell containing three lattice sites. This is the *rhombohedral* lattice R in Table A.2. Thus, trigonal P and hexagonal R refer to the same Bravais lattice.

Table A.1 Parameters for the conventional unit cells of the Bravais lattices

Dimension	Parameters		System	Lattices
1	a			P
2	$a \neq b$	$\gamma \neq 90°$	Oblique	P
	$a \neq b$	$\gamma = 90°$	Rectangular	P, C
	$a = b$	$\gamma = 90°$	Square	P
	$a = b$	$\gamma = 120°$	Hexagonal	P
3	$a \neq b \neq c$	$\alpha \neq \beta \neq \gamma$	Triclinic	P
	$a \neq b \neq c$	$\alpha = \gamma = 90,\ \beta > 90°$	Monoclinic	P, C
	$a \neq b \neq c$	$\alpha = \beta = \gamma = 90°$	Orthorhombic	P, C, I, F
	$a = b \neq c$	$\alpha = \beta = \gamma = 90°$	Tetragonal	P, I
	$a = b \neq c$	$\alpha = \beta = 90°,\ \gamma = 120°$	Hexagonal	P
	$a = b = c$	$\alpha = \beta = \gamma \neq 90°$	Trigonal	P
	$a = b = c$	$\alpha = \beta = \gamma = 90°$	Cubic	P, I, F

Table A.2 Number of lattice sites n in the conventional unit cell and their coordinates (u, v, w) for various lattice types

Lattice type	Symbol	n	(u, v, w)
Primitive	P	1	$(0, 0, 0)$
Base-centered	C	2	$(0, 0, 0)(\frac{1}{2}, \frac{1}{2}, 0)$
Body-centered	I	2	$(0, 0, 0)(\frac{1}{2}, \frac{1}{2}, \frac{1}{2})$
Face-centred	F	4	$(0, 0, 0)(0, \frac{1}{2}, \frac{1}{2})(\frac{1}{2}, 0, \frac{1}{2})(\frac{1}{2}, \frac{1}{2}, 0)$
Rhombohedral	R	3	$(0, 0, 0)(\frac{2}{3}, \frac{1}{3}, \frac{1}{3})(\frac{1}{3}, \frac{2}{3}, \frac{2}{3})$

When the lattice is defined in terms of the conventional unit cell the reciprocal lattice vectors are expressed as

$$\mathbf{K}_h = h\mathbf{a}^* + k\mathbf{b}^* + l\mathbf{c}^* \tag{A.35}$$

where

$$\mathbf{a}^* = \frac{2\pi}{V}\mathbf{b} \wedge \mathbf{c} \tag{A.36}$$

etc., and the Miller indices h, k, l are *integers* the allowed values of which depend on the type of lattice under consideration.

We shall consider in detail only the cubic and hexagonal lattices, for two reasons. First, the majority of crystals that one encounters in practice are either cubic or hexagonal. Second, once one understands these two systems the others present no difficulty.

Cubic lattices

The conventional unit cell for a cubic lattice is a cube of side a and volume $V = a^3$, so that in Cartesian coordinates

$$\mathbf{a} = a(1, 0, 0), \qquad \mathbf{b} = a(0, 1, 0), \qquad \mathbf{c} = a(0, 0, 1) \tag{A.37}$$

The parameter a is usually referred to as the *lattice constant*. The reciprocal lattice is also cubic and

$$\mathbf{a}^* = \frac{2\pi}{a}(1, 0, 0), \qquad \mathbf{b}^* = \frac{2\pi}{a}(0, 1, 0), \qquad \mathbf{c}^* = \frac{2\pi}{a}(0, 0, 1) \quad \text{(A.38)}$$

There are three cubic Bravais lattices: the simple cubic or sc lattice (P), the body-centered cubic or bcc lattice (I), and the face-centered cubic or fcc lattice (F). These lattices and their reciprocals are shown in Fig. A.6 where we also show the primitive translation vectors. Some properties of

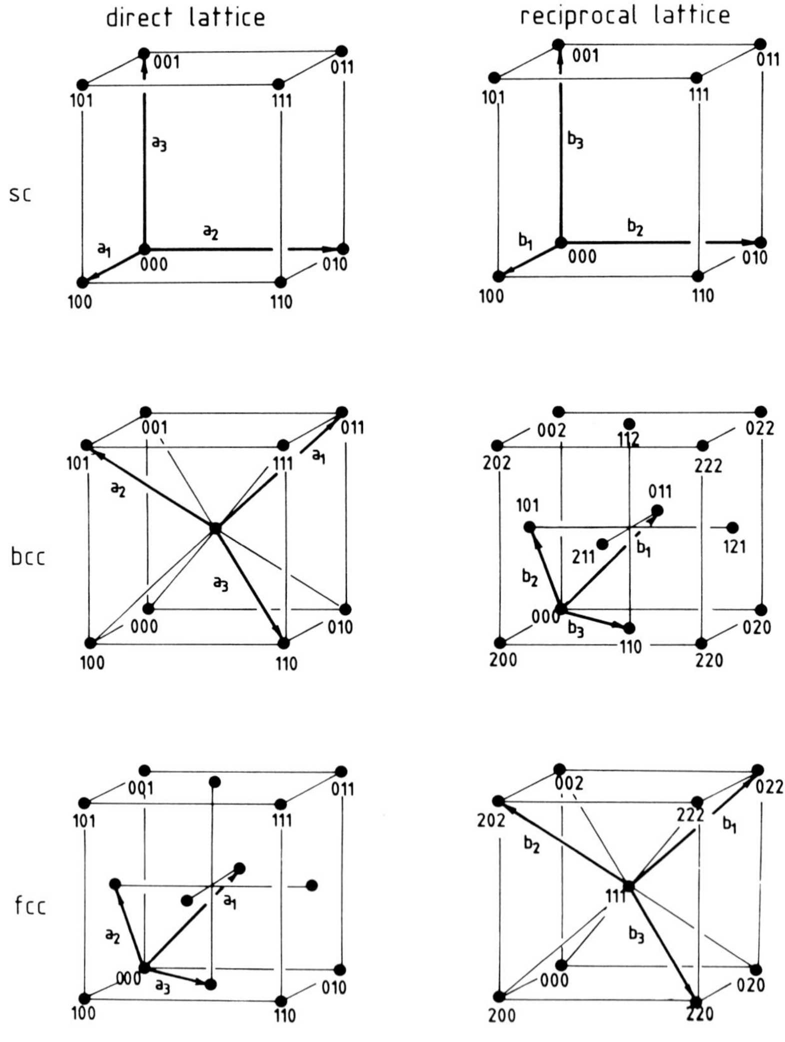

Fig. A.6 Cubic Bravais lattices.

Table A.3 Properties of the cubic Bravais lattices

Quantity	sc	bcc	fcc
$\mathbf{a}_1$	$a(1, 0, 0)$	$a(-\frac{1}{2}, \frac{1}{2}, \frac{1}{2})$	$a(0, \frac{1}{2}, \frac{1}{2})$
$\mathbf{a}_2$	$a(0, 1, 0)$	$a(\frac{1}{2}, -\frac{1}{2}, \frac{1}{2})$	$a(\frac{1}{2}, 0, \frac{1}{2})$
$\mathbf{a}_3$	$a(0, 0, 1)$	$a(\frac{1}{2}, \frac{1}{2}, -\frac{1}{2})$	$a(\frac{1}{2}, \frac{1}{2}, 0)$
$a_1 = a_2 = a_3$	a	$\frac{\sqrt{3}}{2}a$	$\frac{1}{\sqrt{2}}a$
$\alpha_1 = \alpha_2 = \alpha_3$	$90°$	$\cos^{-1}(-\frac{1}{3}) = 109.47°$	$60°$
u	l_1	$\frac{1}{2}(-l_1 + l_2 + l_3)$	$\frac{1}{2}(l_2 + l_3)$
v	l_2	$\frac{1}{2}(l_1 - l_2 + l_3)$	$\frac{1}{2}(l_3 + l_1)$
w	l_3	$\frac{1}{2}(l_1 + l_2 - l_3)$	$\frac{1}{2}(l_1 + l_2)$
$\mathbf{b}_1$	$\frac{2\pi}{a}(1, 0, 0)$	$\frac{2\pi}{a}(0, 1, 1)$	$\frac{2\pi}{a}(-1, 1, 1)$
$\mathbf{b}_2$	$\frac{2\pi}{a}(0, 1, 0)$	$\frac{2\pi}{a}(1, 0, 1)$	$\frac{2\pi}{a}(1, -1, 1)$
$\mathbf{b}_3$	$\frac{2\pi}{a}(0, 0, 1)$	$\frac{2\pi}{a}(1, 1, 0)$	$\frac{2\pi}{a}(1, 1, -1)$
$b_1 = b_2 = b_3$	$\frac{2\pi}{a}$	$\frac{2\sqrt{2}\,\pi}{a}$	$\frac{2\sqrt{3}\,\pi}{a}$
$\beta_1 = \beta_2 = \beta_3$	$90°$	$60°$	$\cos^{-1}(-\frac{1}{3}) = 109.47°$
h	h_1	$h_2 + h_3$	$-h_1 + h_2 + h_3$
k	h_2	$h_3 + h_1$	$h_1 - h_2 + h_3$
l	h_3	$h_1 + h_2$	$h_1 + h_2 - h_3$

the cubic Bravais lattices are also listed in Table A.3. It will be noted that the sc lattice is self-reciprocal while the reciprocal of the bcc lattice is fcc and vice versa.

It is evident from Table A.3 that for the sc lattice the Miller indices h, k, l are unrestricted. For the bcc lattice, on the other hand,

$$h + k + l = 2(h_1 + h_2 + h_3) \tag{A.39}$$

so that h, k, l are restricted to values for which $h + k + l$ is even. For the fcc lattice,

$$h + k = 2h_3, \qquad k + l = 2h_1, \qquad l + h = 2h_2, \tag{A.40}$$

so that the values of h, k, l are either all even or all odd.

In connection with the Bragg plane relation Eq. A.12, note that if n is a common factor of h_1, h_2, h_3 then n is also a common factor of h, k, l. Since the converse is also true, it follows that if n is the largest common factor of h_1, h_2, h_3 then n is also the largest common factor of h, k, l.

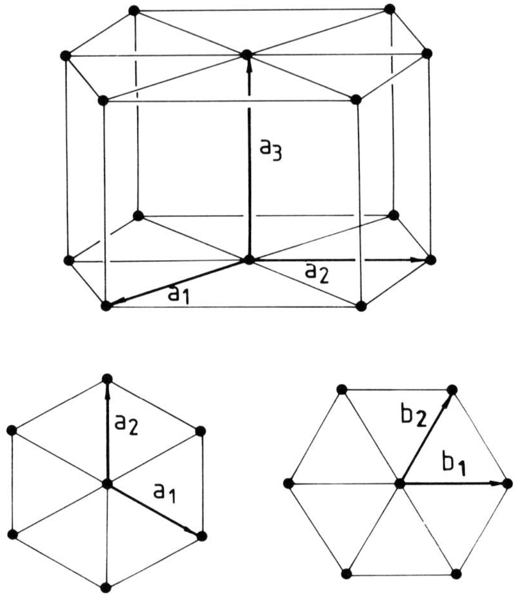

Fig. A.7 Hexagonal lattice.

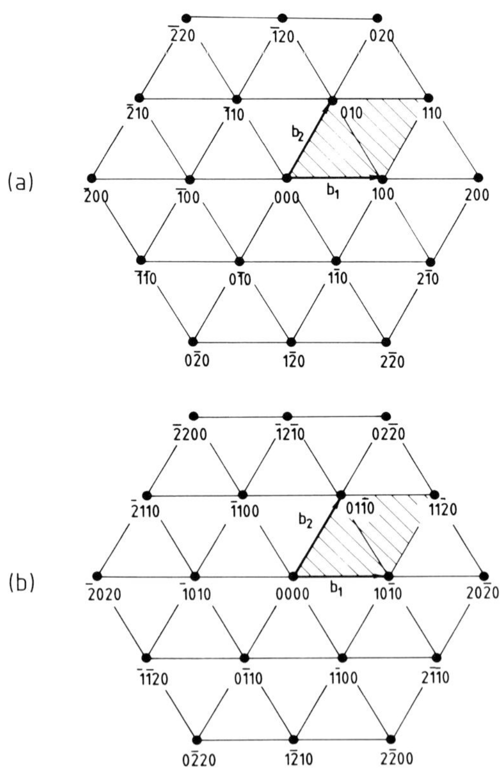

Fig. A.8 The [001] plane of the hexagonal lattice labeled with (a) Miller indices *hkl* and (b) Bravais–Miller indices *hkil*.

Hexagonal lattices

The hexagonal lattice is illuatrated in Fig. A.7 and is characterized by two lattice constants a and c. the conventional unit cell is primitive, so that in Cartesian coordiates

$$\begin{aligned}\mathbf{a} = \mathbf{a}_1 &= a\left(\frac{\sqrt{3}}{2}, -\frac{1}{2}, 0\right)\\ \mathbf{b} = \mathbf{a}_2 &= a(0, 1, 0)\\ \mathbf{c} = \mathbf{a}_3 &= c(0, 0, 1)\end{aligned} \tag{A.41}$$

Hence,

$$\begin{aligned} a_1 = a_2 = a, &\qquad \alpha_1 = \alpha_2 = 90 \text{ degrees}\\ a_3 = c, &\qquad \alpha_3 = 120 \text{ degrees}\end{aligned} \tag{A.42}$$

and the volume of the unit cell is

$$V = V_0 = \frac{\sqrt{3}}{2}a^2c \tag{A.43}$$

The reciprocal lattice is also hexagonal but is rotated by 30 degrees about the c-axis relative to the direct lattice. In particular,

$$\begin{aligned}\mathbf{a}^* = \mathbf{b}_1 &= \frac{4\pi}{\sqrt{3}\,a}(1, 0, 0)\\ \mathbf{b}^* = \mathbf{b}_2 &= \frac{4\pi}{\sqrt{3}\,a}\left(\frac{1}{2}, \frac{\sqrt{3}}{2}, 0\right)\\ \mathbf{c}^* = \mathbf{b}_3 &= \frac{2\pi}{c}(0, 0, 1)\end{aligned} \tag{A.44}$$

Hence,

$$\begin{aligned} b_1 = b_2 = \frac{4\pi}{\sqrt{3}\,a}, &\qquad \beta_1 = \beta_2 = 90 \text{ degrees}\\ b_3 = \frac{2\pi}{c}, &\qquad \beta_3 = 60 \text{ degrees}\end{aligned} \tag{A.45}$$

The reciprocal lattice sites of a hexagonal lattice can be labeled by the Miller indices *hkl* as in Fig. A.8a. However, they are usually labeled instead by the *Bravais–Miller indices hkil,* where $i = -(h + k)$ so that $h + k + i = 0$. The redundant index i serves to display the hexagonal symmetry of the lattice as illustrated in Fig. A.8b. Also, the theoretical expressions for the properties of hexagonal crystals take a more symmetrical form when expressed in terms of the Bravais–Miller indices.

APPENDIX B

CRYSTALS

In our discussion of neutron diffraction in crystals we have assumed that the reader knows all he needs to know about the nature of crystalline materials so that we could focus our attention on the neutron-optical properties of these materials. The purpose of this appendix is to summarize briefly some of the basic concepts, terminology, etc., that were taken for granted. For more details we refer the reader to, for example, the series of books *The Crystalline State* (Bragg, 1962) or the more recent series *Modern Crystallography* (Vainshtein, 1981). Extensive tabulations of crystal structures have been given by Pearson (1958) and Wyckoff (1963). The new edition of the *International Tables for Crystallography* (Hahn, 1983) is also a rich source of crystallographic lore.

Crystal structures

A *crystal* is a solid body in which the atoms have a regular periodic arrangement in space. In the simplest and most common type of crystal the periodicity is characterized by a space lattice in which each unit cell contains the same configuration of atoms as all the others. The position of the νth atom in the lth unit cell can then be expressed as

$$\mathbf{r}_{l\nu} = \mathbf{R}_l + \mathbf{r}_\nu \tag{B.1}$$

where $\mathbf{R}_l$ is the lattice vector giving the position of the lth lattice site and $\mathbf{r}_\nu$ is the position of the νth atom relative to this site. The characteristic property of a *perfect crystal* is that both $\mathbf{r}_\nu$ and the chemical nature of this atom are independent of l, so that all the unit cells are identical.

Crystal structures are defined in terms of the conventional unit cell so that

$$\mathbf{r}_\nu = x\mathbf{a} + y\mathbf{b} + z\mathbf{c} \tag{B.2}$$

To specify the crystal structure one must therefore give (1) the lattice and (2) the chemical species and coordinates (x, y, z) of each atom in the unit cell. A number of common crystal structures are listed in Table B.1. We give only the positions of the atoms in the primitive unit cell. The positions of any remaining atoms in the conventional unit cell can be obtained with the help of Table A.2.

It is instructive to sketch the crystal structures from the information given in Table B.1. This is most easily done in two stages, as illustrated in Fig. B.1. for the diamond structure. First one draws the projections of

Table B.1 Common crystal structures

System	Lattice	Space group	Structure	Chemical formula	(x, y, z)
Cubic	P	O_h^1	Cesium chloride	CsCl	$Cs(0,0,0)Cl(\frac{1}{2},\frac{1}{2},\frac{1}{2})$
			Pervoskite	$CaTiO_3$	$Ca(0,0,0)Ti(\frac{1}{2},\frac{1}{2},\frac{1}{2})$ $0(0,\frac{1}{2},\frac{1}{2})(\frac{1}{2},0,\frac{1}{2})(\frac{1}{2},\frac{1}{2},0)$
	I	O_h^9	Bcc	Na	Na(0, 0, 0)
	F	O_h^5	Fcc	Cu	Cu(0, 0, 0)
			Rock salt	NaCl	$Na(0,0,0)Cl(\frac{1}{2},\frac{1}{2},\frac{1}{2})$
			Fluorite	CaF_2	$Ca(0,0,0)F(\frac{1}{4},\frac{1}{4},\frac{1}{4})(\frac{3}{4},\frac{3}{4},\frac{3}{4})$
			Heusler alloy	Cu_2AlMn	$Al(0,0,0)Mn(\frac{1}{2},\frac{1}{2},\frac{1}{2})$ $Cu(\frac{1}{4},\frac{1}{4},\frac{1}{4})(\frac{3}{4},\frac{3}{4},\frac{3}{4})$
		O_h^7	Diamond	C	$C(0,0,0)(\frac{1}{4},\frac{1}{4},\frac{1}{4})$
		T_d^2	Zinc blende	β-ZnS	$Zn(0,0,0)S(\frac{1}{4},\frac{1}{4},\frac{1}{4})$
Hexagonal	P	D_{6h}^4	Hcp	Mg	$Mg(0,0,0)(\frac{1}{3},\frac{2}{3},\frac{1}{2})$
			Graphite	C	$C(0,0,0)(\frac{1}{3},\frac{2}{3},0)(0,0,\frac{1}{2})(\frac{2}{3},\frac{1}{3},\frac{1}{2})$
			Nickel arsenide	NiAs	$Ni(0,0,0)(0,0,\frac{1}{2})As(\frac{1}{3},\frac{2}{3},\frac{1}{4})(\frac{2}{3},\frac{1}{3},\frac{3}{4})$
		C_{6v}^4	Wurtzite	α-ZnS	$Zn(0,0,0)(\frac{1}{3},\frac{2}{3},\frac{1}{2})$ $S(0,0,u)(\frac{1}{3},\frac{2}{3},\frac{1}{2}+u)u=0.375$

(a)

(b)

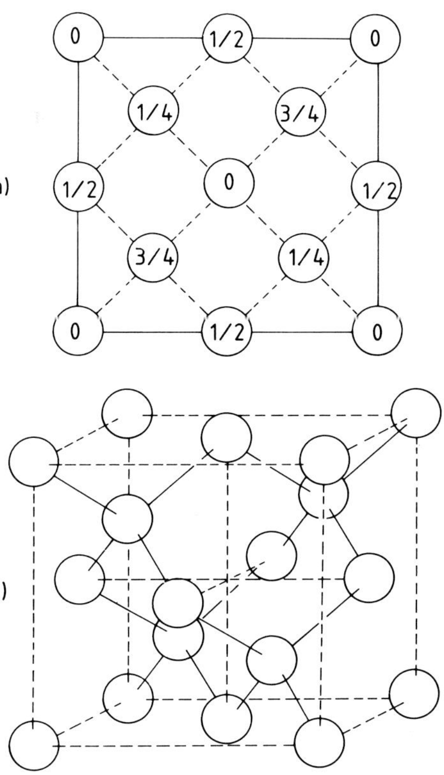

Fig. B.1 Diamond structure.

the atoms on the base of the conventional unit cell, as shown in Fig. B.1a, the numbers indicating the heights of the atoms above the base. With Fig. B.1a in mind one can then draw the three-dimensional view of the crystal structure shown in Fig. B.1b. From this one sees, for example, that in the diamond structure each atom is located at the center of a regular tetrahedron of nearest-neighbor atoms.

Most elements have either the body-centered cubic (bcc), face-centered cubic (fcc), or hexagonal closed-packed (hcp) structure. The first two of these structures are physical realizations of the bcc and fcc lattices in which each lattice site is occupied by an atom. None of the other Bravais lattices are realized in this simple way, so that when these other lattices occur in nature there is always more than one atom in the primitive unit cell. In the hcp structure, for example, there are two atoms per unit cell.

The fcc and hcp structures are shown in Fig. B.2, where it is seen that these are both *close-packed structures* in which each atom has 12 nearest neighbors. The characteristic property of a close-packed structure is that it consists of a stack of hexagonal layers. In the fcc structure the (111) planes are the hexagonal planes and the layer sequence is ABCABC. . . . In the hcp structure the (0001) planes are the hexagonal planes and the layer sequence is ABAB . . . . In the ideal hcp structure

$$\frac{c}{a} = \sqrt{\frac{8}{3}} = 1.633 \tag{B.3}$$

An important difference between the fcc and hcp structures is that the former is centrosymmetric while the latter is not.

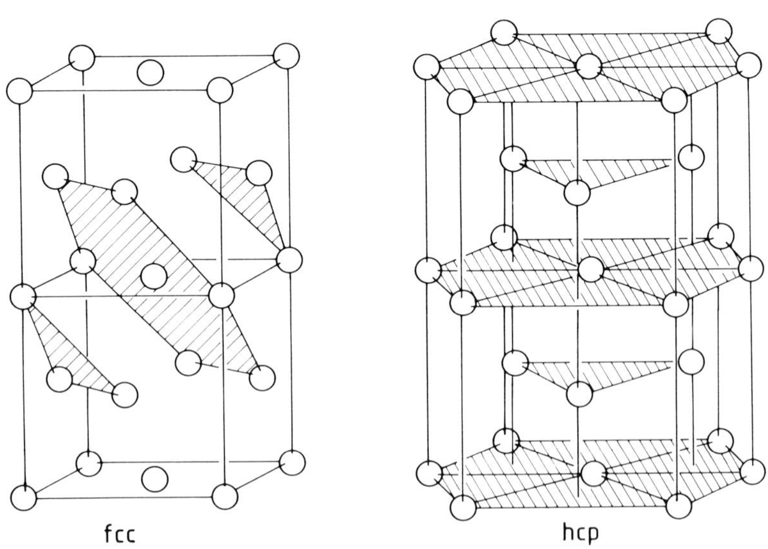

Fig. B.2 Cubic and hexagonal close-packed structures.

Symmetry and coordination

In general, a crystal structure is characterized by its symmetry and coordination. By *symmetry* we mean the space-group operations of translation, rotation and reflection that map the crystal onto itself. By *coordination* we mean, roughly speaking, the topology of the crystal, i.e., the nature and arrangement of the nearest neighbors of each atom in the crystal.

It is clear that different crystals can have the same symmetry but different coordination, and Table B.1 contains many examples of this (e.g., CsCl and $CaTiO_3$). It is perhaps less clear at first sight, but none the less equally true, that different crystals can have the same coordination but different symmetry. The classic example of this is provided by the cubic zinc blende (β-ZnS) and hexagonal wurtzite (α-ZnS) structures, in both of which each atom is surrounded by a tetrahedron of atoms of the other kind. Another example is provided by the cubic NaCl and hexagonal NiAs structures, in both of which each atom is surrounded by an octahedron of atoms of the other kind.

In all but one of the crystal structures listed in Table B.1 the atoms occupy positions that are determined uniquely by the space group. The exception is the wurtzite structure in which the Zn–S distance has a value $u = 0.375$ that can only be determined by experiment. Such a crystal structure is called a *one-parameter structure.* Another important example of a one-parameter structure is the tetragonal rutile structure.

Structural units and bonding

Up to this point we have been considering crystals from a purely structural point of view. From a physical point of view, a crystal is a system of charged particles (nuclei and electrons) interacting primarily through their mutual Coulomb interactions and moving in accordance with the laws of quantum mechanics. The ionic core of an atom, which consists of the nucleus and filled shells of electrons, is essentially the same in a crystal as in the free atom. However, the outer or valence electrons, which bind the ions together in the crystal, are in very different states from those of the free atom.

Crystals can be classified according to the nature of the basic *structural units* and the forces between them.

1. *Covalent forces.* In a crystal like diamond (Fig. B.1) the valence electrons form strong covalent bonds between the neighboring carbon atoms, so that the crystal as a whole is, in effect, one large molecule.
2. *Ionic forces.* In a crystal like NaCl the structural units are the Na^+ and Cl^- ions and the crystal is bound by the Coulomb attraction between neighboring ions.
3. *Metallic forces.* In a metallic crystal like sodium the structural units

are the Na^+ ions that are embedded in the sea of conduction electrons that binds the crystal as a whole.

4. *Van der Waals forces.* In a crystal like argon the structural units are the Ar atoms themselves and the crystal is bound by the weak van der Waals forces between the atoms.

In all of the preceding examples the structural units are atoms or ions. In many crystals the basic structural units are molecules or molecular ions. For example, in a molecular crystal such as methane the structural units are the CH_4 molecules in which the atoms are bound together by strong covalent forces. The molecules themselves are bound together in the crystal by weak van der Waals forces. In an ionic crystal like NH_4Cl the structural units are the NH_4^+ and Cl^- ions. The atoms in the NH_4^+ ions are bound together by strong covalent forces, while the ions themselves are bound together in the crystal by the relatively weaker Coulomb forces.

Thermodynamic phases

In a macroscopic system of atoms that is in thermodynamic equilibrium at some specified temperature and density, the stable equilibrium state is the one for which the Helmholtz free energy,

$$F = U - TS \tag{B.4}$$

is a minimum. The equilibrium state then involves a compromise between minimizing the internal energy U and maximizing the entropy S. At low temperatures T, the free energy is dominated by the internal energy, which usually favors an ordered crystalline state. With increasing temperature, the entropy term eventually becomes dominant and this favors a disordered state such as the liquid phase.

Many molecular liquids have the property that with decreasing temperature the rotational motion of the molecules does not freeze out at the same temperature as the translational motion. In situations (characteristic of globular molecules) in which the translational motion freezes out first, the intermediate phase is called the *plastic crystal* phase. Here the centers of mass of the molecules occupy the sites of a regular crystalline lattice, usually cubic, but their orientations remain largely disordered. In other situations (characteristic of long, rod-like molecules), in which the rotational motion freezes out first, the intermediate phase is called the *liquid crystal* phase. Here the molecules are oriented parallel to one another but their center-of-mass positions remain largely disordered.

Thermal motion in crystals

Up to this point we have ignored the thermal and zero-point motion of the atoms that is always present in any real crystal. In most crystals this motion consists of small-amplitude vibrations about well-defined equilibrium positions. Thus, the instantaneous position of the νth atom in the lth unit cell can be expressed as

$$\mathbf{r}_{l\nu} = \mathbf{R}_l + \mathbf{r}_\nu + \mathbf{u}_{l\nu} \tag{B.5}$$

where $\mathbf{R}_l$ is the lattice vector giving the position of the lth lattice site, $\mathbf{r}_\nu$ is the equilibrium position of the νth atom relative to this site, and $\mathbf{u}_{l\nu}$ is the displacement of this atom from its equilibrium position. By definition,

$$\mathbf{r}_\nu = \langle \mathbf{r}_{l\nu} \rangle - \mathbf{R}_l \tag{B.6}$$

so that

$$\langle \mathbf{u}_{l\nu} \rangle = 0 \tag{B.7}$$

The displacements $\mathbf{u}_{l\nu}$ represent the thermal and zero-point motion of the atoms. In crystals in which the basic structural units are atoms or ions the quantity $\mathbf{u}_{l\nu}$ describes the translational motion of these units. In an ionic crystal such as NH_4Cl the thermal motion of a hydrogen atom is the sum of the translational motion of the center of mass of the NH_4^+ ion plus the librational motion of this ion about its equilibrium orientation. For some purposes it may also be necessary to take the internal vibration of the NH_4^+ ion into account.

The decomposition (B.5) breaks down in the orientationally disordered (plastic) phases of molecular crystals. Here the molecules execute random jumps between different equilibrium orientations, so that some atoms do not have unique equilibrium positions. In this case there is no single vector $\mathbf{r}_\nu$ in terms of which the crystal structure can be specified. Instead, the structure is defined by giving the equilibrium positions of the centers of mass of the molecules plus the parameters that define the molecular orientational distribution function (ODF). In some crystals (such as SF_6) the ODF has the same symmetry as the molecule itself, but in other crystals the ODF may have higher symmetry than the molecules. In CBr_4, for example, the molecules are tetrahedral, while the ODF is cubic. In para-H_2 the molecules are linear, whereas the ODF has complete spherical symmetry.

Magnetic structures

To specify the structure of a magnetic crystal one must give not only the positions of the atoms within the unit cell but also the magnitudes and directions of the atomic magnetic dipole moments (hereafter referred to simply as *spins*). Figure B.3 shows two typical magnetic structures, both of which are bcc as far as the atomic positions are concerned. Figure B.3a represents the structure of a ferromagnet such as iron, in which all the spins are in the same direction, while Fig. B.3b represents the structure of an antiferromagnet such as chromium, in which the spin of each atom is in the opposite direction to that of its nearest neighbors.

The antiferromagnetic structure of chromium is similar to that of CsCl with the up spins corresponding to the Cs^+ ions and the down spins to the

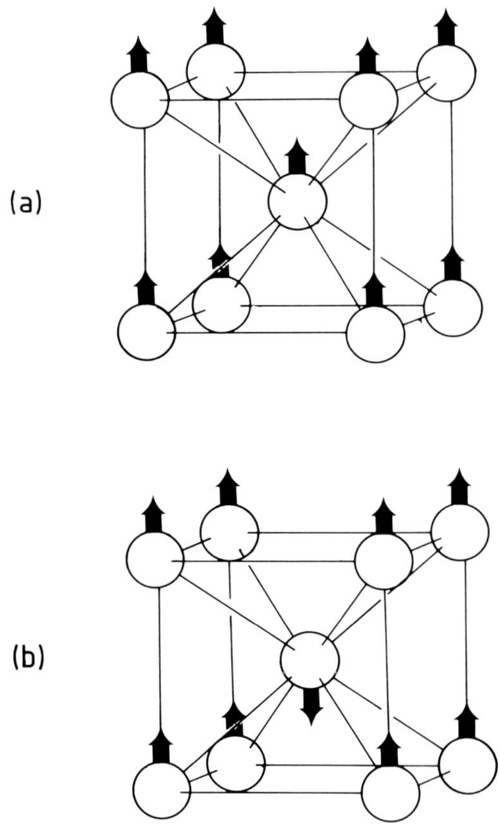

Fig. B.3 Typical structues of (a) a ferromagnet and (b) an antiferromagnet.

Cl^- ions. However, there is also an important difference between these two structures. There is no physical symmetry operation that changes a Cs^+ ion into a Cl^- ion, but there is a symmetry operation, namely time reversal, that changes an up spin into a down spin. If time reversal is included in the set of space group operations, the total number of different space groups in three dimensions is increased from 230 to 1651.

In all the crystals discussed so far the structure is characterized by a single three-dimensional periodicity. However, there are also situations in which the positions and spins of the atoms have multiple periodicities that are not mutually commensurate, so that the structure is not a simple three-dimensional repetition of the same unit cell. The classic example of such an *incommensurate phase* is provided by the helical spin arrangements in rare-earth metals, where the pitch of the helix is not a simple multiple of the atomic lattice constant.

Crystal imperfections

Real crystals always contain imperfections or defects, the nature and extent of which determine the so-called *texture* of the crystal. The diffraction of neutrons by a crystal depends on both its structure and its texture. The texture is determined by the way in which the crystal was originally grown and on the stresses to which it has subsequently been subjected. These stresses include, for example, mechanical stresses such as cold working and thermal stresses such as quenching.

Some common examples of crystal imperfections are listed in Table B.2 according to their dimension. As far as neutron diffraction in crystals is concerned, the most important type of defect is the *dislocation*. Figure B.4 is a schematic illustration of an edge dislocation that occurs when a plane of atoms terminates inside the crystal instead of at the surface. The lattice relaxes around the dislocation so that the residual strain is appreciable only in the immediate neighborhood of the dislocation line, which runs perpendicular to the plane of Fig. B.4.

Dislocations can migrate through the crystal in response to thermal and mechanical stresses. In a carefully annealed single crystal the dislocations tend to form a network of more or less parallel equally spaced lines to as to minimize the overall residual strain in the crystal. This has the effect of partitioning the crystal into a large number of *mosaic blocks,* as illustrated schematically in Fig. B.5. Each such block is essentially a perfect crystal but is slightly displaced and misoriented relative to neighboring blocks. The mean angle of misorientation of the blocks is called the *mosaic spread* and is given approximately by $\eta = 1/N$, where N is the average distance between neighboring dislocations in units of the lattice constant. For a typical mosaic crystal, η is a few minutes of arc ($\simeq 10^{-3}$ radians) so that $N \simeq 10^3$.

Table B.2 Common crystal imperfections

Dimension	Type	Example
0	Point	Vacancy Interstitial Chemical impurity
1	Line	Edge dislocation Screw dislocation
2	Plane	Grain boundary Stacking fault Domain wall
3	Volume	Void Gas bubble Chemical precipitate

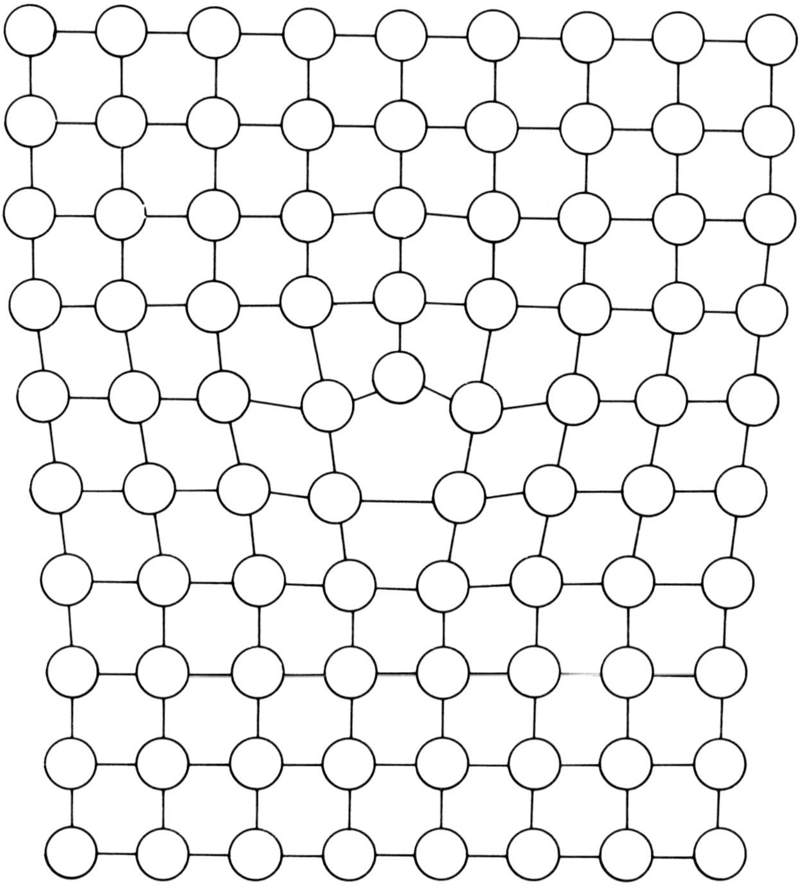

Fig. B.4 Edge dislocation.

The boundary between different mosaic blocks is a region of finite thickness through which there is an essentially continuous change in the orientation of the unit cells, so that there is no discontinuity in the lattice at the boundary. Except in the immediate neighborhood of the dislocation lines, which are separated by distances that are typically thousands of Ångstoms, the arrangement of the atoms in the boundary region is practically the same as elsewhere in the crystal. Thus, a mosaic crystal is not fragmented in the way that, for example, a polycrystal or a powder is.

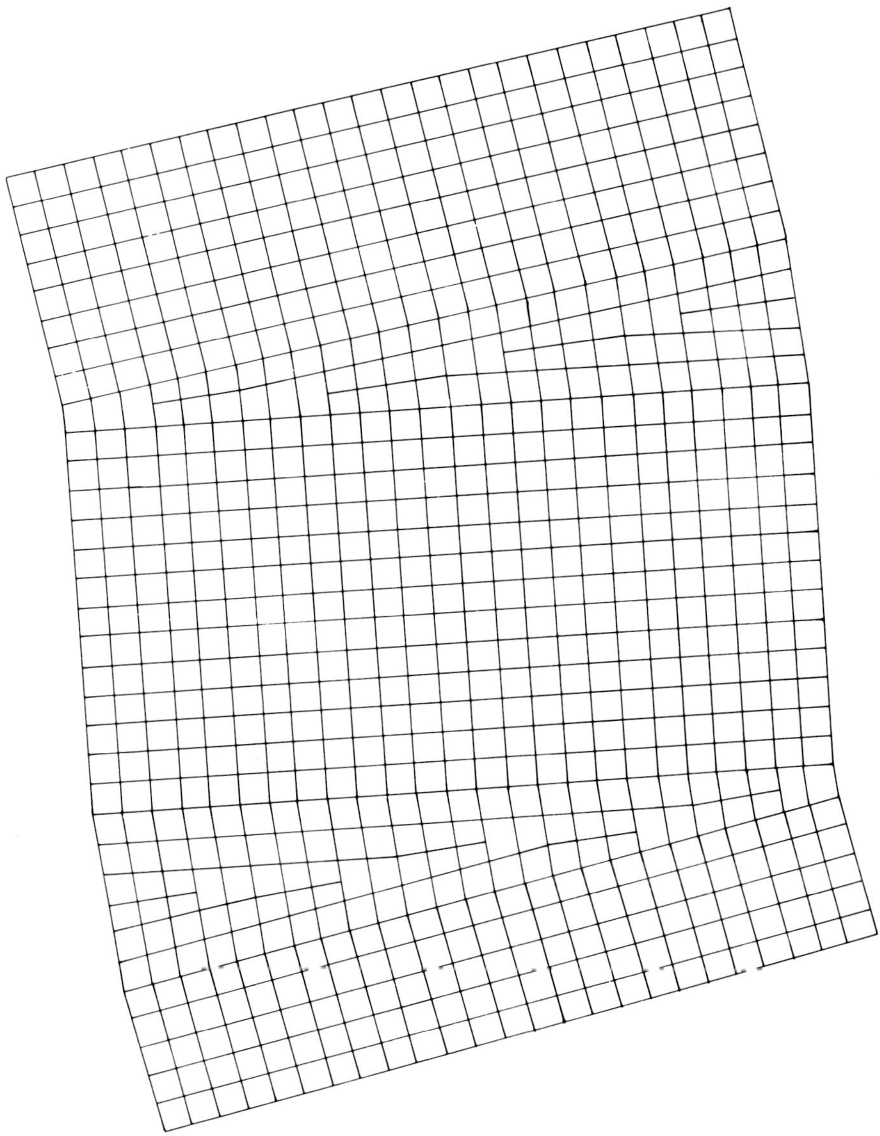

Fig. B.5 Mosaic crystal.

REFERENCES

Abragam, A. and Goldman, M. (1978), *Rep. Prog. Phys.* **41,** 395.

Abragam, A., Bacchella, G. L., Long, C., Meriel, P., Piesvaux, J., and Pinot, M. (1972), *Phys. Rev. Lett.* **28,** 805.

Abragam, A., Bacchella, G. L., Glättli, H., Meriel, P., Pinot, M., and Piesvaux, J. (1973), *Phys. Rev. Lett.* **31,** 776.

Abramowitz, M. and Stegun, I. A. (1965), *Handbook of Mathematical Functions.* National Bureau of Standards, Washington.

Adli, D. G. and Summerfield, G. C. (1984), *Phys. Rev. A* **30,** 119.

Alcock, N. Z. and Hurst, D. G. (1949), *Phys. Rev.* **75,** 1609.

Allen, A. J., Hutchings, M. T., Windsor, C. G., and Andreani, C. (1985). *Adv. Phys.* **34,** 445.

Amaldi, E., D'Agostino, O., Fermi, E., Pontecorvo, B., Rasetti, F., and Segrè, E. (1934), *Ric. Sci.* **5,** 467.

Amaldi, E., D'Agostino, O., Fermi, E., Pontecorvo, B., Rasetti, F., and Segrè, E. (1935), *Proc. Roy. Soc. A* **149,** 522.

Anderson, H. L., Fermi, E., and Marshall, L. (1946), *Phys. Rev.* **70,** 815.

Arif, M., Kaiser, H., Werner, S. A., Cimmino, A., Hamilton, W. A., Klein, A. G., and Opat, G. I. (1985). *Phys. Rev. A* **31,** 1203.

Armstrong, R. L., Morra, R. M., and Powell, B. M. (1985), *Can. J. Phys.* **63,** 988.

Atwood, D. K., Horne, M. A., Shull, C. G., and Arthur, J. (1984). *Phys. Rev. Lett.* **52,** 1673.

Axe, J. D. (1976), Proc. Conf. Neutron Scattering, Gatlinburg, Tennessee, June 6–10, 1976. CONF-760601-P1, p. 353.

Azaroff, L. V. and Buerger, M. J. (1958), *The Powder Method in X-ray Crystallography,* McGraw-Hill, New York.

Bacon, G. E. (1963), *Applications of Neutron Diffraction in Chemistry,* MacMillan, New York.

Bacon, G. E. (1975), Neutron Diffraction, 3d ed, Clarendon Press, Oxford.

Bacon, G. E. and Lowde, R. D. (1948), *Acta Crystallogr.* **1,** 303.

Badurek, G., Rauch, H., Zeilinger, A., Bauspiess, W., and Bonse, U. (1976a), *Phys. Lett.* **56A,** 244.

Badurek, G., Rauch, H., Zeilinger, A., Bauspiess, W., and Bonse, U. (1976b), *Phys. Rev. D* **14,** 1177.

Badurek, G., Rauch, H., Wilfing, A., Bonse, U., and Graeff, W. (1979), *J. Appl. Crystallogr.* **12,** 186.

Badurek, G., Rauch, H., Summhammer, J., Kischko, U., and Zeilinger, A. (1983), *J. Phys. A* **16,** 1133.

Badurek, G., Rauch, H., and Tuppinger, D. (1986), *Phys. Rev. A* **34,** 2600.

Barkan, S., Bieber, E., Burgy, M. T., Ketudat, S., Krohn, V. E., Rice-Evans, P., and Ringo, G. R. (1968), *Rev. Sci. Instrum.* **39,** 101.

Baruchel, J., Guigay, J. P., Mazure-Espejo, C., Schlenker, M., and Schweizer, J. (1982), *J. de Phys.* **43,** C7-101.

Baryshevskii, V. G. and Podgoretskii, M. I. (1965), *Sov. Phys. JETP* **20,** 704.

Batterman, B. W. and Cole, H. (1964), *Rev. Mod. Phys.* **36,** 681.

Bauspiess, W., Bonse, U., Rauch, H., and Treimer, W. (1974), *Z. Phys.* **271,** 177.
Bauspiess, W., Bonse, U., and Rauch, H. (1978), *Nucl. Instrum. Methods* **157,** 495.
Berger, H. (1971), *Ann. Rev. Nucl. Sci.* **21,** 335.
Bernstein, H. J. (1967), *Phys. Rev. Lett.* **18,** 1102.
Bethe, H. A. (1949), *Phys. Rev.* **76,** 38.
Bethe, H. A. and Placzek, G. (1937), *Phys. Rev.* **51,** 450.
Beynon, T. D. and Pink, A. G. (1980), *Nature* **283,** 749.
Bialynicki-Birula, I. and Mycielski, J. (1976), *Ann. Phys. (NY)* **100,** 62.
Blatt, J. M. and Jackson, J. D. (1949), *Phys. Rev.* **76,** 18.
Blaudeck, P. and Lenk, R. (1976), *Phys. Stat. Sol. B* **77,** 601.
Bloch, F. (1936), *Phys. Rev.* **50,** 259.
Bonse, U. and Graeff, W. (1977), in *X-Ray Optics,* editor H. J. Queisser, Springer-Verlag, Berlin; p. 93.
Bonse, U. and Hart, M. (1965), *Appl. Phys. Lett.* **6,** 155.
Bonse, U. and Rauch, H. (editors) (1979), *Neutron Interferometry,* Clarendon Press, Oxford.
Bonse, U. and Rumpf, A. (1986), *Phys. Rev. Lett.* **56,** 2441.
Bonse, U. and Wroblewski, T. (1983), *Phys. Rev. Lett.* **51,** 1401.
Bonse, U. and Wroblewski, T. (1984), *Phys. Rev. D* **30,** 1214.
Born, M. (1926), *Z. Phys.* **38,** 803.
Born, M. and Wolf, E. (1975), *Principles of Optics,* Pergamon, Oxford.
Borrmann, G. (1941), *Phys. Z.* **42,** 157.
Borrmann, G. (1950), *Z. Phys.* **127,** 297.
Bragg, W. H. and Bragg, W. L. (1913), *Proc. Roy. Soc. A* **88,** 428.
Bragg, W. L. (1913), *Proc. Camb. Phil. Soc.* **17,** 43.
Bragg, W. L. (1940), *Proc. Phys. Soc.* **52,** 54.
Bragg, W. L. (editor) (1962), *The Crystalline State,* Bell, London.
Bragg, W. L., James, R. W., and Bosanquet, C. H. (1921a), *Phil. Mag.* **41,** 309.
Bragg, W. L., James, R. W., and Bosanquet, C. H. (1921b), *Phil. Mag.* **42,** 1.
Breit, G. (1947), *Phys. Rev.* **71,** 215.
Breit, G. and Wigner, E. (1936), *Phys. Rev.* **49,** 519.
Breit, G. and Zilsel, P. R. (1947), *Phys. Rev.* **71,** 232.
Brillouin, L. (1930), *J. Phys. Rad.* **1,** 377.
Brockhouse, B. N. (1955), *Phys. Rev.* **99,** 601.
Brockhouse, B. N. (1957), *Phys. Rev.* **106,** 859.
Brockhouse, B. N. (1959), *Rev. Sci. Instrum.* **30,** 136.
Brockhouse, B. N. (1961), in *Inelastic Scattering of Neutrons in Solids and Liquids,* International Atomic Energy Agency, Vienna; p. 113.
Brockhouse, B. N. and Stewart, A. T. (1955), *Phys. Rev.* **100,** 756.
Buerger, M. J. (1960), *Crystal-Structure Analysis.* Wiley, New York.
Burgers, J. M. (1940), *Proc. Phys. Soc.* **52,** 23.
Chadwick, J. (1932), *Nature* **129,** 312.
Chamberlain, O. (1950), *Phys. Rev.* **77,** 305.
Chew, G. F. (1950), *Phys. Rev.* **80,** 196.
Christ, J. and Springer, T. (1962), *Nukleonik* **4,** 23.
Colella, R., Overhauser, A. W., and Werner, S. A. (1975), *Phys. Rev. Lett.* **34,** 1472.
Comsa, G. (1983), *Phys. Rev. Lett.* **51,** 1105.
Cowley, J. M. (1975), *Diffraction Physics,* North-Holland, Amsterdam.
Dabbs, J. W. T., Harvey, J. A., Paya, D., and Horstmann, H. (1965), *Phys. Rev.* **139,** B756.
Dachs, H. (editor) (1978), *Neutron Diffraction,* Springer-Verlag, Berlin.
Darwin, C. G. (1914a), *Phil. Mag.* **27,** 315.

Darwin, C. G. (1914b), *Phil. Mag.* **27,** 675.
Darwin, C. G. (1922), *Phil. Mag.* **43,** 800.
Davison, B. and Sykes, J. B. (1957), *Neutron Transport Theory,* Clarendon Press, Oxford.
Davisson, C. and Germer, L. H. (1927), *Phys. Rev.* **30,** 705.
Debye, P. (1914), *Ann. Phys. (Leipzig)* **43,** 49.
Debye, P. and Scherrer, P. (1917), *Phys. Z.* **27,** 277.
Dederichs, P. H. (1972), *Solid State Phys.* **27,** 135.
Delapalme, A., Schweizer, J., Couderchon, G., and Bathie, R. P. (1971), *Nucl. Instrum. Methods* **95,** 589.
Dickinson, W. C., Passell, L., and Halpern, O. (1962), *Phys. Rev.* **126,** 632.
Dietze, H. D. and Nowak, E. (1981), *Z. Phys. B* **44,** 245.
Doi, K., Minakawa, N., Motohashi, H., and Masaki, N. (1971), *J. Appl. Crystallogr.* **4,** 528.
Dolling, G., Powell, B. M., and Sears, V. F. (1979), *Molec. Phys.* **37,** 1859.
Ebisawa, T., Achiwa, N., Yamada, S., Akiyoshi, T., and Okamoto, S. (1979), *J. Nucl. Sci. Tech.* **16,** 647.
Egelstaff, P. A. and Pease, R. S. (1954), *J. Sci. Instrum.* **31,** 207.
Ekstein, H. (1951), *Phys. Rev.* **83,** 721.
Ekstein, H. (1953), *Phys. Rev.* **89,** 490.
Elsasser, W. M. (1936), *Comptes Rendus* **202,** 1029.
Ewald, P. P. (1913), *Phys. Z.* **14,** 465.
Ewald, P. P. (1916a), *Ann. Phys. (Leipzig)* **49,** 1.
Ewald, P. P. (1916b), *Ann. Phys. (Leipzig)* **49,** 117.
Ewald, P. P. (1917a), *Ann. Phys. (Leipzig)* **54,** 519.
Ewald, P. P. (1917b), *Ann. Phys. (Leipzig)* **54,** 557.
Ewald, P. P. (1921), *Z. Krist.* **56,** 129.
Ewald, P. P. (1940), *Proc. Phys. Soc.* **52,** 167.
Ewald, P. P. (1965), *Rev. Mod. Phys.* **37,** 46.
Ewald, P. P. (1969), *Acta Crystallogr. A* **25,** 103.
Faxén, H. and Holtsmark, J. (1927), *Z. Phys.* **45,** 307.
Feenberg, E. (1932), *Phys. Rev.* **40,** 40.
Fermi, E. (1934), *Nuovo Cim.* **2,** 157.
Fermi, E. (1936), *Ric. Sci.* **7,** 13.
Fermi, E. and Marshall, L. (1946), *Phys. Rev.* **70,** 103.
Fermi, E. and Marshall, L. (1947), *Phys. Rev.* **71,** 666.
Fermi, E. and Zinn, W. H. (1946), *Phys. Rev.* **70,** 103.
Fermi, E., Amaldi, E., Pontecorvo, B., Rasetti, F., and Segrè, E. (1934), *Ric. Sci.* **5,** 282.
Fermi, E., Sturm, W. J., and Sachs, R. G. (1947), *Phys. Rev.* **71,** 589.
Finkelstein, D., Jauch, J. M., Schiminovich, S., and Speiser, D. (1962), *J. Math. Phys.* **3,** 207.
Foldy, L. L. (1945), *Phys. Rev.* **67,** 107.
Foldy, L. L. (1966), in *Preludes in Theoretical Physics,* editors A. De-Shalit, H. Feshbach, and L. Van Hove, North-Holland, Amsterdam; p. 205.
Freund, A. and Forsyth, J. B. (1979), in *Neutron Scattering,* editor G. Kostorz, Academic Press, New York; p. 461.
Friedel, G. (1913), *Comptes Rendus* **157,** 1533.
Friedrich, W., Knipping, P., and Laue, M. (1912), *Sitz. Bay. Akad. Wiss.* 303.
Gähler, R., Kalus, J., and Mampe, W. (1980), *J. Phys. E* **13,** 546.
Gähler, R., Klein, A. G., and Zeilinger, A. (1981), *Phys. Rev. A* **23,** 1611.
Gähler, R., Kalus, J., and Mampe, W. (1982), *Phys. Rev.* D **25,** 2887.
Gibbs, D., Moncton, D. E., D'Amico, K. L., Bohr, J., and Grier, B. H. (1985), *Phys. Rev. Lett.* **55,** 234.
Glättli, H. and Goldman, M. (1986), in *Methods of Experimental Physics,* Vol.

23, Neutron scattering, Part C, editors K. Sköld and D. L. Price, Academic Press, New York; chapter 21.
Goldberger, M. L. and Seitz, F. (1947), *Phys. Rev.* **71,** 294.
Goldberger, M. L. and Watson, K. M. (1964), *Collision Theory,* Wiley, New York.
Goldberger, M. L., Lewis, H. W., and Watson, K. M. (1963), *Phys. Rev.* **132,** 2764.
Goldstein, H. (1950), *Classical Mechanics,* Addison-Wesley, Reading, MA.
Golub, R. and Pendlebury, J. M. (1979), *Rep. Prog. Phys.* **42,** 439.
Graf, A., Rauch, H., and Stern, T. (1979), *Atomkernenergie* **33,** 298.
Greenberger, D. M. (1983), *Rev. Mod. Phys.* **55,** 875.
Greenberger, D. M. and Overhauser, A. W. (1979), *Rev. Mod. Phys.* **51,** 43.
Greenberger, D. M., Atwood, D. K., Arthur, J., Shull, C.G., and Schlenker, M. (1981), *Phys. Rev. Lett.* **47,** 751.
Groshev, I. V., Dvoretsky, V. N., Demidov, A. M., Panin, Yu. N., Lushchikov, V. I., Pokotilovsky, Yu. N., Strelkov, A. V., and Shapiro, F. L. (1971), *Phys. Lett.* **34B,** 293.
Guinier, A. (1939), *Ann. Phys.* (*Paris*) **12,** 161.
Hahn, T. (editor) (1983), *International Tables for Crystallography,* Reidel, Holland.
Halban, H. and Preiswerk, P. (1936), *Comptes Rendus* **203,** 73.
Halpern, O. (1949a), *Phys. Rev.* **75,** 343.
Halpern, O. (1949b), *Phys. Rev.* **76,** 1130.
Halpern, O., Hamermesh, M., and Johnson, M. H. (1941), *Phys. Rev.* **59,** 981.
Hamelin, B. (1976), *Nucl. Instrum. Methods* **135,** 299.
Hamilton, W. A., Klein, A. G., and Opat, G. I. (1983), *Phys. Rev.* A **28,** 3149.
Hawkesworth, M. R. and Walker, J. (1969), *J. Mater. Sci.* **4,** 817.
Hayter, J. B. (1978), in *Neutron Diffraction,* editor H. Dachs, Springer-Verlag, Berlin; p. 41.
Hayter, J. B., Penfold, J., and Williams, W. G. (1978), *J. Phys. E* **11,** 454.
Herrmann, P., Steinhauser, K. A., Gähler, R., and Steyerl, A. (1985), *Phys. Rev. Lett.* **54,** 1969.
Hirsch, P. B. (1956), *Prog. Met. Phys.* **6,** 236.
Hohenberg, P. C. and Platzman, P. M. (1966), *Phys. Rev.* **152,** 198.
Horne, M. A. and Zeilinger, A. (1979), in *Neutron Interferometry,* editors U. Bonse and H. Rauch, Clarendon Press, Oxford; p. 350.
Horne, M. A., Zeilinger, A., Klein, A. G., and Opat, G. I. (1983), *Phys. Rev. A* **28,** 1.
Hughes, D. J. (1954), *Neutron Optics,* Interscience, New York.
Hughes, D. J. and Burgy, M. T. (1951), *Phys. Rev.* **81,** 498.
Hughes, D. J. and Harvey, J. A. (1955), *Neutron Cross Sections,* Brookhaven National Laboratory Report BNL 325.
Hull, A. W. (1917), *Phys. Rev.* **10,** 661.
Ito, Y. and Shull, C. G. (1969), *Phys. Rev.* **185,** 961.
Izyumov, Y. A. and Ozerov, R. P. (1970), *Magnetic Neutron Diffraction.* Plenum, New York.
Jahnke, E. and Emde, F. (1945), *Tables of Functions,* Dover, New York.
James, R. W. (1963), *Solid State Phys.* **15,** 53.
Just, W., Schneider, C. S., Ciszewski, R., and Shull, C. G. (1973), *Phys. Rev.* B **7,** 4142.
Kaiser, H., Rauch, H., Badurek, G., Bauspiess, W., and Bonse, U. (1979), *Z. Phys. A* **291,** 231.
Kaiser, H., Werner, S. A., and George, E. A. (1983), *Phys. Rev. Lett.* **50,** 560.
Kaiser, H., George, E. A., and Werner, S. A. (1984), *Phys. Rev. A* **29,** 2276.
Kaiser, H., Arif, M., Werner, S. A., and Willis, J. O. (1986), *Physica* **136B,** 134.

Kaneno, T. (1960), *Prog. Theor. Phys.* **23,** 17.
Kearney, P. D., Klein, A. G., Opat, G. I., and Gähler, R. (1980), *Nature* **287,** 313.
Kikuta, S., Kohra, K., Minakawa, N., and Doi, K. (1971), *J. Phys. Soc. Japan* **31,** 954.
Kikuta, S., Ishikawa, I., Kohra, K., and Hoshino, S. (1975), *J. Phys. Soc. Japan* **39,** 471.
Kitchens, T. A., Oversluizen, T., and Passell, L. (1974), *Phys. Rev. Lett.* **32,** 791.
Klein, A. G. and Martin, L. J. (1977), *Am. J. Phys.* **45,** 298.
Klein, A. G. and Opat, G. I. (1975), *Phys. Rev. D* **11,** 523.
Klein, A. G. and Opat, G. I. (1976), *Phys. Rev. Lett.* **37,** 238.
Klein, A. G. and Opat, G. I. (1984), *J. de Phys.* **45,** C3-235.
Klein, A. G. and Werner, S. A. (1983), *Rep. Prog. Phys.* **46,** 259.
Klein, A. G. and Zeilinger, A. (1984), *J. de Phys,* **45,** C3-239.
Klein, A. G., Martin, L. J., and Opat, G. I. (1977), *Am. J. Phys.* **45,** 295.
Klein, A. G., Kearney, P. D., Opat, G. I., Cimmino, A., and Gähler, R. (1981a), *Phys. Rev. Lett.* **46,** 959.
Klein, A. G., Kearney, P. D., Opat, G. I., and Gähler, R. (1981b), *Phys. Lett.* **83A**, 71.
Klein, A. G., Opat, G. I., Cimmino, A., Zeilinger, A., Treimer, W., and Gähler, R. (1981c), *Phys. Rev. Lett.* **46,** 1551.
Klein, A. G., Opat, G. I., and Hamilton, W. A. (1983), *Phys. Rev. Lett.* **50,** 563.
Knowles, J. W. (1956), *Acta Crystallogr.* **9,** 61.
Koester, L. (1965), *Z. Phys.* **182,** 328.
Koester, L. (1976), *Phys. Rev. D* **14,** 907.
Koester, L. (1977), *Springer Tracts Mod. Phys.* **80,** 1.
Koester, L. and Nistler, W. (1971), *Phys. Rev. Lett.* **27,** 956.
Koester, L. and Nistler, W. (1975), *Z. Phys. A* **272,** 189.
Koester, L. and Ungerer, H. (1969), *Z. Phys.* **219,** 300.
Koester, L., Nistler, W., and Waschkowski, W. (1976), *Phys. Rev. Lett.* **36,** 1021.
Koester, L., Rauch, H., Herkens, M., and Schröder, K. (1981), *Summary of Neutron Scattering Lengths,* Jülich report no. Jül-1755.
Koester, L., Knopf, K., Waschkowski, W., and Klüver, A. (1985), *Z. Phys. A* **320,** 661.
Koester, L., Waschkowski, W., and Klüver, A. (1986), *Physica* **137B,** 282.
Kostorz, G. (editor) (1979), *Neutron Scattering,* Academic Press, New York; p. 227.
Kurz, H. and Rauch, H. (1969), *Z. Phys.* **220,** 419.
Lamb, W. E. (1937), *Phys. Rev.* **51,** 187.
Lamb, W. E. (1939), *Phys. Rev.* **55,** 190.
Landau, L. D. and Lifshitz, E. M. (1958), *Quantum Mechanics,* Addison-Wesley, Reading, MA.
Landkammer, F. J. (1966), *Z. Phys.* **189,** 113.
Laue, M. (1931), *Ergeb. Exakten Naturwiss.* **10,** 133.
Lax, M. (1951), *Rev. Mod. Phys.* **23,** 287.
Lax, M. (1952), *Phys. Rev.* **85,** 621.
Lenk, R. (1975), *Phys. Stat. Sol. B* **69,** 369.
Lippmann, B. A. (1950), *Phys. Rev.* **79,** 481.
Lippmann, B. A. and Schwinger, J. (1950), *Phys. Rev.* **79,** 469.
Lovesey, S. W. (1984), *Theory of Neutron Scattering from Condensed Matter,* Clarendon Press, Oxford.
Lushchikov, V. I. (1977), *Physics Today* **30**(6), 42.
Lushchikov, V. I., Pokotilovskii, Yu. N., Strelkov, A. V., and Shapiro, F. L. (1969), *JETP Lett.* **9,** 23.

Lynn, J. E. (1968), *Theory of Neutron Resonance Reactions,* Clarendon Press, Oxford.

Lynn, J. W., Kjems, J. K., Passell, L., Saxena, A. M., and Schoenborn, B. P. (1976), *J. Appl. Crystallogr.* **9,** 454.

MacEwen, S. R., Faber, J., and Turner, A. P. L. (1983), *Acta Metall.* **31,** 657.

Maier-Leibnitz, H. (1962), *Z. Angew. Phys.* **14,** 738.

Maier-Leibnitz, H. (1972), in *Neutron Inelastic Scattering,* International Atomic Energy Agency, Vienna; p. 681.

Maier-Leibnitz, H. and Springer, T. (1962), *Z. Phys.* **167,** 386.

Maier-Leibnitz, H. and Springer, T. (1963), *J. Nucl. Energy* **17,** 217.

Majkrzak, C. F. (1984), *Appl. Opt.* **23,** 3524.

Majkrzak, C. F. and Passell, L. (1985), *Acta Crystallogr. A* **41,** 41.

Malgrange, C., Petroff, J. F., Sauvage, M., Zarka A., and Englander, M. (1976), *Phil. Mag.* **33,** 743.

Mandel, L. and Wolf, E. (1965), *Rev. Mod. Phys.* **37,** 231.

McReynolds, A. W. (1951), *Phys. Rev.* **83,** 172.

Mendiratta, S. K. (1976), *Phys. Rev. B* **14,** 155.

Messiah, A. (1964), *Quantum Mechanics,* North-Holland, Amsterdam.

Mezei, F. (1976), *Commun. Phys.* **1,** 81.

Mezei, F. and Dagleish, P. A. (1977), *Commun. Phys.* **2,** 41.

Mitchell, D. P. and Powers, P. N. (1936), *Phys. Rev.* **50,** 486.

Moon, R. M. and Shull, C. G. (1964), *Acta Crystallogr.* **17,** 805.

Moon, R. M., Riste, T., and Koehler, W. C. (1969), *Phys. Rev.* **181,** 920.

Mughabghab, S. F. (1984). *Neutron Cross Sections,* Vol. 1, Part B, Academic Press, New York.

Mughabghab, S. F., Divadeenam, M., and Holden, N. E. (1981), *Neutron Cross Sections,* Vol. 1, Part A, Academic Press, New York.

Nathans, R. and Pickart, S. J. (1963), in *Magnetism,* Vol. III, editors G. T. Rado and H. Suhl, Academic, New York; p. 211.

Nathans, R., Shull, C. G., Shirane, G., and Andresen, A. (1959), *J. Phys. Chem. Solids* **10,** 138.

Newton, R. G. (1976), *Am. J. Phys.* **44,** 639.

Nieman, H. F., Tennant, D. C., and Dolling, G. (1980), *Rev. Sci. Instrum.* **51,** 1299.

Nowak, E. (1982a), *Z. Phys. B* **45,** 265.

Nowak, E. (1982b), *Z. Phys. B* **49,** 1.

Overhauser, A. W. and Colella, R. (1974), *Phys. Rev. Lett.* **33,** 1237.

Page, L. A. (1975), *Phys. Rev. Lett.* **35,** 543.

Patterson, A. L. (1939), *Phys. Rev.* **56,** 972.

Pearson, W. B. (1958), *A Handbook of Lattice Spacings and Structures of Metals and Alloys,* Pergamon, New York.

Peres, A. (1979), *Phys. Rev. Lett.* **42,** 683.

Peres, A. (1983), *Am. J. Phys.* **51,** 947.

Peshkin, M. and Ringo, G. R. (1971), *Am. J. Phys.* **39,** 324.

Peterson, S. W. and Levy, H. A. (1951), *J. Chem. Phys.* **19,** 1416.

Peterson, S. W. and Smith, H. G. (1961), *Phys. Rev. Lett.* **6,** 7.

Petrascheck, D. and Folk, R. (1976), *Phys. Stat. Sol. A* **36,** 147.

Pinsker, Z. G. (1978), *Dynamical Scattering of X-rays in Crystals,* Springer-Verlag, Berlin.

Plummer, J. P. and Summerfield, G. C. (1963), *Phys. Rev.* **131,** 1153.

Rauch, H. and Petrascheck, D. (1978), in *Neutron Diffraction,* editor H. Dachs, Springer-Verlag, Berlin; p. 303.

Rauch, H. and Seidl, E. (1987), *Nucl. Instrum. Methods A* **255,** 32.

Rauch, H. and Suda, M. (1974), *Phys. Stat. Sol. A* **25,** 495.

Rauch, H., Treimer, W., and Bonse, U. (1974), *Phys. Lett.* **47A,** 369.

Rauch, H., Zeilinger, A., Badurek, G., Wilfing, A., Bauspiess, W., and Bonse, U. (1975), *Phys. Lett.* **54A,** 425.

Rauch, H., Seidl, E., Zeilinger, A., Bauspiess, W., and Bonse, U. (1978), *J. Appl. Phys.* **49,** 2731.

Rayleigh, Lord. (1878), *The Theory of Sound,* reprinted by Dover, New York, 1945.

Rayleigh, Lord. (1914), *Proc. Roy. Soc. A* **90,** 219.

Renninger, M. (1937), *Z. Phys.* **106,** 141.

Saxena, A. M. and Schoenborn, B. P. (1977), *Acta Crystallogr. A* **33,** 805.

Schärpf, O. (1982), in *Neutron Scattering*—1981, editor J. Faber, AIP Conf. Proc. No. 89, American Institute of Physics, New York; p. 182.

Scheckenhofer, H. and Steyerl, A. (1977), *Phys. Rev. Lett.* **39,** 1310.

Schelten, J. and Mika, K. (1979), *Nucl. Instrum. Methods* **160,** 287.

Schermer, R. I. (1963), *Phys. Rev.* **130,** 1907.

Schlenker, M. and Shull, C. G. (1973), *J. Appl. Phys.* **44,** 4181.

Schmatz, W., Springer, T., Schelten, J., and Ibel, K. (1974), *J. Appl. Crystallogr.* **7,** 96.

Schneider, C. S. (1973), *Rev. Sci. Instrum.* **44,** 1594.

Schneider, C. S. (1976), *Acta Crystallogr. A* **32,** 375.

Schneider, C. S. and Shull, C. G. (1971), *Phys. Rev. B* **3,** 830.

Schoenborn, B. P., Caspar, D. L. D., and Kammerer, O. F. (1974), *J. Appl. Crystallogr.* **7,** 508.

Schütz, G., Steyerl, A., and Mampe, W. (1980), *Phys. Rev. Lett.* **44,** 1400.

Schwinger, J. and Teller, E. (1937), *Phys. Rev.* **52,** 286.

Sears, V. F. (1969), *Phys. Rev.* **185,** 200.

Sears, V. F. (1975), *Adv. Phys.* **24,** 1.

Sears, V. F. (1977), *Acta Crystallogr.* A **33,** 373.

Sears, V. F. (1978), *Can. J. Phys.* **56,** 1261.

Sears, V. F. (1982a), *Phys. Rep.* **82,** 1.

Sears, V. F. (1982b), *Phys. Rev. D* **25,** 2023.

Sears, V. F. (1983a), *Nucl. Instrum. Methods* **213,** 561.

Sears, V. F. (1983b), *Acta Crystallogr. A* **39,** 601.

Sears, V. F. (1984a), *J. Appl. Crystallogr.* **17,** 226.

Sears, V. F. (1984b), *Phys. Rev. B* **30,** 44.

Sears, V. F. (1985a), *Z. Phys. A* **321,** 443.

Sears, V. F. (1985b), *Phys. Rev. A* **32,** 2524.

Sears, V. F. (1985c), *Nucl. Instrum. Methods B* **12,** 430.

Sears, V. F. (1986a), *Phys. Rep.* **141,** 281.

Sears, V. F. (1986b), in *Methods of Experimental Physics,* Vol. 23, Neutron scattering, Part A, editors K. Sköld and D. L. Price, Academic Press, New York; p. 521.

Sears, V. F. (1988), *Physica* B **151,** 156.

Sears, V. F. and Khanna, F. C. (1972), *Phys. Rev. Lett.* **29,** 549.

Segrè, E. (editor) (1962), *Enrico Fermi, Collected Papers,* University of Chicago Press, Chicago, IL.

Sherwood, J. E., Stephenson, T. E., and Bernstein, S. (1954), *Phys. Rev.* **96,** 1546.

Shimony, A. (1979), *Phys. Rev. A* **20,** 394.

Shore, B. W. (1967), *Rev. Mod. Phys.* **39,** 439.

Shull, C. G. (1968), *Phys. Rev. Lett.* **21,** 1585.

Shull, C. G. (1969), *Phys. Rev.* **179,** 752.

Shull, C. G. (1973), *J. Appl. Crystallogr.* **6,** 257.

Shull, C. G. and Ferrier, R. P. (1963), *Phys. Rev. Lett.* **10,** 295.

Shull, C. G. and Oberteuffer, J. A. (1972), *Phys. Rev. Lett.* **29,** 871.

Shull, C. G. and Shaw, W. M. (1973), *Z. Naturforsch.* **28a,** 657.

Shull, C. G. and Siegel, S. (1949), *Phys. Rev.* **75,** 1008.

Shull, C. G. and Smart, J. S. (1949), *Phys. Rev.* **76,** 1256.

Shull, C. G. and Wollan, E. O. (1951), *Phys. Rev.* **81,** 527.

Shull, C. G., Wollan, E. O., Morton, G. A., and Davidson, W. L. (1948), *Phys. Rev.* **73,** 842.

Shull, C. G., Wollan, E. O., and Koehler, W. C. (1951), *Phys. Rev.* **84,** 912.

Shull, C. G., Atwood, D. K., Arthur, J. and Horne, M. A. (1980a), *Phys. Rev. Lett.* **44,** 765.

Shull, C. G., Zeilinger, A., Squires, G. L., Horne, M. A., Atwood, D. K., and Arthur, J. (1980b), *Phys. Rev. Lett.* **44,** 1715.

Sippel, D. and Eichhorn, F. (1968), *Acta Crystallogr. A* **24,** 237.

Sippel, D., Kleinstück, K., and Schulze, G. E. R. (1962), *Phys. Stat. Sol.* **2,** K104.

Sippel, D., Kleinstück, K., and Schulze, G. E. R. (1964), *Phys. Lett.* **8,** 241.

Sippel, D., Kleinstück, K., and Schulze, G. E. R. (1965), *Phys. Lett.* **14,** 174.

Somenkov, V. A., Shilstein, S. S., Belova, N. E., and Utemisov, K. (1978), *Solid State Commun.* **25,** 593.

Squires, G. L. (1978), *Thermal Neutron Scattering,* Cambridge University Press, Cambridge.

Stassis, C. and Oberteuffer, J. A. (1974), *Phys. Rev. B* **10,** 5192.

Staudenmann, J. L., Werner, S. A., Colella, R., and Overhauser, A. W. (1980), *Phys. Rev. A* **21,** 1419.

Steyerl, A. (1969), *Phys. Lett.* **29B,** 33.

Steyerl, A. (1977), *Springer Tracts Mod. Phys.* **80,** 57.

Steyerl, A. and Malik, S. S. (1986), *Physica* **137B,** 270.

Summerfield, G. C. (1964), *Ann. Phys. (NY)* **26,** 72.

Summhammer, J., Badurek, G., Rauch, H., and Kischko, U. (1982), *Phys. Lett.* **90A,** 110.

Summhammer, J., Badurek, G., Rauch, H., Kischko, U., and Zeilinger, A. (1983), *Phys. Rev. A* **27,** 2523.

Trammell, G. T. (1962), *Phys. Rev.* **126,** 1045.

Trammell, G. T. and Chalk, J. D. (1966), *Phys. Rev.* **141,** 815.

Treimer, W., Fiedler, W., and Rückert, B. (1985), *Phys. Lett.* **110A,** 173.

Triftshäuser, W. (1965), *Z. Phys.* **186,** 23.

Vainshtein, B. K. (editor) (1981), *Modern Crystallography,* Springer-Verlag, Berlin.

Van Bueren, H. G. (1961), *Imperfections in Crystals,* North-Holland, Amsterdam.

Van Hove, L. (1954), *Phys. Rev.* **95,** 249.

Van Kranendonk, J. and Sipe, J. E. (1977), *Prog. Opt.* **15,** 245.

Vineyard, G. H. (1954), *Phys. Rev.* **96,** 93.

Von Der Hardt, P. and Röttger, H. (1981), *Neutron Radiography Handbook,* Reidel, Dordrecht, Holland.

Wagenfeld, H. (1968), *Acta Crystallogr A* **24,** 170.

Waller, I. (1923), *Z. Phys.* **17,** 198.

Warner, M. and Gubernatis, J. E. (1985), *Phys. Rev. B* **32,** 6347.

Weiss, R. J. (1951), *Phys. Rev.* **83,** 379.

Werner, S. A. (1980), *Phys. Rev. B* **21,** 1774.

Werner, S. A. and Klein, A. G. (1986), in *Methods of Experimental Physics,* Vol. 23, Neutron Scattering, Part A, editors K. Sköld and D. L. Price, Academic Press, New York; p. 259.

Werner, S. A., Colella, R., Overhauser, A. W., and Eagen, C. F. (1975), *Phys. Rev. Lett.* **35,** 1053.

Werner, S. A., Staudenmann, J. L., and Colella, R. (1979), *Phys. Rev. Lett.* **42,** 1103.

Whitaker, M. D. and Beyer, H. G. (1939), *Phys. Rev.* **55,** 1101.

Wick, G. C. (1937), *Phys. Z.* **38,** 689.
Willis, B. T. M. (editor) (1970), *Thermal Neutron Diffraction,* Oxford University Press, Oxford.
Wollan, E. O. and Shull, C. G. (1948), *Phys. Rev.* **73,** 830.
Wollan, E. O., Shull, C. G., and Marney, M. C. (1948), *Phys. Rev.* **73,** 527.
Word, R. E. and Trammell, G. T. (1981), *Phys. Rev. B* **24,** 2430.
Word, R. E. and Werner, S. A. (1982), *Phys. Rev. B* **26,** 4190.
Wyckoff, R. W. G. (1963), *Crystal Structures,* Interscience, New York.
Zachariasen, W. H. (1945), *Theory of X-ray Diffraction in Crystals,* Wiley, New York.
Zachariasen, W. H. (1967), *Acta Crystallogr.* **23,** 558.
Zeilinger, A. (1981), *Am. J. Phys.* **49,** 882.
Zeilinger, A. and Beatty, T. J. (1983), *Phys. Rev. B* **27,** 7239.
Zeilinger, A. and Shull, C. G. (1979), *Phys. Rev. B* **19,** 3957.
Zeilinger, A., Shull, C. G., Horne, M. A., and Finkelstein, K. D. (1986), *Phys. Rev. Lett.* **57,** 3089.
Zeldovich, Ya. B. (1959), *Sov. Phys. JETP* **9,** 1389.
Zilsel, P. R., Darling, B. T., and Breit, G. (1947), *Phys. Rev.* **72,** 576.
Zinn, W. H. (1947), *Phys. Rev.* **71,** 752.

INDEX